SILENT TRAVELERS

Germs, Genes, and the "Immigrant Menace"

ALAN M. KRAUT

BasicBooks
A Division of HarperCollins*Publishers*

Designed by Craig Winer

Library of Congress Cataloging-in-Publication Data
Kraut, Alan M.
Silent travelers: germs, genes, and the "immigrant menace" / Alan M. Kraut.
p. cm.
Includes bibliographical references and index.
ISBN 0-465-07823-0
1. Immigrants—Health and hygiene—Government policy—United States. 2. Medical policy—United States. 3. Immigrants—Medical examinations—United States. 4. United States—Emigration and immigration—Government policy. I. Title.
RA448.5.I44K73 1994 93-34572
614.4'08'69—dc20 CIP

95 96 97 ◆/RRD 9 8 7 6 5 4 3 2

To My Wife,
Deborah Aviva Kraut,
Debby's Due

Contents

Preface

Appropriately enough, this book's beginnings occurred on the streets of that great incubator of new Americans, Manhattan's Lower East Side. One broiling day in July 1979, I led a small group of undergraduates up and down its famous streets: Hester, Essex, Delancey, Grand, Ludlow, Rivington—home of Jewish immigrants at the turn of the century. Then over to Mulberry and Elizabeth, for the Italians. On to Mott for the Chinese. The walking tour was part of an American University summer institute on the immigrant experience. These narrow streets with their tall, crumbling tenements defined America to generations of immigrants who lived, worked, and raised families in the apartments and lofts above. To me and my class, it appeared that these same streets were bustling afresh with newly arrived Chinese, Southeast Asians, and Latin Americans, although remnants of earlier groups remained, especially individuals too poor or too old to have joined the flight to suburbia, or at least to have moved uptown.

By afternoon my class was melting before my eyes, getting an all too authentic impression of life without air conditioning. As we walked and chatted, we passed an elderly man with a Yiddish-language newspaper folded under his arm. His curved spine, pallor, and general appearance hinted at a hard life in unhealthy conditions, perhaps a life spent leaning over a sewing machine in the garment trades. My students stared at him, then quickly averted their eyes, possibly in deference to the contrast of their well-groomed, healthy, tanned good looks and his. One student blurted out an unsympathetic description, then stopped, stunned when she caught my frigid glare. The son of a factory worker and grandson of a tailor, I was two generations removed from these streets. My grandparents had escaped to the Bronx. Now I groped for words to explain the harsh reality over which these students had stumbled.

What book could I recommend that would explain as fully as I wished how much matters of health and disease had shaped the lives of the men, women, and children who met America in these streets? How could I explain why so many Jewish immigrants began and ended conversations with references to health—"Be well!" "Stay well!" "Live and be well!"—sometimes in Yiddish, also in English or Russian or Polish? I flipped through my mental bibliography cards searching for a book on the linkage of health, disease, medicine, and immigration in the United States that no one had yet written.

A second nudge toward writing this book came one morning in March 1983 when, listening to the radio over breakfast, I learned that the Centers for Disease control thought Haitians a high risk for AIDS. A whole national group? But why? How will the stigma affect those trying to flee the Duvalier regime, I wondered. How many times in American history have other groups been branded as bearers of disease? Someone ought to tell that story, I recall saying to my wife. But I was already involved in another project.

I thought again about the relationship of immigration to public health as I carefully adjusted my hardhat and picked my way through the rubble on the floor of Ellis Island's Great Hall in the mid-1980s. I was touring the ruins of America's famous immigration depot as one of the eleven historians on the History Committee of the Statue of Liberty–Ellis Island Commission (later, a committee of the SoL/ELI Foundation), advising and making recommendations on the restoration and interpretation of the monument. When we passed the spot where medical inspections occurred, I was reminded that a physician was one of the first Americans an immigrant encountered on Ellis.

On September 9, 1990, Ellis Island was reopened to the public, the largest historic restoration in American history. Guests gathered in Battery Park to wait in line for boats that would ferry them to festivities on the island, even as immigrants had waited for boats to shuttle them in the opposite direction. AIDS activists hoisted signs protesting the HIV testing of all immigrants and the exclusion of those who tested positive. The protesters' voices echoed the pain of yesterday's immigrants, fearing lest their own bodies keep them from America's opportunities. Clearly, this was a story that needed telling.

No book would ever have been written without the gracious assistance of many. Beyond the cliché, though, there are some long overdue thank-yous.

Robert J. T. Joy, M.D., and Dale C. Smith, Ph.D., of the Uniformed University of the Health Sciences in Bethesda are much more than colleagues and friends; they have been my teachers in the history of medicine ever since a happy coincidence led me to their doorstep. I listened to lectures, participated in seminars, and read books they recommended. When I lunched with them, it was always with a pen in hand and a note card next to my sandwich.

Bob and Dale introduced me to other helpful scholars too numerous to mention. Those at the Johns Hopkins History of Medicine Seminar of the Institute of the History of Medicine under Gert Brieger's leadership heard an early presentation and always made me welcome.

Kathleen Neils Conzen, Robert J. T. Joy, Judith Walzer Leavitt, Allan J. Lichtman, James W. Mooney, Dale C. Smith, and Jon L. Wakelyn read and offered helpful critiques of all or parts of the manuscript.

At archives and libraries in New York, Philadelphia, Chicago, San Francisco, Denver, Boston, Minneapolis/St. Paul, and Washington, D.C., I benefited from the cooperation and kindness of countless scholars, archivists, and librarians. Special thanks to Jeanne Abrams of the Rocky Mountain Jewish Historical Society in Denver, Beth Horrocks and Martin Levin of the American Philosophical Society in Philadelphia, the late Ira Berlin of the Northwest Memorial Hospital Archives in Chicago, Joseph Anderson of the Balch Institute for Ethnic Studies in Philadelphia, Dorothy Levenson of Montefiore Hospital in the Bronx, Joel Wurl of the Immigration History Research Center of the University of Minnesota, Adele Lerner of the New York Hospital Archives, Nancy Zinn, Special Collections Librarian at the University of California, San Francisco, former archivist James Mann and current archivist Daniel May of the Metropolitan Life Insurance Archives in New York, Fred Miller, former head of the Urban Archives at Temple University, David Klaasen of the Social Welfare Policy Archives at the University of Minnesota, and Richard Strassberg, Director and Archivist of the Labor-Management Documentation Center at the Catherwood Library of the New York State School of Industrial and Labor Relations at Cornell University. At the National Library of Medicine, I received help from former NLM archivist Peter Hirtle, former History Division chief and now historian of the Public Health Service, John Parascandola, Deputy Chief Philip Teigen, and Historian James Cassedy; at the U.S. Food and Drug Administration, John Swann and Suzanne White; and in the Public Health Service, Director of the Bureau of Health Professions, Fitzhugh Mullan and Thomas Bornemann of the Refugee Health Project. Independent scholar and collector William Helfand generously shared images from his collection.

Colleagues on the Statue of Liberty–Ellis Island project offered information and advice. In addition to Kathy Conzen, immigration historians Roger Daniels, Victor R. Greene, Moses Rischin, and Rudolph J. Vecoli pointed me toward sources. Phyllis Montgomery, who directed research for the interpretive material on Ellis Island, and her associates, Fred Wasserman and Mary-Angela Hardwick, generously shared materials and information.

Other scholars sent me copies of papers they had written or photocopies of documents. My sincere thanks to Randolph H. Boehm, Gretchen Condran, Pete Daniels, David M. Emmons, Gerald Grob, Howard Markel, George

Pozzetta, David Rosner, Paul Rozin, Susan Cotts Watkins, and many others. Saul Benison's advice and bibliographical references made our convention breakfasts a special treat. Suzanne Michael of the Committee on Family Health of the New York Immigrant Health Task Force kept me up-to-date on the contemporary immigrant health scene.

In 1987, my office mate, British historian Janet Oppenheim, and I gave papers together on the history of medicine at an American Historical Association meeting. Her suggestions and warm friendship have been constants. Long-time friend and colleague Scott Parker shared books and articles about psychological testing of immigrants and throughout the project asked me questions that caused me to tighten and refine my arguments. James P. May of the Washington College of Law offered advice and assisted my research on several early nineteenth-century cases dealing with conflicts of state and federal jurisdiction.

Throughout the project I benefited from conversation with my friend and colleague James W. Mooney. We had long and valuable discusions about hegemony, social control, and many other topics. Always, his thoughtful approach to American intellectual and cultural history assisted me in sharpening my own thinking. His practiced editorial eye and high professional standards saved my page proofs from a myriad of minor errors and inconsistencies, when my own eyesight and attention span were at low ebb.

My project received generous financial support. At the American University, course releases and stipends sustained me, including a Sabbatical Support Award in 1987–88. Former University Provost Milton Greenberg, former Dean of Faculties Frederic Jacobs, and interim Provost Ann Ferren have been especially kind to me and my project over the years. College of Arts and Sciences Dean Betty Bennett has always been an enthusiastic supporter of my research endeavors, as have Robert Beisner and Roger Brown, who chaired the History Department at different times during the project.

A Rockefeller Foundation Fellowship in the Humanities in 1987–88 sent me to the Francis G. Wood Institute for the History of Medicine at the College of Physicians of Philadelphia, then directed by Diana Long. College of Physicians Librarian Thomas Horrocks offered advice on sources, and Mutter Museum Director Gretchen Worden, always a kind friend, has continued to be helpful, even assisting me in finding some of the photographs in this volume. At Thursday afternoon brown-bag seminars, I benefited from the expertise of Philadelphia's medical history community, including Charles E. Rosenberg, Rosemary Stevens, Barbara Bates, Joan Lynaugh, Karen Buhler-Wilkerson, Russell Maulitz, Steven Peitzman, and Janet Tighe, among others. Historians Janet Golden and Eric Schneider shared their learning, their pizza, and their evenings at the Spectrum, where history and Seventy-Sixer basketball mixed. Barbara Traister, my fellow Rockefeller colleague that year, was a wonderful friend and sounding board for ideas.

More time free of teaching duties was made possible by a Senior Postdoctoral Fellowship in the Medical Sciences Division of the Smithsonian Institution's Museum of American History and an Interpretive Research Grant from the National Endowment for the Humanities. At the Smithsonian, Curator Ray Kondratas offered advice and criticism. Michael Harris and Judy Chelnick made certain I found materials I needed. At the Endowment, Dr. Daniel Jones not only read preliminary drafts of grant applications but was a constant source of encouragement.

I have enjoyed the assistance of E. J. Prior, Kelly King, and Christopher Canavan, my excellent work-study students. American University interlibrary loan librarian Eileen Gloster has always been gracious and efficient in getting me materials from other institutions. University Librarian Patricia Wand and librarians Mary Mintz, Diana Vogelsong, George Arnold, and Elizabeth Nibley have been especially helpful. History graduate students Jennifer Altenhofel and Joseph Ryan generously shared documents they found while doing their own projects.

During the course of my work, two of my graduate students earned their Ph.D.'s—Drs. Sarah Larson and Melissa Kirkpatrick. I am proud of their achievements and grateful for their help and support. Treasured friends, they taught me even as I taught them. Melissa periodically helped me overcome computer illiteracy while offering valued comments on the entire manuscript.

Sarah Larson has taken a blue pencil—really a hot pink or green felt-tip—to drafts of each chapter more than once. She is a superb developmental editor with rare sensitivity to the nuances of historical argument. Her criticisms are on target and are always delivered with a care that bespeaks an understanding of the subtext as well as the text. She also assisted in selecting and captioning photographs.

Steven Fraser is a friend as well as a fine editor. He had confidence in me and this project from our first discussion. His candid, intelligent, and incisive comments on early manuscript drafts helped shape the book. I share the high regard for him that is all but unanimous among our fellow historians. Jane Judge has been a pleasure to work with as she attentively guided the manuscript through the production process.

Family is traditionally acknowledged last, perhaps to underscore the importance of their contributions. My mother-in-law, ethnomusicologist Irene Heskes, shared the sheet music from *Menshen-Fresser* and translated the lyrics from the Yiddish. My father-in-law, Jacob Heskes, CPA retired, computerized my data on medical exclusions from government reports and loyally read drafts, gently advising me that this book would probably not up my tax bracket. My own parents, Jeanette and Harry Kraut, are no longer living, but parts of chapters six and seven bear their imprint, reminding me how often I write my family history as I tell the American story.

And now for the wife. This book bears a long overdue dedication to Deborah Aviva Kraut, my partner of over twenty years and friend for longer, who was enthusiastic about this study from the onset. She has worked most of her career in health care facilities and enjoyed this project in which our professional paths crossed. She reads all chapter drafts and we argue when she doesn't like my anecdotes, but her judgments are invariably correct. She encouraged me to spend a year commuting to Philadelphia, never complaining when my research took me afar. If the places were fun, such as San Francisco and Chicago, she followed, finding great places for dinner and architectural wonders I ought to notice. She forced me to look up, take my hands off the keyboard, and enjoy a life off the page. What more can I ask?

When I wrote the acknowledgments for my first book in 1981, my daughter, Julia Rose Kraut, sat atop my desk strapped into her infant chair. Now she is preparing for her Bat Mitzvah. I have loved being her father more than anything else in the world. On one of our New York walks, she insisted we visit the building where the Triangle Shirtwaist fire occurred. As she carefully read the commemorative plaque and looked up to see how far the victims had jumped to escape the flames, I knew whose daughter she must be. Happy Bat Mitzvah, Julia.

Introduction: The Double Helix of Health and Fear

Easter weekend, 1992, a recurring drama in the annals of American immigration history and the history of public health was replayed. This time the scene was an old airplane hangar at the United States Naval Base at Guantánamo Bay, Cuba. Lately, American officials had been processing Haitian refugees in the hangar before flying them to the United States, so there was anticipation in the air when American servicemen gathered 140 refugees together. Those assembled eagerly expected that they, too, were headed for America. But not this time.

Instead, a double dose of grim news from American military officials dashed the Haitians' hopes. The refugees were being denied admission to the United States on health grounds. Test results had indicated they were HIV-positive. AIDS, not America, loomed in their future.

As the first infected refugees in the group heard the horrible news, word spread. A woman shrieked. Then, an interpreter told a reporter, "All hell broke loose. The whole place just became chaotic." Grief-stricken and desperate, the Haitians began "throwing things, screaming, knocking over cots." Military police restored order, but neither they nor a Haitian-American priest who sought to pacify the crowd with prayer could calm the fears or relieve the agony of those who had just learned their ghastly fate.[1] In defense of the public's health, the United States government had yanked the welcome mat from beneath the Haitians' feet. It would be a year and a half before a federal judge permanently enjoined the United States government from detaining the Haitians at Guantánamo Bay, observing that their illness hardly warranted "the kind of indefinite detention usually reserved for spies and murderers."[2]

The Haitians' tragedy echoed the aspirations and apprehensions, the hopes and fears, of earlier arrivals. Slightly more than a century before the April 1992 Guantánamo melee, the United States had opened its flagship immigration depot on Ellis Island in New York harbor. Between January 1, 1892, and November 29, 1954, over twelve million immigrants were inspected and interrogated and finally told whether or not they were fit enough for America. Fannie Kligerman was among those examined in the years to come.

She, too, encountered uniformed individuals astride America's threshold. They were not soldiers, but the young Russian girl thought they were. Seventy-five years later, an elderly Kligerman recalled her eyes being examined and remembered the uniform: "And who examined us? A soldier. This I remember, too."[3] It was a Public Health and Marine Hospital Service physician, not a military officer, but to Fannie Kligerman, who was eventually admitted, all that mattered was that a uniformed sentry stood between her and a new life.

Just as continual replenishment of the American population through immigration has shaped the politics, economics, and culture of American society, ethnic pluralism has defined American medical culture and its approach to public health.[4] Medicine has been an important instrument employed by native-born Americans to assimilate immigrants into American society in a manner that would most effectively preserve the established order's cultural preferences and priorities. Immigrants have engaged in a love/hate relationship with America's public health priorities, at times resisting them as an unwelcome intrusion in their lives and at other times embracing them as a path to upward mobility.[5]

The reciprocal influence of immigration and public health in the United States stands at a busy cultural and social intersection, where at least four significant themes converge to shape the relationship.[6] The first theme is the relationship among health, disease, and nativism, those prejudices and policies that express opposition to the foreign-born.[7] The medicalization of preexisting nativist prejudices occurs when the justification for excluding members of a particular group includes charges that they constitute a health menace and may endanger their hosts. While some members of an immigrant group may or may not have a contagious disease that can cause others to become sick, the entire group is stigmatized by medicalized nativism, each newcomer being reduced from "a whole and usual person to a tainted, discounted one," because of association with disease in the minds of the native-born.[8] Thus, there is a fear of contamination from the foreign-born.[9]

In the early years of America's awareness of AIDS, the Haitian immigrant community suffered the stigma of association with the disease. In 1985, a twenty-nine-year-old social worker from Haiti named Marcelle Fortune was denied a place to live when a Miami landlord refused to rent to her on the grounds that his tenants feared a Haitian immigrant would spread AIDS throughout the building. Fortune, an employee of Florida's Department of Health and Rehabilitative Services, did not have AIDS.[10] However, over two years earlier the Centers for Disease Control (CDC) in Atlanta had sent a tremor of trepidation throughout Miami and other areas of Haitian concentration by announcing that those immigrants had been classified in a special high-risk category for the deadly contagious disease. Consistent with essayist Susan Sontag's contention that the stigma of disease can become a metaphor for already marginalized individuals, culturally defining them even further from society's mainstream, Marcelle Fortune, a black Haitian immigrant, found herself stigmatized, caught in the maelstrom of panic over a disease with which she was not afflicted.[11] By April 1985, when the CDC finally dropped Haitians from its high-risk list, the psychological, financial, and social damage had been done. Tens of thousands of Haitians in the United States and in Haiti, most no more infected with AIDS than Marcelle Fortune, had lost jobs, housing, or educational opportunities. Though refugees from Haiti were still being granted asylum in the United States, the welcome for political refugees fleeing the oppressive Jean-Claude Duvalier regime turned frigid in most American communities. Haitian refugee children, already struggling to adjust to life in America, found themselves ostracized as public health menaces even on the school playground.[12]

It was old wine in new bottles. In ancient times, the presence of strangers seemed so perilous to the native-born of a locale that such travelers were routinely shunned or harmed with socially sanctioned violence in the communities they entered. The Bible warned against such behavior. God cautions the Israelites to welcome strangers and to guarantee their safety: "And if a stranger sojourn with thee in your land, ye shall not vex him. But the stranger that dwelleth with you shall be unto you as one born among you, and thou shalt love him as thyself; for ye were strangers in the land of Egypt."[13] Despite this admonition, immigrants often have been despised and feared as harbingers of disease, debility, or even death to the native-born.

The biblical injunction has hardly been irrelevant in the history of the United States, a nation priding itself on its diversity. Even as Marcelle Fortune experienced the stigma of association with a disease she did not have, in 1916 newspapers and some public health officials smeared Italian immigrants residing along the East Coast by blaming them for the polio epidemic that terrorized Americans during a long and fearful summer. Earlier still, in 1900, public officials branded the Chinese in San Francisco health menaces after an outbreak of bubonic plague, just as in the previous century some nativists associated Irish immigrants with a cholera outbreak that plagued newcomers and their American-born neighbors.

During America's peak immigration period between 1890 and the 1920s, exclusions never exceeded 3.0 percent and averaged less than 1 percent. However, an increasing percentage of those debarred were excluded for medical reasons, reaching an unprecedented 69 percent of those excluded by 1916. The conundrum of what kind of physical specimens to admit to the United States of America and how to maximize the physical and mental potential of new arrivals has been a crucial chapter of both immigration history and the history of public health in the United States.

The advances and perceived limitations of scientific medicine constitute a second theme. In the most recent wave of immigration that began in the 1970s and heightened during the two subsequent decades, scientific medicine has played a significant role. For example, epidemiological researchers identified the HIV virus and developed a blood test for its detection, with the subsequent exclusion of all immigrants testing positive.

The present role of scientific discovery in the linkage between immigration and public health echoes an earlier era.[14] After the 1832 cholera epidemic, sanitarians contended that dirt, not effluvia from decaying organic matter, was the origin of disease. Some nativists believed that the Irish were inherently diseased. Increasingly common, though, was the view that the Irish were merely filthy and unmindful of public hygiene, and therefore, were culpable for epidemic diseases.[15] By contrast, Germans, who arrived in large numbers at the same time as the Irish, were either omitted from blame, or praised for their clean homes and orderly life habits.

By the close of the nineteenth century, the squabble among nativists, politicians, and public health bureaucrats over admission and exclusion was most affected by the development of medical bacteriology by Louis Pasteur in France and Robert Koch in Germany during the 1880s. Germ theory and subsequent significant advances in the medical sciences that enhanced physicians' ability to test for the presence of pathogenic organisms magnified scientific medicine's value in excluding infected newcomers who could potentially infect others or, at the very least, find themselves unable to work and lead productive, self-sufficient lives. However, improvements in medicine also provided a foundation for hope that with proper medical care and health education, American society could become a level playing field upon which immigrants and natives could compete equally for political power, cultural influence, and material resources.

By 1900 nativists baldly claimed in scientific medicine a weapon that white Anglo-Saxon Protestant civilization could use to defend itself against the intrusion of those it regarded as of inferior breed. Nativists trumpeted data suggesting that certain diseases were more prevalent among the newcomers than among the native-born. They also boasted that the general vitality and robustness of Americans marked a standard well beyond that of new arrivals, adding that it was unlikely the foreign-born could ever prove the physical equals of America's pioneering breed. Lest that breed suffer dilution in the descendants of biologically imprudent liaisons, exclusion seemed the best solution. Other Americans, some equally troubled by the health and unimposing appearance of the newcomers, advocated assimilation rather than exclusion.

Assimilationists favored a continuation of immigration, though they differed over whether or not some restriction might still be necessary. They contended that newcomers could receive care and education in health and hygiene that would transform life-styles. Had not Franz Boas demonstrated that even physical appearances could change through altered habits of nutrition and life-style? Health education in schools and preventive care in communities could remake outsiders into insiders, changing cacophonous, foreign-speaking voices of unhealthy, undersized individuals into robust, conforming, disciplined, healthy members of the American choir, capable of making their voices heard and their presence felt in the normal competitive din of American life.

These assimilationists, often progressive reformers and social workers, were not for the most part consciously sinister or cynical advocates of their own class interests.[16] Still, many assimilationists of the Progressive bent shared a patrician confidence in social stewardship that obligated them to protect both the public health and newcomers from the chaos of social drift, as Walter Lippmann called it, by engaging the findings of medical science as the foundation of what was intended to be a benevolent social control.[17] There can be little doubt that the acculturation, if not complete assimilation, of the foreign-born was their goal. They hoped to achieve cultural uniformity humanely, relying largely, though not exclusively, upon persuasion rather than coercion, inclusion rather than exclusion.[18]

Not all Americans were equally optimistic about medicine's prospects as a device for social control.[19] Eugenicists, for example, argued that only exclusion and careful education in right choices of human breeding could dilute the potentially disastrous impact that the inferior nature of the foreign-born could have on the future of American physical and mental vitality.[20]

Politicians were generally enthusiastic about scientific medicine because they could refer to the opinions of medical experts as justification for public health policies toward newcomers, whatever those policies might be. Public health bureaucrats seized on scientific medicine as the means of keeping natives and newcomers alike healthy and able to withstand the rigors of an industrializing, urbanizing nation. With more and more immigrants needed to fuel America's industrial economy at the turn of the century, pressure mounted to admit rather than exclude new arrivals. The demand for solutions to the public health problems and health care needs of newcomers escalated. To paraphrase Karl Marx, immigration became a locomotive of change in regard to American medicine and public health.

A third theme, then, is the institutional response of national, state, and local governments to both the qualitative and quantitative dimensions of immigration as it affected and was affected by public health considerations. Throughout American history, natives have gazed upon newcomers with a mind to determine whether or not they looked well or ill, strong or weak, alert or dull, sane or insane, attractive or homely; in short, whether or not they were desirable as neighbors, fit for America. Those deemed fit, which often as not has meant similar in appearance and origins to those already here, were allowed to remain. Those judged unfit, either physically or mentally, were excluded, lest they enfeeble those already here. Unhealthy, then, became a convenient metaphor for excludable, legislation the remedy, and public health bureaucrats—first state, later federal—the instruments of cure.

Codifying and regulating issues of public health, generally, always required a balancing act, delicately weighing individual rights and public welfare. Prior to the Civil War and in the years immediately after, state and municipal governments performed the regulatory stunt when it was done at all. However, by the 1890s, bickering over public health that had been confined to the floors of state legislatures and the smoke-filled backrooms of city halls increasingly echoed in Washington's corridors as well.[21]

The impact of immigration upon public health was unmistakably felt in the public sector even as it was by private public interest groups dedicated to social reform and uplifting the poor and foreign-born. There is ample evidence that the increasing numbers of foreign-born in the industrial work force and the educational system persuaded lawmakers and lobbyists alike to use institutional means to secure healthier work conditions and to make children's health a responsibility of the public schools.

Public policies, indeed all measures affecting immigrants' health and hygiene, required negotiation between newcomers and their hosts. Even if English novelist C. S. Lewis was correct in his general observation that "Man's power over Nature turns out to be a power exercised by some men over other men with Nature as its instrument," immigrants to the United States have hardly accepted passively efforts to subordinate them in the name of public health.[22]

Thus, a fourth and final theme treats immigrant and refugee response to differences between themselves and the native-born in matters of therapy, health care, disease prevention, and hygiene—all areas that impinge upon public health.

Immigrants do not automatically abandon traditional systems of healing for Western health care upon arrival.[23] In 1988, the *Washington Post* reported that Asian immigrants were turning to the proprietors of an Asian herb shop for medicine rather than consulting a physician. The *Post* disclosed that Danny Ung, owner of the China Herbs store in Falls Church, Virginia, was selling fellow Asians deer antler shavings for poor blood circulation and an array of ground sea horses, ginseng roots, and extra-strength "Tiger balm" for other ailments such as headaches, stomachaches, and muscle pains. At the Liberty Department Store in nearby Arlington, John Hoang, a Vietnamese immigrant, was doing a booming business among recent immigrants in Chinese, Indonesian, and Korean herbs and oils. Although neither was qualified to diagnose illness by Western standards, Ung and Hoang testified to a high volume of trade.[24]

In recent years, American physicians and pharmacists have acknowledged that alternative drugs and systems of therapy imported by newly arrived immigrants may have some healing value, and that, even if these therapies do not heal by Western standards, they may be essential for the patient's welfare. Hmong refugees from northeastern Laos, many of whom settled in the Minneapolis–St. Paul, Minnesota, area during the 1980s, vigorously resisted altering their traditional system of health care. By late 1985, ten thousand Hmong lived in the community and were locked in a cultural tug-of-war with the medical community. The Hmong preferred the ministrations of ritual healers, shamans, to Western physicians. They recoiled from modern therapeutics in favor of herbal medicine and amulets. Physicians, especially those at hospitals and clinics where the Hmong were guided by social workers for medical care, sought to understand and respect rather than dismiss patient expectations and preferences.

In Minnesota, social workers encouraged a blending of traditional and scientific therapeutics when it was safe to do so. Physicians in St. Paul voiced little objection when children suffering from middle ear infections drank a shaman's herbal remedy. However, these doctors then expected the shaman to recommend to the patient the physician's antimicrobial prescription. Cooperation and cultural literacy have generally resulted in better health care for these newcomers. Still, blood drawing and surgery often do not get the support of the shamans. At times negotiations break down and degenerate into cultural warfare between physician and shaman at a patient's bedside.[25]

And what happens when physicians are asked to take second place to traditional healers? That was the case in Newington, Connecticut, when Juliet

Cheng was charged with neglect and her seven-year-old daughter, Shirley, was temporarily taken into custody by the state's Department of Children and Youth Services. Ms. Cheng had rejected a physician's recommendation of surgical repair on her daughter's knees, hips, and left ankle for severe juvenile rheumatoid arthritis. Instead, Cheng wanted to take the child home to China for physical therapy and traditional herbal treatment. Explaining her side of the story to the press, Cheng said, "Shirley is very upset in the hospital. Every time she sees me she asks me to take her out of her chair. She doesn't want the surgery. I just want her back so we can return to China." A lawyer for Connecticut Legal Services placed the case in larger context, observing that it was an example of "the medical establishment deciding they have the truth of the matter and totally disparaging alternate forms of medicine."[26]

Eventually, it became a federal case. Two out of three physicians approved by a federal judge to examine Shirley recommended against surgery. One of the physicians, Dr. Thomas J. A. Lehman, chief of the division of pediatric rheumatology at the Hospital for Special Surgery in Manhattan, explained his conclusion that surgery was not immediately necessary by referring to the value of cultural sensitivity in affecting therapy. "If you do something where you need the cooperation of the entire family for the child to get better, when it's against the family's wishes your probability of success is vastly reduced." He continued, "Ms. Cheng accepts the fact that Shirley will probably eventually need surgery, but wants to continue working on the therapy to try and get Shirley better without surgery."[27]

As in Juliet Cheng's case, newcomers in earlier eras—Irish, Chinese, Italians, eastern European Jews, Latin Americans—were hardly passive pawns in the discourse over their health, choice of health care, or public health regulations imposed upon them. Earlier in the century, the Chinese Six Companies sought out legal counsel to prevent San Francisco officials from imposing a quarantine on all of Chinatown when a case of bubonic plague was found there. Sometimes misunderstandings caused riots, as there were on New York's Lower East Side in 1906 when angry eastern European Jewish and Italian parents stormed several public schools to protest what they thought were bloody murders of their children and what turned out to be adenoidectomies. The surgery, designed to improve their children's health, had required prior consent, which parents had willingly given, though they may not have known that curettage would be required to ease their youngsters' nasal breathing.

More often, though, the cultural battles were not in the law courts or the streets but in homes and hospitals and workplaces where newcomers and the native-born wrestled over their cultural differences in defining wellness, disease, cleanliness, health care, and religion's role in healing the sick. Health and disease are socially constructed concepts; so also is the definition of appropriate care. Did the family of an eastern European Jewish immigrant child who suffered a burn send for a female folk healer, who would rub the burn with pulverized cooked lima beans, or a physician, who might recommend hospitalization? As with today, traditional healers and allopathic physicians of one hundred years ago often stood glaring at the bedside of a sick immigrant child, arguing that only his or her definition of the condition and therapy could save the patient.

Newcomers have never left unopposed the imposition upon them of New World concepts of wellness and illness. For many, better health as defined in American terms seemed to require relinquishing cherished customs, traditions, and trusted therapies on the altar of assimilation. Overwhelmingly, members of the various ethnic groups turned to their own institutions, to their own physicians and care givers, especially in the initial years after arrival, in an effort to obtain health care that did not require them to abandon their cultural identities.

There is a tension that pervades this book. On the one hand, immigrants can be and, from time to time, have been the bearers of diseases harmful or even fatal to the native-born population. Officials responsible for the public health have had to deal forthrightly with this threat. On the other hand, preexisting ethnic prejudices and public hysteria in the face of disease often created a wholly false linkage between illness and specific immigrant groups. Public health officials have often found themselves charged responding to this fear, too. It is not this book's purpose to castigate those sincere in their concern for the public health, however misguided their approach or uninformed their science may have been. It is rather to explore the complexity of their arguments and to shed light upon other arguments, those grounded in nativist prejudices, that may have been motivated by something less than altruistic concern with the health and welfare of the population.

The anxious questions asked of yesterday's immigrants on Ellis Island and the anguished screams of today's newcomers at Guantánamo Bay suggest an all too familiar continuity. The double helix of health and fear remains encoded in American society and culture, reappearing in patterns fresh but familiar. It is the task of this volume to decipher that double helix while preserving, intact, the voices of immigrants, medical professionals, and worried public alike.

1

"The Breath of Other People Killed Them": First Encounters

Just two years shy of the five-hundredth anniversary of Christopher Columbus's arrival in the New World, the medical implications of the Columbian Exchange made headlines. Even as North American politicians were angling for prominent spots in Columbus Day parades and European leaders, especially those in Spain and Italy, were conspiring to outdo each other in expressions of national pride for having "discovered" America, Indians were dying from the encounter of Old World and New much as they had centuries before.

The Yanomami, a nine-thousand-member tribe that lives on both sides of Brazil's border with Venezuela, considered by anthropologists to be the last major Stone Age tribe of the Amazon, were dying of white man's diseases, especially malaria and tuberculosis.[1] For five centuries, the curare-tipped arrows of the Yanomami had protected the tribe's health, excluding outsiders and their illnesses. However, the lure of gold, diamonds, and other resources kept the prospectors, or *garimpeiros,* coming. In 1990, a member of a French medical team, Médecins du Monde, told journalists, "Practically every Indian we examined has malaria, some also have tuberculosis, and one of every two children is undernourished." Making the analogy to another suffering people, he continued, "They look like African children: Their hair is turning blond, their stomachs are bloated, their arms and legs are spindly." The Yanomami lacked natural immunities to the strains of malaria contracted from prospectors. "Malaria is like AIDS for them," Dr. Ivone Menegola told a reporter.[2] The tribe was in imminent danger of extinction, devastated by the gold rush that began in 1987. The tribe's mortality rate had climbed by 1990 to approximately 15 percent, a figure difficult to calculate because the Yanomami destroy all trace of the dead and they do not count. Anything above two is *wahoro,* many.

With many men confined to their hammocks suffering from influenza, chicken pox, dysentery, measles, and gonorrhea, as well as malaria and tuberculosis, there were food shortages. Pollution newcomers brought to forests and streams further reduced food supplies. Game fled the forests, scared by the noise of dredging pumps and airplanes. But the loudest sounds were the accusations of ecologists, the Indians' advocates, against government officials whom they accused of inaction, of overlooking the obliteration of the Yanomami so that mineral exploitation could proceed unhindered. When the government finally acted to bar prospectors, the damage to the food supply could not be repaired. Emaciated Yanomami trekked to prospector campsites to beg for food. It was then the ecologists' turn to come under attack for trying to preserve a native population as if it existed in a glass jar, rather than supplying food and gradually helping the tribe to adjust to the diet and conditions of the outside world.

As for the Yanomami, their reaction to the outsiders from a different place and time underwent a transformation, according to Roman Catholic missionaries. Once the enemy, white men were culturally recast as good and contact courted to procure "flashlights, knives and axes." Loincloths and bare feet were abandoned for shorts and flip-flop shower sandals. Sneakers and digital watches became highly desired objects that Yanomami all but venerated. The encounter could not be undone. The last transaction in the Columbian exchange was concluding.[3]

Historically, quarantine, or the separation of one group from another, has been the most common mechanism for halting the spread of disease. But use of such a preventive presupposes awareness of risk. Certainly, Native Americans who encountered Christopher Columbus, the first Italian to arrive in North America, saw neither him nor his men as health menaces, any more than the Yanomami did when they spotted their first *garimpeiro.*[4] Why should they? Native Americans, or Indians, as Columbus and later arrivals would call them, certainly had experience with sickness and injury. However, there is no evidence that they had suffered from the epidemic diseases that ravaged European communities such as the lethal collaboration of rats and fleas that had spread plague across fourteenth-century Europe, annihilating approximately a third of Europe's population.[5] Indeed, to many of the Europeans, their hosts appeared more robust and healthier than themselves or those left behind in Europe.[6] Smallpox and measles were not in evidence and there is no indication that malaria, yellow fever, typhoid, or typhus had preceded Europeans to the Americas either.[7]

Even the presence of the same bacteria need not affect two diverse populations in different parts of the globe the same way. Syphilis may be a case in point. By the time of the Columbian Quincentenary, scholars still disagreed over whether the scourge was another of Europe's gifts to the New World or perhaps had traveled in the other direction. New biochemical methods, especially the analysis of DNA and immunoglobulins from human skeletal tissue, fueled historical debate.[8] Some investigators took a pre-Columbian stance, arguing that while syphilis was at times confused with leprosy in Europe, the venereal disease was present in Europe as an epidemic disease well before 1492 and, therefore, could not have come from America.

Others, advocates of the Columbian theory, have argued that syphilis was not endemic to Europe and that venereal syphilis did not exist in Europe until explorers returned with it. As evidence that the disease was endemic to the New World, scholars suggest that the permanent disfigurement of bones in many pre-Columbian Native American skeletons resulted from advanced syphilis. However, not everyone is convinced.[9]

Complications in the dispute abound. In 1992 there were reports that some pre-Columbian English skeletons unearthed showed treponemal damage, suggesting that treponema—perhaps endemic syphilis—was in Europe before the Italian explorer ever set sail.[10] If syphilis was endemic and venereal on both sides of the Atlantic prior to 1492, it may never have figured into the Columbian exchange at all.[11] Other speculation suggests that while the arrival of syphilis in Europe had little to do with Native Americans, it was an unanticipated by-product of African trade. Evidence is far from conclusive.[12]

Whether or not syphilis made the transatlantic journey, other epidemic diseases traveling westward found Native Americans unprotected by natural immunities.[13] The encounter with Columbus cost the Indians their health almost immediately, perhaps from animal rather than human viral infection. Beginning in 1494, an estimated 1,100,000 Indians were reduced to 10,000 on the island of Hispaniola by the ravages of disease, which historians were once quite certain was smallpox. However, fresh examination of epidemiological evidence in accounts by Christopher Columbus and his physician, Alvarez Chanca, suggest that the horrendous mortality was the result of a swine influenza epidemic.[14] Eight hogs taken on board ship at La Gomera, Canary Islands, and in contact with the crew and all those who traded with the Columbus expedition on Santo Domingo appear to have spread infection.[15] The admiral himself was stricken and took five months to recover. American pre-Columbian Indians had few if any domestic animals. Thus, they were first exposed to domestic animal viruses only when Columbus landed with horses, cows, sheep, goats, and pigs at La Isabela, in Santo Domingo. Native Americans had no immunities to animal viruses; farm animals became agents of death.[16]

It was not long though before North American Indians were being felled by human-borne diseases, as well.[17] Awed Europeans looked on as disease ravaged the natives of the New World. Indians north of Mexico who may have numbered as high as seven or eight million (five million in the area of the modern coterminous United States) prior to contact with Europeans were most devastated by smallpox, but typhus, yellow fever, diphtheria, influenza, and dysentery all contributed to the decimation of tribes by as much as 50 to 90 percent of the population after just one contact with European immigrants.[18] Illnesses that Europeans already regarded as childhood diseases—measles, mumps, chicken pox, and whooping cough—wreaked havoc on Native American communities that lacked natural immunities.[19]

Coastal plain tribes were all but wiped out by the time of the American Revolution. Eastern Massachusetts tribes had begun to see their population dwindle with the first European contact. The Iroquois suffered several epidemic waves in the seventeenth and eighteenth centuries. In 1759, the Catawba of North Carolina were halved in number by a single epidemic. In 1738, the Cherokee lost half their population to smallpox. The Creeks, too, suffered devastation from the disease in the seventeenth century. Earlier, Spaniard Solorzano Pereira, witnessing the death of hundreds of Native Americans, wrote with some justification that "the breath of other people killed them."[20]

Pereira may have been more correct than he knew. In dealing with these new diseases, Native Americans were as much at a disadvantage culturally as they were biologically. At times the deadly breath that killed their kin was their own. Notions of isolation and quarantine were not foreign to Native Americans. However, Indians had no way of knowing when separation from the victims of these unfamiliar ailments was absolutely necessary and ought to take priority over other cultural imperatives that demanded close personal contact with the sick.

A French missionary among the Hurons in the 1820s, Father Gabriel Sagard, observed that "Sometimes the medicine-man orders one of the sick people to leave town and encamp in the woods or in some other place apart, so that he may practice upon him there during the night his devilish contrivances." Sagard was quite certain that quarantine was the motivation because "this is only done for those who are infected with some unclean or dangerous disease, and such persons only, and no others, do they force to isolate themselves from the community until they are completely cured." The missionary, perhaps despairing of European practices inadequate to prevent the spread of infection, praised such customs as "laudable and most excellent custom and ordinance, which indeed ought to be adopted in every country."[21]

Native Americans also exercised quarantine when they isolated menstruating women, fled villages when smallpox erupted, and compelled wounded warriors returning from battle to wait in a holding area before returning to their homes. In the latter case, braves had to wait "four moons . . . as in the case of their women after travail; and they keep them strictly separate, lest the impurity of one should prevent the cure of the other."[22]

However, some tribes' response to disease often only accelerated the spread of germs passed on through close human contact. To aid and comfort a sick relative, Native Americans often clustered around the ailing one. Quarantine of the sick would have seemed cruel, an undeserved ostracism, but might have slowed the progress of some diseases' scythelike movement through village populations. Instead, culturally determined behavior patterns of how the well must treat the sick made those not yet felled vectors of decimation.[23] This was an irony because the Indians were hardly without medical knowledge.

Native Americans had a sophisticated and effective arsenal of therapies for diseases and other ailments with which they were familiar. Many Europeans acquired a distinct admiration for Indian skill in tending the sick. The Indian doctor was a figure respected by whites as well as Native Americans, and if concoctions of herbs, sweatings, and rubbings were accompanied by non-Christian attention to charms, amulets, and incantations, few of the sick argued with those who were making them well. In 1670, a German immigrant surgeon who was exploring Virginia's Blue Ridge Mountains, John Lederer, owed his life to an Indian with healing skills. While asleep, Lederer was stung by a "Mountain-spider" and "had not an Indian suckt out the poyson, I had died: for receiving the hurt at the tip of one of my fingers, the venome shot up immediately into my shoulder, and so inflamed my side, that it is not possible to express my torment." The Indian used "a small dose of Snakeroot-powder, which I took in a little water; and then making a kinde of Plaister of the same, applied it neer to the part affected: when he had done so, he swallowed some by way of Antidote himselfe, and suckt my fingers end so violently, that I felt the venome retire back from my side into my shoulder, and from thence down my arm: having thus sucked half a score time, and spit as often, I was eased of all my pain, and perfectly recovered."[24]

Americans and epidemic illness in colonial America. (Courtesy of Chicago Historical Society)

Europeans also derived much benefit from Indian botanical drugs such as alum root, sassafras, or yerba santa.[25] More than two hundred indigenous drugs used by one or more Indian tribes have been in *The Pharmacopeia of the United States of America* for different periods since its first publication in 1820, or in the *National Formulary* since its inception in 1888. Unfortunately, traditional Indian therapies were often not potent in curing the new diseases from abroad. The sweating of fever victims, for example, only worsened the condition of smallpox sufferers.

A popular therapy for a host of illnesses was a good long session in a steaming sweatlodge, followed by the bracing contrast of a naked dip into a frigid river or an icy snowbank. But such therapy shocked the body and fueled the fevers of smallpox that would soon consume its victims. In cases where colonists succeeded in persuading tribes to abandon such therapy, they succeeded in saving some of the lives that were being lost to the illnesses they themselves had brought across the ocean.[26] However, even after the Indians recognized the impotence of traditional medicine in the face of new diseases, large-scale suffering continued.

As in the case of Brazil's Yanomami, when almost everyone in an Indian village would get sick at the same time, there were few individuals left still healthy enough to make fires, haul water, cook food, and offer care and comfort to the sick. Likewise, there was little time to prepare for the onset of illness. In the absence of a systematic notion of quarantine to preserve community health, the very messenger who brought word of smallpox in another village might also carry the disease.[27]

Every aspect of tribal life was affected by the epidemic diseases that cut down Indian populations. After weeks or months the pestilence would abate, leaving the remnant of a people to reconstruct life as best they could. Like wars, epidemic diseases bisected history. Tribal storytellers talked of life before an especially memorable epidemic as if it was separated by a kind of biological abyss from how life proceeded afterward. Kinship networks were disrupted. The elderly who carried in their heads ancient histories, cures, and crafts were often wiped out quickly, taking with them generations of a tribe's collective understanding of the world and itself. Describing the experience of the Etiwan tribe in early eighteenth-century South Carolina, an observer wrote, "They keep their Festivals and can tell but little of the reasons: their Old Men are dead."[28]

In all societies, healers who cannot heal are scorned. During the great epidemics of the seventeenth and eighteenth centuries, shamans and conjurers and medicine men, anyone who had claimed special powers or expertise, saw their authority diminished and respect for their ways curbed, although priests were still needed for the burials that now occurred with greater frequency.

Survivors often merged with remnants of other tribes, further diluting tribal rituals and lines of authority. Every merger required a process of assimilation not unlike that which characterizes patterns of international migration. Ways of speech, dress, government, and recreation were different. So, too, were marriage patterns. Even standards of honesty varied. Stealing from one's own tribe was a disgrace, but stealing from strangers, especially Europeans, was no moral offense at all. As one scholar describes the process, "When different Piedmont peoples came together . . . they had to redefine the meaning of 'stranger.'"[29]

Not surprisingly, refugees in flight from disease eased the strain of adjustment by merging with tribes most like themselves. Plains tribes sought out others that shared a familiar life-style, as did peoples in the hill country. However, language in treaties with whites stipulating separate and distinct names of merged tribes and distinctive burial plot patterns suggests that remnant groups did not forget who they were and sought to preserve their identities across generations. Tribes that had migrated and merged later tried to move back to familiar land where ancestors were buried or a sacred stone could be found. The fact that some sought return to the site where their village once stood is testimony to the powerful loss Native Americans felt as disease diminished their numbers and shattered their ability to retain tribal independence.[30]

Indian populations dwindled, then, because rates of mortality so far exceeded birthrates, leaving the remnants of once robust populations to evolve strategies for survival. As one historian has so aptly put it, "Old World pathogens served as the shock troops of the European invasion, softening up the enemy before the battalions of busy farmers waded ashore."[31] Whatever designs Europeans may later have had upon Indian lands and whatever excesses, even atrocities, Europeans committed on behalf of their own greed or religious convictions, the unintentional devastation of "virgin soil" epidemics exceeded them all.[32] At times, death claimed the hosts before the newly arrived, uninvited guests had even left their vessels, as in 1616 when a shipborne plague killed Indians living in New England's coastal villages. The Indians "died in heapes, as they lay in their houses." Because so few were left untouched by the disease, they were "not able to bury the dead, they were left for Crowes, Kites, and vermin to pray upon." To an English settler, it seemed "a new found Golgotha."[33]

The religious metaphor is more than coincidence. Not unlike some contemporary Christian fundamentalists who interpret the AIDS epidemic as a scourge designed by God to punish and purge homosexuals from society, both Indians and Europeans sought to comprehend the decimation in spiritual terms and perhaps even to take some reassurance in the belief that Indian health could not have been protected by human precautions because disease was an instrument of a higher will.

Indians well understood that fractures, dislocations, wounds, skin irritations, abrasions, and animal bites were the result of rational causes, human or natural. They were treated accordingly and often quite successfully. However, persistent internal ailments, where the cause was not readily visible, were frequently perceived as supernatural in origin. When their usual therapies did not cure, Indians turned to shamans who used incantations, amulets, prayers, dances, and noises produced by rattles or drums to purge disease. The remedy of choice depended upon the supernatural cause. These could include "sorcery, taboo violation, disease-object intrusion [imagined entry of a worm or snake into the body], spirit intrusion, and soul loss," or causes more specific to a particular tribal culture. Iroquois believed unfulfilled dreams or desires could trigger disease.[34] Other tribes attributed illness to spiritual offenses committed against nature such as the disrespect shown fire when it is extinguished with urine, spit, or offal. Many tribes also saw in witches and their evil spells a significant cause of disease.[35] And at least one scholar has argued that tribes involved in the fur trade saw new diseases and the inability of traditional cures and curers to cope with them as evidence of a shattered relationship with other life forms. These Indians blamed sickness on a conspiracy against them of the game animals they hunted and with whom they believed they had once lived in harmony. These tribes believed that only extensive hunting and slaughter of such creatures could stop diseases that were decimating their numbers.[36]

In 1639, when the Cherokee in South Carolina were struck hard by smallpox, believed by some to have been spread by African slaves or "Guinea-men" (as eighteenth-century trader James Adair called them) and later by unintentionally infected goods sold to the Indians by English merchants, Cherokee medicine men claimed to see a divine spirit behind the catastrophe. Not unlike later fundamentalist Christians, Cherokee healers saw a pattern of sexual immorality and taboo violation at the root of the problem. Cherokee society, the medicine men believed, had been contaminated by a rash of "unlawful copulation" by young married couples in the tribe who had disobeyed "in every thicket" certain ancient marital customs concerning cohabitation. Available accounts suggest a considerable number of suicides among the offenders, not as a belated public health measure or shame for their spiritual transgression, but because of the disgust (and perhaps guilt) inspired by the sight of their own pockmarked visages in the hand mirrors that, ironically, they had acquired in trade from Europeans.[37]

Significantly, Europeans also imposed upon these biological catastrophes a supernatural, or as they preferred, theological interpretation with self-serving political implications. As American studies scholar David Stannard has observed, Franciscan and Dominican priests in Latin America debated the theological meaning of the Indians' mass devastation by disease. Some comprehended the pestilence as aimed at the heathens for sinning and failing to accept Jesus Christ, while others lamented that such deaths were God's way of punishing the Spanish, by depriving them of their slaves.[38]

The vulnerability of the Indians to diseases against which white settlers had immunities led many faithful European Christians in North America to the conclusion that the suffering of the savage was a confirmation of God's approval of European colonization, a divine clearing of the path for the Lord's "chosen flock." The success with which missionaries spread Christianity suggests that the lesson was not lost on Native Americans, though even conversion did not slow the devastation of disease. Puritan minister Increase Mather wrote that when the colonists implored the Lord to clear the Indians from their way, God did not hesitate, "For it is known that the Indians were distressed with famine, multitudes of them perishing for want of bread; and the Lord sent Sicknesses amongst them, that Travellers have seen many dead Indians up and down in the woods that were by famine and sickness brought unto that untimely end. Yea the Indians themselves have testified, that more amongst them have been cut off by the sword of the Lord in these respects than by the sword of the English."[39]

During the 1616–17 epidemic that diminished the Native American population of New England, Increase Mather's son, Cotton Mather, a minister and sometime medical adviser, explained that the pestilence had been divinely provided to rid the forests "of those pernicious creatures to make room for better growth." Mather's confidence was grounded in the observation that despite close contact with tribe members, Europeans never so much as "felt their heads to ache."[40]

The English colonizing North America's Atlantic coast were not without compassion for their heathen neighbors, who were suffering God's judgment upon them. Plymouth Governor William Bradford recalled that in 1634 a smallpox epidemic ravaged local Indians, who feared the disease "more than the plague." Smallpox, or variola, is an infectious disease evidenced by fever and skin eruptions, which leave a scar that has given the disease its popular name. During this period, smallpox was often confused with syphilis, which became known as "large pox," as compared with the smaller pox of variola. Temperatures of more than 103 degrees were not uncommon, accompanied by a quick pulse, headache, vomiting, and pains in the loins and back; three days was the duration of these symptoms. On day three or four came the eruptions, often first on the forehead or scalp, but soon spreading over the body. Dark crimson spots rapidly developed into pimples that later burst. It is a common virus, the pathophysiology of which is still not fully understood. In 1987, ten years after the eradication of the disease, experts debated whether or not to destroy the last remaining samples of the virus that had ruined so many lives, samples "stored in little vials kept in freezer lockers at minus 94 degrees Fahrenheit, waiting only for some hypothetical future use."[41] Plans coordinated by the World Health Organization call for the eventual destruction of the variola virus in the 1990s after the genetic code of the virus has been broken by DNA sequencing.

Over three centuries before its eradication, William Bradford witnessed the horror of smallpox. In his history of Plymouth, the colonial governor recounted how Native Americans fell into a "lamentable condition as they lie on their hard mats, the pox breaking and mattering and running into one another, their skin cleaving by reason thereof to the mats they lie on. When they turn them, a whole side will flay off at once as it were, and they will be all of a gore blood, most fearful to behold. And then they being very sore, what with cold and other distempers, they die like rotten sheep." So weakened, the tribe members could not care for each other or even bury their dead. Too weak to gather kindling wood, their last hours were warmed by fires fueled with wooden eating utensils and their bows and arrows. Burning with fever, they crawled to water on all fours, dying on the way.[42]

With pride, Bradford recounted how the English braved the risk of infection to care for these heathen who appeared so susceptible to smallpox compared with themselves. He self-righteously credited correct belief with providing the immunity that these non-Christians did not share, "But by the marvelous goodness and providence of God, not one of the English was so much as sick or in the least measure tainted with the disease," despite close contact with diseased Indians, including the ministrations for which the survivors expressed much gratitude, according to Bradford.[43]

Since the early eighteenth century, well before Edward Jenner's inoculations with cowpox (vaccination), the inoculation of healthy individuals with matter from those infected with smallpox to prevent the disease (variolation) had existed as a medical procedure in Western Europe and among some non-Western groups even longer. Whether or not God was on the Puritans' side, European settlers may have owed their smallpox resistance to some practical precautions.[44]

The extent to which Europeans understood how disease could be spread and intentionally disseminated disease among the Native Americans to wipe them from desired lands remains unclear. There is some indication that European settlers may have sold Indians blankets that the settlers believed to be infected with smallpox. How often this primitive biological warfare was undertaken and with what rate of success cannot be estimated. It seems unlikely, though, that Europeans were knowledgeable enough in techniques of disseminating contagion to wreak nearly as widespread damage as did the uncalculated contact of two peoples with different patterns of immunity and insufficient understanding of disease to take effective steps to protect community health.[45]

European colonists never had sufficient control over diseases to be able to manipulate them to their own political ends. However, as historian Gary Nash has observed in his study of Indians, African slaves, and Europeans in early America, the slaves were inoculated against smallpox by their owners because they were valuable property. For Indians, on the other hand, inoculation was never even an issue.[46]

Native Americans were hardly alone in suffering the ravages of "virgin soil epidemics" triggered by the absence of acquired immunities. Newly arrived European immigrants as well suffered from diseases endemic to a locale and to which they lacked natural immunities. While Puritan families settled in New England's temperate climate and largely healthy environment, unmarried male indentured servants who contracted to work off their debt in the Chesapeake region in the mid-seventeenth century were often felled by malaria or the many other endemic diseases of the damp bay area. High mortality rates brought still more indentured servants to replace those who had died; these newcomers, too, lacked proper immunities. Only the gradual increase of a native-born population resistant to local disease strains and the switch to black slave labor immunized in Africa ended the terrible cycle of immigration and death among European colonists around the Chesapeake.[47]

The sons and daughters of European settlers might not have grasped the mechanics of "virgin soil" epidemics, but they certainly could see that recent immigrants had a much higher incidence of illness. Americans' understanding of disease and its prevention in the eighteenth century explains much about their reticence to extend an unqualified welcome to newcomers. The great majority of European-Americans believed the origins of disease to be an invisible poisonous vapor arising from organic matter in an advanced state of putrefaction. This miasmic theory can be traced to age-old observations that malaria was contracted by individuals in swampy, low-lying areas.[48]

A similarly odoriferous miasma seemed to emanate from the impoverished sections of towns and villages, where the poor resided. It was suggested that early urban dwellers and their families, often the most visible victims of both endemic and epidemic diseases, might experience an improvement in their overall health if the miasmas that enveloped them could somehow be purged. But not until the middle of the nineteenth century would American public health reformers and sanitarians seek to conquer disease through cleanliness, purifying water supplies, efficiently disposing of waste, ventilating tenements, and ensuring the purity and freshness of milk, bread, and other foods.[49]

A second position accepted by many at the end of the eighteenth century was that specific contagia were the exclusive causes of infections and epidemic diseases. But the machinations of these contagia, and their prevention, remained a mystery. Only with the discovery of germ theory at the end of the nineteenth century would it become clear how a specific microorganism can cause a specific disease in another living being.

A third position synthesizing miasma and contagion theories was dubbed "contingent contagionism" by historian of medicine George Rosen in his *History of Public Health.*[50] This nineteenth-century view maintained that disease was the result of contagia, whether specific or nonspecific, but that these contagia could not cause disease unless other conditions prevailed that either nourished the contagia or contributed to their transmission. Thus, a specific virus might cause a specific disease, but unless climatic conditions favored breeding, the virus was not infectious. The filthy, damp basements or sewer systems in cities were perfect places for some contagia to multiply.

This melding of the miasmatic theory with a growing fear of contagia served to bind the cause of disease directly to the sufferer. If immigrants chose to live in unhealthy surroundings and if they carried with them illnesses that would flourish in those conditions, then American communities must take steps to protect themselves, concluded many of the native-born. The solution was often selected exclusion by statute and legislated quarantine.

In 1700, Massachusetts Colony passed a detailed immigration law that selectively excluded the sick or physically disabled. From a required list of passengers kept by each vessel's master, authorities could determine, "No lame, impotent, or infirm persons, incapable of maintaining themselves, should be received without first giving security that the town in which they settled would not be charged with their support." Should the security be defaulted upon, the ship's master was required to carry such persons back to the port of embarkation. Prefiguring controversies over illegal aliens two centuries in the future, many colonists objected to the common practice of allowing such unwanted individuals to go ashore at shallow areas near the coast where there were no officials present to read lists of passengers.[51] By 1722, the law was amended to give the selectmen of any town, even towns not possessing a formal port of entry, the authority to demand passenger lists and security bonds and the right to refuse landing of undesirables. Though "the intent was only to prevent the importation of poor, vicious, and infirm persons," the law appears to have interfered sufficiently with commerce to require further changes. An amended act still required presentation of a passenger list but allowed entry to all immigrants possessing fifty pounds beyond the worth of their household goods and clothing, as well as to all able-bodied farmers, seamen, carpenters, laborers, and indentured servants, provided they were not individuals of "vicious habits."[52] The final revision of this colonial selective exclusionary legislation was in 1756, when a law was passed expressly forbidding admission to sick, impotent, or infirm persons "from foreign ports or other colonies," without the approval of a town's selectmen and with the usual security posted by the ship's captain. Massachusetts's busy ports rendered the colony and later the state especially sensitive to immigration and its hazards to the public welfare.[53]

After the American Revolution, reservations about immigration were codified in the various states under the new federal Constitution. No state ever entirely shut its doors to the foreign-born. Still, the hardships of life in the new country and the chaos of the Napoleonic Wars kept the number of immigrants to a quarter of a million between 1790 and 1820. In the early years of the nineteenth century's second decade, annual arrivals numbered about ten to twelve thousand; later in the decade, twenty thousand came each year.[54]

Neither the United States government nor the governments of individual states had a well-organized, systematic procedure for inspecting and selectively excluding unhealthy individuals, whether physically unfit or mentally enfeebled. Quarantine remained the first line of defense against disease from abroad as it had since the Middle Ages, when it was inaugurated by the Venetians. In July 1377, the municipal council of Ragusa on the Dalmatian coast had mandated a thirty-day period of isolation for those coming from places known to have experienced the plague. Eventually the period was extended to forty days, the origin of the term *quarantine*, which was derived from *quarantenaria*.[55] Likewise, in 1808, the Boston Board of Health had sought to reconcile the fear of diseases, especially smallpox and yellow fever, with the need for a free flow of commerce. It required that every year from May through October, all arriving ships from the West Indies, the Mediterranean, and other tropical ports be quarantined for three full days or until twenty-five days had elapsed since departure, whichever was longer. Other ports less busy than Boston's were generally only quarantined when there was known to be an outbreak of a contagious disease. Individuals sidestepped the pesky quarantine regulations by disembarking prior to their ship's entry into Boston harbor. Those caught were prosecuted.[56] Penalties varied but were frequently severe. To prevent violations of the quarantine, Rhode Island's law provided for the death penalty "without benefit of clergy."[57]

New York had been the main port under the British and remained so after independence. With the exception of a raging yellow fever epidemic in 1795 and a few milder outbreaks of disease such as measles in 1788 and influenza in 1789, it was a remarkably healthy place at the end of the eighteenth century. New York's colonial quarantine laws were reenacted in 1784. Bedloe's Island was the quarantine station with a port physician appointed. He was charged with inspecting all incoming vessels that he suspected of carrying infection and reporting conditions aboard to the governor or the mayor, who would then decide whether a ship should be quarantined and for how long. Immigrants quarantined for any length of time in the port of New York were generally sent to the Marine Hospital at Red Hook.[58] Because shipmasters sought to maximize profits by cramming into their holds as many immigrants as possible, illness and even death was quite common. Typhus and diarrhea often ravaged tightly packed passengers to the New World. Friedrich Kapp, one of New York's commissioners of emigration, wrote in 1870 that "the lower deck of an emigrant vessel as late as 1819, was no better than that of a slaver or a coolie ship; the passengers were just as crowded, and just as little thought of, as those unfortunate beings from Africa or China."[59] Kapp explained that the steerage deck was only five or six feet high at most, while the lower or orlop deck used for sleeping accommodations "was not much better than a black hole, too bad to shelter cattle."[60] Consequently, there was "a high mortality." According to Kapp, "Ten deaths among one hundred passengers was nothing extraordinary; twenty per cent was not unheard of; and there were cases of 400 out of 1,200 passengers being buried before the ships left port."[61] In New York, the Marine Hospital filled to capacity when several ships docked at the same time.

The severity of shipboard epidemics of typhus, cholera, and smallpox moved to action a Congress cautious, even reluctant, to enter legislative waters traditionally patrolled by colonial and later state legislatures. The Act of March 2, 1819, "An act regulating passenger ships and vessels" (3 Stat. 488), limited the number of passengers a ship might carry to two for every five tons of the ship. It also initiated the collection of vital data on the foreign-born, requiring masters of arriving ships to furnish "a list or manifest of all passengers taken on board," which included each passenger's and crew member's name and national origin. Now port officials would know who was coming and from where.[62]

Later passenger acts improved shipboard health by specifying the amount of space per passenger that must be allowed and the ventilation, "privy or watercloset" facilities, and cooking accommodations that must be provided passengers.[63] The Act of March 3, 1855 (10 Stat. 715), substituted for earlier acts a nineteen-provision code for the health and safety of ship passengers from abroad.[64] The federal government had acted, but there was little political support for expanding federal powers over immigration. It was left to the states to consult ship manifests and ship physicians, and to worry about who was coming and in what condition.

There is ample evidence that in the late eighteenth and early nineteenth centuries, the inexplicable devastation of epidemic diseases caused Americans to stigmatize whole nations of the foreign-born as disease carriers. At times, whether or not immigrants were to blame for an epidemic even became enmeshed in domestic politics, with the origin of deadly fevers fueled by partisan fervor.

A century before Walter Reed learned that a female mosquito, *Aedes aegypti*, was the vector that passed yellow fever to human beings and that mosquito control was a public health imperative, Philadelphians blamed the Germans. By 1793, Philadelphians had enjoyed a thirty-year respite from yellow fever epidemics, after an outbreak in the earliest days of the city when William

Penn was alive. Then some called it the "Barbados distemper," believing its origins to be in the Caribbean and its arrival an unwelcome by-product of commerce and migration.[65] In 1793, native Philadelphians renamed it "Palatine fever," because so many immigrant victims of the disease originated in that region of the German states. Philadelphians responded with the few public health measures available: quarantine, appointment of a port physician to ensure compliance, and construction of a new pesthouse on Fisher's Island in the Delaware River just below the city, a place officially named Province or State Island, but more popularly referred to as Mud Island.[66]

Quarantine precautions were no match for yellow fever in 1793. Only the frost ended the epidemic. The "Palatine Fever" had passed, but scars inflicted during the crisis suggest a secondary infection—nativism.

Beginning with sociologist Erving Goffman, social scientists have identified the abomination of the body as the most essential form of stigma. Bodies that are stigmatized by association with disease seem especially threatening because disease-causing contagion cannot be seen by the naked eye or easily eluded. There appears to be no way to shun a disease other than shunning the person who has already become its victim. Knowing that the stigmatized victim is from another place brings with it the reassurance that one's own body and surroundings are inherently healthy and would remain so were it not for the presence of the stranger.[67] Eighteenth-century Philadelphians could not accept that yellow fever was a product of the city's own deadly miasmas and instead traced the scourge to imported contagion. In 1793, while some Philadelphians were content to blame the Germans, party politicians looked to Haitian immigrants. The yellow fever epidemic became a broader political conflict.

In early August 1793, over two thousand French refugees from the black slave revolution in Haiti arrived. There were blacks and whites among the newcomers, but unlike the French royalists who had emigrated after the revolution of 1789, many of the white Frenchmen in the group were radicals and moderates. They departed only when their lives were at risk after black slaves assumed leadership of the revolution. Jeffersonian Republicans were often sympathetic to these Frenchmen. The pro-English Federalists generally were not.

Just as a later generation of Haitians in the 1980s would be stigmatized by being classified as a high-risk group for AIDS, these Haitians were held responsible for the yellow fever epidemic by Federalists who deplored the political affinity and potential electoral support that these newcomers might offer the Jeffersonians. As historian of medicine Martin Pernick observed, "In general, Republican physicians, including the refugee doctors, believed the fever to be local. The 'importationists' were almost all nonpartisans or Federalists."[68] Pernick also notes the irony that both sides were correct. As Walter

Reed's research would later demonstrate, "a yellow fever epidemic requires both locally bred mosquitoes and an initial pool of infected persons, such as the exiled Haitians." Mosquitoes picked up the infection from the Haitians and carried it to uninfected Philadelphians, whom they bit.

As the epidemic worsened, those physicians who attributed it to domestic causes denounced the unsanitary conditions and unhealthy climate of Philadelphia, while their opponents denounced the policy of admitting disease-bearing refugees. The eminent physician and American revolutionary Dr. Benjamin Rush, a loyal Republican, regarded yellow fever as the product of poisonous "miasmata" from local swamps and "effluvia" from the filthy docks of the port. Neither Rush nor most other Philadelphia physicians allowed politics to dictate their science, but there can be little doubt that politics and yellow fever epidemiology were deeply entwined. Philip Freneau, recruited by the Jeffersonians, was not a physician but an articulate satirist who spoofed the human comedy in verse, managing to plug the localist epidemiological view:

Doctors raving and disputing,
Death's pale army still recruiting—
What a pother
One with t'other
Some a-writing, some a-shooting.

Nature's poisons here collected,
Water, earth, and air infected—
O' what a pity
Such a City,
Was in such a place erected![69]

Freneau left little doubt that a city where "Nature's poisons" were collected was the culprit in the yellow fever epidemic, not the foreign-born. However, there was ample coverage for other points of view. One of Freneau's fellow Republicans, newspaper editor Benjamin Franklin Bache of the *General Advertiser*, believed that the disease was imported, but from the British West Indies rather than from Haiti. Enraging Federalist Anglophiles, Bache referred to yellow fever as "a present from the English."[70] Fiery Republican newspaper editor Mathew Carey of Vermont, an Irish immigrant himself, believed the disease to be of foreign origin and allowed that the scourge might even have come from his own homeland.

The dispute was more than an abstract disagreement on epidemiology. It was used to justify immigration policies that were highly partisan in motive. Federalists cited the threat to the public health from abroad in their demands for the quarantine or even complete exclusion of the French radicals hoping to escape Haiti, believing that by so doing they were depriving their Jeffersonian opponents of allies and potential votes. Meanwhile, Republicans saw the Federalist enthusiasm for curbs on immigration and commerce with Haiti as a plot to shatter Republican merchants' valuable trade with the West Indies.

There was even the charge of a politically motivated cover-up. Dr. William Currie, a Federalist who believed the disease to be of foreign origins, charged that yellow fever had come to the United States aboard a French privateer, *Sans Culotte*, which had been captured as a prize in July 1798 during the height of political tensions with France. Currie and his colleagues charged the French and port physician, Dr. James Hutchinson, with concealing illness discovered among the crew. Federalists, perhaps realizing the potential for a political bonanza, encouraged antagonism toward the French in Philadelphia. They called for boycotting of French physicians, while rumors spread that other new arrivals were contaminating wells, using a kind of germ warfare preparatory to a French invasion of the United States. The possibility of mob violence against French refugees caught in the middle heightened as Federalists and Republicans engaged in partisan bloodletting.

The political controversy surrounding yellow fever also extended to the broader issue of public health precautions. Federalist Levi Hollingsworth secured the agreement of the College of Physicians that the city was as healthy as could be, a pronouncement that has appeared ludicrous to historians, who have observed that the city lacked a sewage system, a fresh water supply, and an institutionalized system for regular disposal of garbage. However, in fairness to Philadelphia's medical elite, their city was no exception; most other cities in the early republic were lacking in one or more of these essentials. Thomas Jefferson, defending the theory that the epidemic was a domestic product and taking the opportunity to echo his general suspicion of urban centers as detrimental to the perpetuation of democracy, confidently predicted that "yellow fever will discourage the growth of great cities in our nation."[71]

The brilliant author of the Declaration of Independence was wrong. Great cities, crowded with workers, many of them foreign-born, were to be a significant factor in the nation's history. Dr. Benjamin Rush was perhaps a better prognosticator when he made the general observation on the events of his own day that, "Loathsome and dangerous diseases have been considered by all nations as of foreign extraction."[72]

Whatever the origins of yellow fever, for many the reason why Philadelphia was devastated remained as much a matter of theology as the questions surrounding smallpox had been for the Puritans of Massachusetts Bay in an earlier era. Numerous Philadelphians believed that epidemics generally were expressions of divine displeasure. Therefore, they were not surprised that cholera and typhoid epidemics seemed to disproportionately affect the poor, many of whom were immigrants already regarded by the native-born as lazy and morally suspect.

Yellow fever was different, being spread, as one scholar wittily commented, by mosquitoes who observed "republican egalitarianism" in choosing whom to bite.[73] Still, the religious rationale for the epidemic was recast in political terms. While Republicans suggested that God was punishing the pride and conceit of domestic leaders, their opponents found the French to be morally culpable. Ministers inclined to the Federalist view, such as the Rev. William Smith, denounced from the pulpit French political refugees from Haiti, whose "wild principles and restless conduct" suggested an impatience of "all rule of authority."[74]

The French refugees from Haiti vehemently denied both the charges of immorality and claims that they were vectors of yellow fever in Philadelphia. One physician among the newcomers was especially vocal. Dr. Jean Deveze, thirty-nine, had learned medicine at Bordeaux and at the age of twenty-two emigrated to the Caribbean, settling in the Northern Province of Santo Domingo at Cap François, where he caught and survived yellow fever. He then returned to Paris for more study. By 1778, he was back in the Caribbean as chief surgeon for the national troops of the Northern Province at Cap François. There he founded his own hospital, which he administered even as he served the French military, catching yellow fever a second time and again surviving. When the revolution turned bloody, he barely escaped with his life. His possessions seized by British privateers who had intercepted his ship, he managed to reach Philadelphia with little more than his life just as the fever that had twice felled him was erupting all over the city. Deveze considered the fever to be contagious and used what some called the "French cure" of quinine and stimulants. At the urging of Stephen Girard, a French merchant in Philadelphia, Deveze treated the sick during the epidemic. He dismissed as ridiculous the notion that French refugees were guilty of infecting Philadelphia with yellow fever or that the disease was a punishment for any immorality. He denounced such charges as based in popular ignorance and the partisan interests of those who knew better.[75]

The frosts of November ended the yellow fever epidemic but not the debate over its cause and possible public health precautions to prevent future pestilence from plaguing Philadelphians. At the beginning of the nineteenth century, most Philadelphians firmly believed that disease came from abroad, spread by the bodies and/or breath of immigrants. Therefore, high on the list of public health precautions was strengthening the city's quarantine capabilities. State laws in 1806 and 1818 created a Board of Health for the port and city of Philadelphia and granted the governor authority to appoint board members and staff. Among the steps that the board took was the creation of a quarantine station, the Lazaretto, on the Delaware River about ten miles south of Philadelphia. It was an extensive facility, occupying ten acres and including living quarters, large kitchens, stables, and a three-story, two-winged hospital building.[76]

During the summer months, when it was believed epidemics were most likely to occur, the Lazaretto physician examined passengers and seamen on all arriving vessels, while the quarantine master inspected each ship and its cargo for signs that the ship posed a health menace to Philadelphians. At any sign of sickness, the two officers would detain the ship and remove the sick to the hospital. Only then would the vessel be issued a health certificate. During the rest of the year, October to May, the port physician, who had a desk at the Board of Health's city office, bore the responsibility for detaining, inspecting, and certifying incoming ships.

The Philadelphia Lazaretto and similar quarantine stations stemmed from the belief that whatever other public health measures were taken to purify the water, remove the sewage, and collect garbage, precautions were incomplete without provision to stop disease from abroad. Even as many western Europeans and Americans were abandoning their belief in miasmas as the origin of disease and concentrating upon dirty, unhygienic conditions as the source of sickness, the foreign-born continued to be perceived as the most significant public health menace. It was argued that, after all, it was not the wealthy and successful of other nations who chose to emigrate, but those who clung to the bottom rungs of society's ladder and had already been living lives of physical deprivation and moral degradation.

In the decades following the American Revolution, as immigration accelerated in response to economic expansion, the equation of disease and the foreign-born increasingly resulted in the stigmatization of one ethnic group or another. Eager for new labor, Americans nonetheless remained wary of the "breath of other people." Struggling to understand the origins of disease, the nation simultaneously sought to develop government mechanisms to stop the spread of disease. Reliance on quarantine, coupled with an innate distrust of foreigners, suggested to the native-born that regulation of immigration was crucial to safeguarding the health of the nation. Such regulation more and more was considered the province of state—and eventually federal—government. But formalizing exclusion and restriction served as a self-fulfilling prophesy, codifying the connection between immigrant and illness. Disease and difference—this was the lesson learned in the early years of European settlement of the New World.

2

"A Scourge, a Rod in the Hand of God": Epidemics and the Irish Mid-Century

A silent traveler from abroad, cholera sailed into New York and other East Coast ports in 1832 even as an increasingly large number of foreign-born were arriving. Physicians were unable to offer clear explanations of the scourge and few were sufficiently confident to publicly dismiss fears that the pestilence was somehow a divine punishment being visited upon these newcomers, one from which many innocent native-born victims would suffer, as well. Whether or not stigmatization of the foreign-born satisfied an immediate demand for answers during the cholera epidemic, it raised long-term questions about the origins of disease, the adequacy of public health institutions to protect urban populations being suddenly and rapidly swelled by the foreign-born, and immigrants' need for health care institutions tailored to their unique cultural perspectives.

Caused by the bacterium *Vibrio cholerae,* cholera is an acute infection with symptoms of diarrhea, vomiting, muscular cramps, dehydration, and collapse.

It is contracted by the ingestion of water or foods contaminated by the excrement of infected persons. The incubation period is seventy-two hours, with a crisis several days after the initial symptoms are experienced. If the dehydration and chemical imbalances are not treated, the weakened victim suffers collapse and death. Modern medicine can help. Physicians use antibiotics to destroy the bacteria and shorten the illness's duration, but rehydration remains the most important measure. Cholera is not endemic to the United States, and public health precautions, especially with respect to sanitation, are generally the best preventive. Such public health measures were uneven at best in antebellum America when cholera made its appearance.

The cholera epidemic of 1832 had originated in Asia. Quarantines and other precautions proved inadequate to the task of prevention. After claiming its victims in Europe, the disease spread westward across the Atlantic. In East Coast cities, large numbers of foreign-born were struck down, especially in the poorest quarters, where immigrants from Ireland were concentrated. Many Americans perceived a link between these two unwelcome guests, cholera and the Irish.[1]

Why the Irish? In part, because so many of them headed for the United States. Between 1820 and 1830, America's busiest port, New York, received 92,884 newcomers—81,827 of them from England and Ireland and another 7,729 from Germany.[2] Later, between 1840 and 1860, hard times would send an additional 1.8 million Irish hurtling toward North America. The Irish were readily visible because they tended to concentrate in cities such as Boston, Providence, and especially New York. A great majority who arrived in New York City migrated elsewhere within three to four years, only to be replaced by others. By 1860, 203,740 foreign-born Irish lived in New York City; one of every four New Yorkers was born in Ireland.[3]

In American cities the native-born, largely of Protestant English ancestry, reviled the newcomers, sometimes for their Irishness, at other times for their Catholicism, and increasingly after 1832 for cholera.[4] Such collective aggression, stoked by the intensity of epidemics, was hardly unprecedented. In medieval and early modern Europe, Jews had been popular scapegoats blamed for the onset of disease.[5] While such libels were far less common in the United States than they had been in Europe, the very openness, even rootlessness of American society and life nurtured fears of subversion when sudden, unsettling changes in community life occurred.[6] These anxieties were fed by a general belief in the unsullied quality of the American continent, an Eden of pure air and water that had nurtured a pioneering race of remarkable physical fortitude who had cleared the forests and founded the republic.[7] As one historian of medicine has observed, our national consciousness of a heroic tradition—what some scholars have described as an American exceptionalism—extended beyond political institutions to the very health and fitness of the population. Thus, health was "regarded as indigenous to our soil, disease as an odious alien."[8] When the 1832 epidemic struck, Americans did not hesitate to blame the victims for their disease.

Living in run-down shanties and tenements, Irish immigrants felled during the 1832 cholera epidemic were believed by many of the native-born to have died of individual vices typical of their group, a divinely determined punishment that might be spread to those undeserving of such retribution. And what were these vices? Intemperance and a lack of cleanliness were the character flaws most responsible for the sickness suffered by the "low Irish" who were crammed into city slums, according to a New York Board of Health report.[9] Irish-haters such as journalist Hezekiah Niles, editor of the *Niles Weekly Register*, blamed the Irish for taking the jobs of native-born workers, adding to the public's burdens by cramming the urban almshouses with their destitute, opening saloons to lure thirsty and gullible workers, and perverting local politics with their smooth-talking ward healers.[10] Was cholera bred in bars and political clubhouses? Anti-Irish nativists were only too willing to answer with a grimace and a nod.

That so many of those Irish stricken in New York were Roman Catholic did not escape public notice. How could it? A Second Great Awakening was sweeping across the states of the Northeast and Midwest, arousing Protestant sensibilities; by 1832 evangelical preachers had already penetrated the mountains of Kentucky and deep into the South as well. Charles Grandison Finney and other revivalists scorched the earth of western New York with the heat of their fire-and-brimstone sermons, leaving behind a "burned-over district" whose embers glowed for decades afterward, lighting the way for abolitionists, temperance advocates, and women's suffragists.[11] Whether or not they were drawn to reform movements, some graduates of revivalism, such as historian Daniel Walker Howe's American Whigs, saw Protestantism as part of a "cultural matrix" identified with political freedom. Conversely, the Church of Rome seemed no friend of free institutions and individual liberties. Whigs and others who equated the Catholic Church with rigid hierarchy and oppression were apprehensive that Irish Catholic immigrants would be the papacy's footsoldiers, aiding and abetting the Roman Catholic Church's intended counterreformation, its subversion of American Protestantism.[12] Authors of pulp fiction were already enjoying a brisk trade, depicting Catholics and especially Catholic clergy as morally impure.[13]

Not all Protestant clergymen agreed on the details of cholera's social construction as the Irish Catholic disease. Some saw the disease as a literal result of human sinfulness. In a sermon preached on a day of "public fasting, humiliation, and prayer," one minister declared that cholera was "not *caused* by intemperance and filth, in themselves, but it is a *scourge,* a rod in the hand of God."[14] Others, of a more liberal bent, suggested that cholera was not a scourge sent by God with the intention of punishing people but rather a disease that existed in nature triggered by violations of natural law. Drinking or eating to excess or failing to keep oneself clean were failures to conform in body and soul to the divine plan, a plan understood to be distinctly Protestant in nature.[15]

Most native-born Americans did not suggest that the Irish were inherently flawed. Rather, the influence of Roman Catholic priests in the Irish community was attributed by many American Protestants to environmental factors such as poverty, oppression, and misgovernment. Catholicism thrived, Protestants thought, because Irish minds were limited in their imagination by a lack of education and economic independence. The church merely benefited from Irish ignorance and desperation. Similarly, the Irish inclination toward drink was seen as a moral failing derived from the need to escape the misery of an impoverished condition. In short, Irish immigrants were often regarded not as moral profligates, the wages of whose sins were disease, but victims of bad circumstances, whose shortcomings could be cured.[16] In one of his nativist tracts, the sire of the telegraph, Samuel F. B. Morse, despaired that so many of the newcomers were Catholics, who would "obey their priests as demigods, from the habit of their whole lives" and were unlikely to be disabused of their illusions simply because of their arrival in the United States.[17]

Non-Irish immigrants, especially the numerous German immigrants, some of whom were also Catholic, appear not to have shared in the blame for cholera. In at least one instance, German immigrants turned the tables and blamed municipal authorities for conditions in their neighborhoods that bred disease. In 1854, while the Irish on the New York side of the Hudson River were fending off nativist characterizations of them as disease breeders, German immigrants on the New Jersey side were casting blame on their hosts. Over 10,000 German immigrants lived in the Hill section of Newark, many falling victim to cholera. According to a newspaper report, a "Committee of Germans" cited stagnant pools of water in their community, the result of poor street construction, as incubators of illness. Seizing the offensive, the Germans claimed anti-immigrant prejudice as the reason for municipal neglect. A fuming member of the German delegation claimed he heard a Newark street commissioner express indifference as to whether or not the "Dutchmen" died in "one heap." Such sentiments made the indignant Germans feel that, aside from the value of their votes and taxes, newcomers were no more than "d——d Dutch" to their neighbors.[18]

The famine of the mid-1840s loosed yet more Irish Catholics across the Atlantic. Those leaving from the English port of Liverpool were required to stand a medical inspection conducted by British government physicians. A doctor who made the crossing many times on American ships wrote a letter to the *New York Times* in 1851 describing at best a perfunctory exam, one that bordered upon charade. Each passenger "inspected" by the doctors was required to stick out his or her tongue, while overburdened physicians barely glanced up from their papers. Passengers infested with vermin or in the early stages of smallpox were often overlooked. On busy days when several ships were departing, as many as a thousand travelers might pass before a team of two doctors who, standing behind a small window, asked each emigrant whether he or she was well, glanced at tongues, and stamped tickets speedily.[19] Complaints did reach the ears of British emigration commissioners urging more thorough inspections to prevent shipboard disease. However, the commissioners dismissed these complaints, arguing that most shipboard disease was the result of changes of diet, seasickness, fear, and other such conditions that compromised the body, making it vulnerable to disease.[20]

Examining physicians freely admitted that emigrants were not stripped and that, beyond taking pulses and examining tongues, only hands were closely scrutinized in the search for symptoms of serious ailments. Rejections were for typhus fever, childhood illnesses, or infectious eye diseases such as ophthalmia. The physicians defended their exams, noting that they differed little from those given army inductees and convicts, where prisoners were lined up in the jail yard as a doctor walked before them in search of those who might be ill. When asked if immigrants suffering from consumption were rejected, a physician said no, because it was not an infectious disease. He was not a fool, merely as mistaken as other physicians of his day.[21]

Medical inspection of emigrants certainly cannot be characterized as thorough, but it would be a mistake to dismiss such exams as egregiously superficial. By the 1840s and 1850s, the state of medical knowledge, especially that relating to infectious disease, remained thin. Diagnostic techniques were few and highly impressionistic. Acute observation and the physician's judgment were still the most reliable instruments at hand.[22] Equally significant might be the method of physician payment: a pound for every hundred persons inspected. By 1854, Liverpool had three full-time positions and had switched from piecework to a fixed annual salary of 400 pounds.[23]

A second cholera epidemic struck in 1849. It appeared to spread north from Philadelphia to New York and Boston. Although the figures were skewed by wealthier, native-born Americans who tended to avoid hospitals when ill, a tabulation of hospital statistics for six cities (Cincinnati, New York, Buffalo, Brooklyn, Boston, and New Orleans) compiled by historian Charles Rosenberg reveals that 4,309 of 5,301 patients during the height of the 1849 cholera epidemic were listed as foreign-born. More than 40 percent of those dying of cholera in New York were of Irish birth.[24] As in 1832, the connection seemed obvious to many. Were not the Irish homes the poorest in drainage, least well ventilated, and most crowded? And was not poverty a product of moral shortcomings?

Fewer native-born Americans in 1849 than in 1832 viewed cholera as a direct act of retribution upon the sinful and spiritually unworthy Irish Catholics. While many continued to see a social dimension to the disease, the public was less interested in deciphering God's judgment and more directed at eliminating the filthy conditions that some blamed for the disease and blocking the entry of those already diseased. As Americans gradually abandoned their belief in miasmas for the sanitarian theory, it seemed more important than ever before to exclude aliens whose baggage included disease. Increasingly, port quarantine procedures seemed insufficient and medical inspection of individual arrivals a desirable complement to quarantine.

In May 1855, an old fort at the tip of Manhattan, Castle Garden, was leased, renovated, and reopened on August 1 as a state immigration reception center: New York's Emigration Landing Depot. Quarantine procedures would now be supplemented by the individual examination of immigrants. Every vessel bringing immigrants had to anchor at the quarantine station, six miles below New York City. There, a New York State emigration officer boarded to ascertain a count of passengers, deaths during the voyage, the degree and kind of illnesses suffered during the trip, and the overall cleanliness of the vessel. A report was sent to the general agent and superintendent at Castle Garden and the boarding officer remained on the ship as it steamed up the bay to ensure that no one from the ship contacted anyone on shore prior to the authorized disembarkation of emigrants. This prohibition was also designed to inhibit the activity of runners and swindlers who accosted arriving immigrants with schemes and scams.

Once the ship was anchored near the depot, a New York City policeman detailed to Castle Garden assumed authority. Passengers were transferred to the jurisdiction of the Landing Department. After a customs inspection of luggage, immigrants and their belongings boarded barges and tugs that took them to the Castle Garden pier. There, passengers were examined by a state medical officer to discover "if any sick have passed the Health authorities at Quarantine (who are thereupon transferred by steamer to the hospitals on Ward's or Blackwell's Island), and likewise to select all subject to special bonds under the law—as blind persons, cripples, lunatics, or any others who are likely to become a future charge." After examination, the migrants were directed into a rotunda in the center of the depot, where they registered their names, nationalities, former places of residence, and intended destination.[25]

By the mid-1850s, inspection procedures similar to those at Castle Garden were instituted by all states with busy ports of arrival. Fear of infectious disease from abroad had triggered entirely new processes and procedures in public health policy. Still, many Americans continued to rant against the threat of foreign-borne disease. A nativist physician, Dr. Samuel Busey, wrote in 1856, "In the cities, those direful and pestilential diseases, ship fever, yellow fever, and small pox, are almost exclusively confined to the filthy alleys, lanes, and streets, and low, damp, filthy and ill-ventilated haunts, which are exclusively tenanted by foreigners."[26] Others, as fearful of immigrants as Busey, unsuccessfully pressed the federal government from 1859 to 1866 to at least assume quarantine responsibilities, even if medical inspection of individual immigrants remained a state prerogative. Republicans preferring that health matters be under state jurisdiction, and not subsumed under federal authority over interstate commerce, joined Democrats, who remained suspicious of any federal intrusions upon states rights even after the Civil War, in the losing coalition.[27]

As for the Irish, not until 1866, when a third major cholera epidemic hit the East Coast of the United States, did they escape condemnation as health menaces, though many Americans continued to regard the Irish presence in American cities as an unfortunate contribution to urban degradation. By then most physicians were sufficiently confident of a link between dirt and disease that most Americans accepted their construction of cholera as a social problem rather than a moral one.[28] As historian Charles Rosenberg has observed, by 1866 Americans had learned that "the soul could be made healthy while

THE KIND OF "ASSISTED EMIGRANT" WE CAN NOT AFFORD TO ADMIT.

Members of the New York Municipal Board of Health (foreground) armed with a bottle of carbolic acid to fight off cholera arriving in the port of New York, 1883. (*Puck*)

the body which housed it was diseased. Chloride of lime, not fasting, brought deliverance from cholera; the cure for pauperism lay in education and housing, not prayers and exhortation."[29] Consequently, much of the social energy once channeled into revival meetings was now poured into a fresh institutional direction, the creation of permanent boards of health to tackle health problems with moral implications. In New York City, more urban dwellers turned to their spanking new Metropolitan Board of Health than to the pulpit for aid and explanations. Similar structures arose in other states and cities. No longer would governments scramble to react to health crises already in progress. Future epidemics would be met by preexisting public health agencies prepared with data and expertise. Boards of health now shared in the moral authority once reserved for the church alone because they appeared to be laboring toward the same end, a moral and spiritual health that could not be separated from the body's well-being. Moral passion was not absent or wasted during the 1866 epidemic, merely redirected.

Attention now turned from the subversive influence of Catholicism to the need for clean water supplies, effective drainage and waste removal, and new regulatory powers for health officials. In Louisiana, where the Irish had also been blamed for disease, a State Board of Health, the first state board, was created in 1855, largely in response to yellow fever epidemics. Quarantine supervision in New Orleans was its main responsibility.[30] However, other boards such as Massachusetts's in 1869 took a broader perspective on public health. While there is little evidence that the relative mildness of the 1866 cholera epidemic was the direct result of Board of Health plans or precautions, New Yorkers of the era thought otherwise, singing the praises of public health and their new administrative creation. Immigration and the health conditions in immigrant neighborhoods remained matters of concern for state and municipal boards of health, but discussions of God's disapproval of newcomers' morals was generally not high on the new state and municipal boards' agendas. Community disapproval seemed sufficient.[31]

Though there is real doubt that the Irish were any more susceptible than other immigrants to illness, it is certainly true that their plight more often led to institutionalization where they could be counted. Natives of Ireland made up 53.9 percent of New York City's foreign-born population in 1855 but 85.0 percent of all patients of foreign birth at the city's Bellevue Hospital.[32] New York officials juggled statistical data in an effort to appreciate the dimension of the problems facing newcomers and their hosts. However, distortions in the data could shed false light on the extent of the plight of the Irish, as was the case with mental illness.

If not an epidemic disease such as cholera, mental illness nevertheless seemed to native-born Americans every bit as pervasive among the Irish. In the mid–nineteenth century, insanity was defined by aberrant behavior. Discussions of etiology were protean and quite general in character, reflecting broad social perceptions and prevailing religious, moral, and cultural values. Most physicians believed that insanity grew out of a violation of physical, mental, and moral laws that, when properly understood and obeyed, resulted in the highest development of the white Anglo-Saxon Protestant race, indeed of all civilization. Abnormality in the population might be traced to the increasing rigors of civilization, the immigration of those already on the road to degeneracy, an inherited predisposition to mental illness, or external elements—an endless list that included alcohol, sexual excesses, improper nutrition, grief, or anxiety.[33]

According to published data of the era, Irish immigrants appeared to have the highest rates of illness, generally, and the highest rates of insanity, specifically. From 1849 to 1859, three-fourths of the admissions to New York City's lunatic asylum on Blackwell's Island were immigrants; two-thirds of these were Irish. Trying to account for such skewed figures, the asylum's resident physician speculated that, "Either the ratio of insane is very much less among the natives, or they are kept at their homes. Probably the first supposition is true, and this may arise in part from the shipment of the insane from Europe during a lucid interval."[34] It is more likely the second supposition was on point. Clearly, the native-born preferred taking care of sick relatives at home, which left a preponderance of foreign-born in existing institutions. In 1854, physician Edward Jarvis calculated that, although only 43 percent of the poor insane ended up in a state institution, almost everyone from among the foreigners did.[35] The native-born, especially those who were more affluent, sought to avoid exposing loved ones to the poor conditions in mental hospitals, where they might associate with society's bottom rung, a category that always included the foreign-born.[36]

Although there is little doubt that Irish immigrants in New York City and elsewhere were institutionalized while the native-born deranged were treated at home, rates of institutionalization were not disproportionate to the size of the Irish immigrant population. Improved techniques of statistical analysis suggest that mid–nineteenth-century data on the admissions of immigrants to mental hospitals were greatly distorted.[37]

Data inaccuracies notwithstanding, the popular perception at midcentury was that Irish immigrants had a greater propensity to mental illness than their hosts. Patterns of aberrant behavior diagnosed as insanity in the 1840s seemed quite common among newly arrived immigrants. The governors of the New York Alms House attributed the condition to "privations on shipboard," or "the changes incident to arriving in a strange land," and "to want of sufficient nourishment." At least some of these conditions were judged to be temporary. In 1854, thirty-five of the hundred patients admitted to New York's Lunatic Asylum and chargeable to the state's commissioners of emigration had been in New York City less than one year. Many of these were diagnosed as temporarily deranged and were later pronounced recovered and released.[38]

Almshouse physicians in port cities reported that both Irish men and women were particularly susceptible after immigration to schizophrenia. One New York almshouse physician, perhaps looking through the gender kaleidoscope of his era, claimed that young women arrivals seemed especially prone to mental illness. He attributed it to "the combined moral and physical influences of their leaving the homes of their childhood, their coming almost destitute to a strange land, and often after great suffering."[39] Irish women appeared to be more vulnerable to mental disease than women of other groups if judged by the large number found in hospitals and asylums. However, the high rate of institutionalization may be accounted for by the fact that many Irish women emigrated alone and had no one to care for them in time of illness. A New York physician treating patients on Blackwell's Island commented upon the large number of Irish women there, but he attributed their illnesses to environmental strains rather than to a predisposition to mental instability. Other observers suggested that Irish women, who often married late and came to America alone in search of work as domestics or teachers, seemed more homesick for the parents and siblings they left behind than did Irish males. (Interestingly, more recent scholarship suggests that the males may have been even more vulnerable than the females.[40]) Still, well into the twentieth century, over two-thirds of institutionalized Irish insane were women and the Irish continued to be the largest immigrant group in American insane asylums.[41]

Once institutionalized, the Irish suffered from anti-Catholicism on the part of many Protestants in charge of mental institutions. The trusting relationship between physician and patient necessary for therapy was disrupted by unbridgeable ethnic differences and prejudice. Common religion, values, and culture were absent and even sympathetic physicians found themselves at a loss. One physician confessed his inability to approach his foreign-born patients "in the proper way." He puzzled, "Modes of address like those used in our intercourse with our own people, generally fall upon their ears like an unknown tongue, or are comprehended just enough to render the whole misunderstood, and thereby excite feelings very different from such as we intended." Another superintendent regretted that his personnel could not easily gain the trust of immigrant patients, "for they seem jealous of our motives" and embarrassed by language difficulties, together significant obstacles to successful therapy and recovery.[42] Historian David Rothman argues that the very foreignness of the immigrant mental patient contributed in part to undermining "the original postulates of the asylum movement" that sought to "re-create both in and out of the asylum, a well ordered, balanced, harmonious, and ultimately homogeneous community." Homogeneity was impossible in an ethnically pluralist nation. However, dangers posed by the mentally ill, perhaps even criminally insane, from abroad lent support for "straight-forward incarceration" as legitimate use for the asylum in protecting society from these less than desirable newcomers.[43]

Some professionals believed that the inner strength of those who undertook a journey to a foreign land predisposed the mentally ill among them to benefit from treatment. But again, cultural imperative influenced medical assessment. Germans were praised for possessing "a healthy and elastic mental constitution." They seemed "docile and affectionate" to their doctors and quick to express gratitude. The reverse was said of the Irish, who were described as less intelligent, resourceful, refined, or self-possessed than the native-born or patients of other groups. Ralph L. Parsons, superintendent of the New York City Lunatic Asylum on Blackwell's Island, described many of the Irish he saw as "persons of exceptionally bad habits." He believed the Irish patients to be of a "low order of intelligence, and very many of them have imperfectly developed brains. When such persons become insane, I am inclined to think that the prognosis is peculiarly unfavorable."[44]

In 1854, an Irish female was described as "noisy and troublesome," having nymphomania and exposing herself, being "vulgar and obscene" as well as "noisy and destructive."[45] The seeming prevalence of insanity among the Irish contributed to an evolving racial nativism. As historian John Higham and others have noted, concepts of race remained vague prior to the Civil War, and the obvious physical similarities between the Irish immigrants and native-born Anglo-Americans hardly encouraged serious scientific argument for the two as distinctive races.[46] However, opponents of continued Irish immigration argued that it was more than the economic circumstances engendered by the famine that made the Irish undesirable additions to the republic; hereditary characteristics defined the Irish as beings of an inferior order.[47]

While few actually read the studies of European ethnologists, many Americans were interested in the study of humankind. The popular press was filled with speculation on the value of various techniques to measure and define national characteristics with scientific precision, often referring to nationality as "race." Not surprisingly, Irish immigrants were a topic of much interest, and their features were subjected to frequent phrenological and physiognomical analysis. An 1851 *Harper's Magazine* article discussing two Irish newcomers described the "Celtic physiognomy" as "distinctly marked—the small and somewhat upturned nose, the black tint of the skin; the eyes now looking gray, now black; the freckled cheek, and sandy hair. Beard and whiskers covered half the face, and short, square-shouldered bodies were bent forward with eager impatience."[48] Similar descriptions appeared frequently. Faces were often portrayed as simianlike with extended cheekbones, upturned nose, and protruding teeth. Newspaper and magazine caricatures reflected these stereotypes or traced out derogatory phrenological charts of the head. Late in the nineteenth century, renowned cartoonist Thomas Nast continued to depict the Irishman and ape as closely related.

The Irish were also portrayed as short and stocky, the very physique pseudoscientists associated with people who were not very active but somewhat "slothful" or "lazy." Many Irish were depicted as having coarse red hair, precisely the kind thought to indicate an "excitable," "sociable," or "gushing" personal manner. If Irish were ruddy-complexioned, this was seen as a sign that they were given to raw, unrestrained passions and self-indulgence. Those with dark eyes could be expected to be lustful or excessively sensuous. According to phrenologists in the 1850s, such individuals would not be contented with indoor or sedentary labor but would gravitate toward outdoor occupations because they required "a great amount of air and exercise."[49] That many of the Irish who arrived midcentury worked on the docks or the railroads seemed to confirm their "scientific" profile. It didn't occur to readers of Orson and Lorenzo Fowler's *New Illustrated Self-Instructor in Phrenology and Physiology* (1859) or James Redfield's *Outline of a New System of Physiognomy* (1850) that the authors might have ascribed a certain personality profile to those physical characteristics already part of the Irish stereotype.

Irish response to nativist charges that they were health menaces was neither collective nor systematic. But many seemed sensitive to charges of cultural or social inferiority; they seemed determined to demonstrate that the Irish were already fully capable of holding their own on the level American playing field. As historian Kerby Miller suggests, well-off Irishmen who had achieved or been born to some measure of wealth and social standing sought to impose their social values upon the Irish immigrant masses as a means of ensuring their own influence and status. They often forged alliances with the Catholic Church in America. Both hoped to impose social discipline upon impoverished Irish Catholic immigrants by encouraging "industry, thrift, sobriety, self-control, and domestic purity," habits that would cement newcomers to the Church in America and shape citizens willing to be led by those who already considered themselves to be their group's stewards.[50] Insistence upon high standards of personal health and hygiene among the Irish immigrant masses and provision for their medical care, then, was one means of securing the desired conformity.

Prominent Irish physicians in New York City, committed to encouraging American middle-class behaviors among the newcomers, responded by taking special care of their own. One of these was Dr. William James MacNeven, who arrived in the United States on July 4, 1805. MacNeven's background was atypical of Irish newcomers in almost every respect.[51] In 1763, he was born to wealth, the son of landed Irish Catholic aristocracy. He was raised in

Prague by an uncle who was titled nobility and a physician. William, too, became a physician, completing his studies in Vienna and returning to Ireland, where he sought a cure for his countrymen's ills in politics. He joined the United Irishmen, was elected to a seat in parliament, and served until 1798, when he was arrested for his role in the Irish rebellion. After his release in 1802, he emigrated with two other Irish radicals, Thomas Addis Emmet, Jr., and William Sampson.

MacNeven settled in New York City, where he taught medical sciences and established himself as a respected member of New York's literary society. He continued to work for Irish nationalism, but unlike many Irish exiles who were too busy with the issues to touch the lives of individual sufferers, MacNeven used his medical skills to help his impoverished fellow immigrants. He was known to treat the poor for no fee.[52] He often attended sick or dying immigrants who were stricken during their journey to America and were in desperate need of care when they reached New York. He offered individuals instruction in how to live a healthy life in America and created an emigrant aid society that sought to relieve unhealthy urban congestion by colonizing newcomers on the western frontier.

William MacNeven's efforts on behalf of the Irish poor and his advocacy of standards of health and hygiene drew the approval of those of his own class—both Irish and non-Irish alike—and may in part account for his immense popularity. When he died in 1841, he was regarded by many as the "recognized leader of the Catholic Irish community in New York."[53] MacNeven looked upon the cholera epidemic's victims with compassion but also regarded them as newcomers whose habits must be changed if they were to become healthy wage earners and acceptable to non-Irish compatriots.

The clamor for Irish immigrants to assimilate, to become more "American," was heard and answered at the altar where most of them worshipped. The Roman Catholic Church set out not only to assist stricken immigrants but to do so in a manner consistent with the faith, as it historically had done in other countries.[54] Nuns and priests responded promptly to the 1832 epidemic. The Sisters of Charity nursed cholera victims in New York and other cities. Father Felix Varela, a native of Cuba who became much beloved among New York's Irish Catholics, "lived in the hospitals" during the epidemic.[55] Even so, attacks on the Catholic clergy by Protestant newspapers were not uncommon. During the 1849 cholera epidemic, the distinguished editor of the *New York Herald*, James Gordon Bennett, chided New York's equally well-known Catholic Bishop John Hughes for vacationing at Saratoga Springs while his clergy labored to comfort those dying of cholera.[56] But neither the Church nor Hughes was indifferent to the suffering among newcomers or to the ways in which opponents of the Catholic Church sought to capitalize upon it. Their response was institutional, the building of a Catholic hospital in New York City.

In 1830, New York's Bishop John Dubois wrote to the secretary of the Association for the Propagation of the Faith in Lyons, describing his diocese and its needs. One of the most urgent was "a hospital in New York, where a number of the emigrants who daily arrive and suffer for want of attention might regain both corporal and spiritual health." He regretted that newcomers in need of medical care were being treated in a hospital three miles outside the city, one "administered by Protestants." At such a distance he could not adequately care for the spiritual wants of "the seven hundred Catholic sick who are in that institution [probably Bellevue]."[57] Four years later, Bishop Dubois called for the creation of a Catholic hospital to "afford our poor emigrants, particularly from Ireland, the necessary relief, attendance in sickness and spiritual comfort, amidst the disease of a climate new to them."[58] Dubois was one of many Catholic priests troubled by the Protestant ethos of American public institutions. In private and public institutions operated by Protestants, priests were frequently denied access to Catholic patients. Protestant clergy converted the elderly on their deathbeds and conducted Protestant services in juvenile wards to capture young minds and hearts.[59] The bishop of New York well understood that maintaining a hospital would provide an institutional identification for Catholic immigrants and would serve as a foundation from which to battle Protestants for newcomers' souls.[60]

In 1847, Bishop John Hughes published a pastoral letter attacking the intrusion of Protestant clergy upon Catholic public almshouse and hospital patients and called for the establishment of a Catholic hospital in New York to care for immigrants.[61] Examples of the warfare for souls were plentiful. In 1846, a priest was refused opportunity to administer last rites to a dying man in a Fitchberg, Massachusetts, almshouse by a superintendent who insisted that he would not allow practice of any "paddy superstition" in a house for which he was responsible. The priest eventually won out, but not before he boldly declared to his antagonist, "I'll allow no infidel or bigot to debar me from the exercise of my office."[62] Whether or not the Catholic source of this anecdote reported the exchange precisely, evidence abounds of these bedside battles for souls.

In early 1848, a young, illiterate Irish Catholic girl, Ellen Duffy, published an affidavit in which she claimed that, while living in the House of Industry, a private, nonsectarian benevolent institution to help the poor find employment, she took sick and was denied the ministrations of a Catholic priest after specifically requesting them. She claimed that one of the assistant managers of the institution, a Miss Balch, had told her, "Ellen, you do not want any Catholic clergyman, you want my minister." Father George McCloskey confirmed that he had been summoned by one of Ellen's friends and denied access to her until she was moved to Bellevue Hospital. Even his efforts to secure the intervention of a New York Alderman had come to nought.[63]

In 1849, St. Vincent's Hospital opened its doors in lower Manhattan, on East Thirteenth Street. The first published report on the institution left little doubt about why it was founded. "Although a Catholic institution, its doors are ever opened to the afflicted of all denominations who seek admission, and who may be attended during their illness by their own ministers, if desirable."[64] The hospital was staffed by the Sisters of Charity, under the direction of Sister Mary Angela Hughes.[65]

By 1849, the Sisters were well experienced with hospitals and understood the enormous potential for linking cures for the body to those that their faith offered the soul. Indeed, some argue that the Sisters' spiritual concerns both for themselves and their patients fueled the fires of their zealous efforts to offer medical care and comfort to beleaguered newcomers.[66] In 1828, five of them had started the first Catholic hospital in the United States in St. Louis.[67] Later, in 1834, the governors of Charity Hospital in New Orleans, a public institution, asked for the sisters to work there because the understaffed hospital was sinking under the burden of hurricane and fire victims, as well as numerous leprosy cases. In New York, the Sisters were already well known for their ministrations to cholera patients during the 1832 epidemic.

In 1849, St. Vincent's Hospital had a thirty-bed capacity; by 1861, the hospital had moved to West Eleventh Street and its facilities had been expanded to accommodate 150 patients. At the turn of the century, it was treating in excess of thirty-five hundred patients annually.[68] Although the hospital was open to all the city's Catholics, the inability of the Sisters of Charity to speak German contributed to its becoming a very largely Irish Catholic institution, especially after the German Sisters of the Poor St. Francis founded St. Francis Hospital on New York's Lower East Side in 1865. By the turn of the century, even as immigrants from southern and eastern Europe were pouring into lower Manhattan, over a quarter of St. Vincent's patients remained Irish.[69]

St. Vincent's quickly captured the hearts of those it was founded to serve. Typical was the patient whom surgeon Van Buren described as "a true Irishman." Responding to an inquiry about the discomfort he was feeling after the removal of a cancer from his lip and cheek, this patient answered, "Sure I suffered, but did not our Blessed Savior suffer more for me on the cross? I thank God Almighty that He has been pleased to send me here, where I have so well got rid of it!"[70]

St. Vincent's Hospital not only allowed patients to draw upon the comforts of their faith in their suffering, it enhanced patients' dignity by charging modest fees rather than setting up as an almshouse. In 1858, a physician in Manhattan's New York Hospital observed that working-class Irish preferred St. Vincent's to his own institution and seemed not to mind paying their way. Still, the Sisters of Charity well understood the charitable needs of the community they served. As St. Vincent's annual report noted in 1858, "Although the terms for board and medical attendance are three dollars per week, nevertheless many patients are received free."[71]

There were never enough resources. The report lamented, "Daily are they [the Sisters of Charity] obliged, with pain and sorrow to turn away some poor afflicted one whose means will not allow him to pay even a part of the small sum asked for admission. It is impossible for the Sisters to accommodate all the poor patients who apply for aid and medical treatment. Often are they obliged even to furnish some of them with articles of clothing, besides receiving them as free patients."[72] As always, the church appealed to the philanthropy of its flock, encouraging Catholic New Yorkers to purchase "yearly subscriptions which would secure free beds, and thus afford relief to many of the poor suffering creatures."[73] By the turn of the century, the needs were still great. Of the 3,681 patients treated at the hospital, 1,161 were total charity cases, 1,255 were receiving partial assistance, and 1,265, a third, were paying for their care.[74]

The needs of the New York community were not atypical. In antebellum Philadelphia, too, the arrival of immigrants stirred clergymen, laity, and physicians to join together to establish denominational hospitals that would synthesize care of the body and the soul. There, too, the absence of public support required that paying patients and charitable contributions substitute for tax dollars. However, Philadelphia's Roman Catholic community, like those in other major cities, was not monolithic but crosshatched by ethnic differences that resulted in the founding of separate institutions: St. Joseph's Hospital for the Irish and St. Mary's Hospital for the Germans.[75]

In 1848, Father Felix Barbelin, the Jesuit pastor of St. Joseph's Church, gathered a group of laity and physicians sympathetic to the Irish, including Dr. William Horner, a convert to Catholicism and the dean of medicine at the University of Pennsylvania. Out of the meeting came a new organization, "St. Joseph's Society for the Relief of Distressed Immigrants from Ireland and for the Establishment of a Hospital."[76] The cholera epidemic that menaced East Coast cities in 1848 hastened efforts, but it was largely the compassion for fellow Irish that stirred both passions and philanthropy. In the *Catholic Herald* there was a call for a hospital that would "relieve much of the present distress among the poor immigrants," many of whom arrived "in the most destitute condition—debilitated by disease incident to a long voyage, in a crowded vessel, and unable to make any provision for themselves or their helpless families." In support of its call for contributions, the *Herald* made two points. First, the care of newly arrived Irish Catholic immigrants in a Catholic institution would maintain their allegiance to the church, rather than alienate them so much that they might "become revilers of her sacred doctrines—and objects of disgrace to this community, or worse yet, Protestant converts." Second, by making newcomers well rather than simply comforting them in their demise, the church was helping to keep Irish families off charity roles, promoting family unity in the face of immigration's trials, and perhaps even boosting the Catholic community's reputation for social responsibility in the eyes of Protestants, some of whom hoped that well-off Catholics would exert the same moral stewardship over newcomers that prominent Protestants exercised over their own flock.[77]

Opened in 1849, St. Joseph's was too small to help much with the cholera epidemic, and the lack of funding from charitable sources soon drove the institution's overseers to rechart their course. Rather than become an alternative to the almshouse for the Irish Catholic immigrant indigent, St. Joseph's would mitigate "calamities of disease, of accidents, and other bodily indispositions" of the worthy or "meritorious" poor, a designation that included "deserving journeymen, domestics in families, and operatives of various kinds" who worked and may have had some savings, but whose means were still insufficient to provide hospital room and board, nursing, and medicines.[78] Charging three dollars per week for the care of a sick patient, two dollars less than the five dollars estimated cost that other institutions charged, permitted St. Joseph's to survive but not to offer care to more than one indigent patient for every five who could afford to pay, and that was after the initial debt for establishing the hospital had been paid off.[79] Impoverished Irish arrivals seeking free medical care from their own were being asked to take their place in a line that promised to move slowly.

There appears to have been no shortage of the worthy poor even among the Irish who were St. Joseph's most frequent patrons. In 1850, 83 percent of St. Joseph's patients listed Ireland as their place of birth and the proportion of Irish remained high, double the percentage of Irish admitted to the secular Pennsylvania Hospital until 1900. Many of these sons and daughters of Ireland must have been not only worthy but working and, therefore, able to pay for hospital care because in 1850, 78 percent of St. Joseph's inpatients paid for their board; indeed, from 50 to 80 percent of St. Joseph's patients were paying customers for health care over the next five decades. That figure was one of the highest in Philadelphia. Why? Aside from the fact that an increasing number of patients at St. Joseph's and other well-run hospitals were getting well and were soon able to return to their families rather than dying of illness or infection, appealing hospital facilities, especially the availability of private rooms, increased St. Joseph's advantage over other institutions. In 1859, the hospital could boast eleven such rooms at six dollars per week, a number that increased to sixty-four rooms by 1876. Nicely appointed, some even had connecting private baths. What better statement of an immigrant's respectability, even modest success, than to be able to afford privacy and comfort as one suffered the trials of illness in a distant land. Who were these worthy poor and how could they afford such rooms? Hospital annual reports suggest that many were domestics, perhaps the beneficiaries of employers' generosity, employers who recognized the value of healthy and grateful help in the home. Others were referred by priests and physicians or Catholic lay groups. The Society of St. Vincent de Paul extended a helping hand to hard-working newcomers who needed only their health to sustain their self-sufficiency.[80] However, Irish immigrants often had severe health problems, including chronic ailments and needs for surgery well beyond those with which most physicians were trained to cope. An increasing number of Irish—one in three by the turn of the century—came to the hospital for surgery. In the operating room, competent surgeons were more important than the religious or ethnic flavor of the institution. Once out of surgery, though, patients' recovery was aided or inhibited by their surroundings. The calm, reverent atmosphere of a Catholic hospital, supervised by English-speaking nuns—many of them as Irish as their patients—and supported by priests offering the sacraments, might have a medical impact as profound as a surgeon's skills.[81]

The arrival of millions of immigrants, and especially the Irish, did truly transform the face of American immigration and the profile of public health in the mid–nineteenth century. Vague fears of strangers coalesced into specific stigmatization of a particular group—the Irish. Preexisting anti-Catholic sentiments and the sheer size of the migration sent shock waves of apprehension through native-born Americans, especially in port cities such as New York where the presence of the Irish was most visible and where sick or disabled newcomers most strained existing medical facilities. Those responsible for the public health responded to the twin threat of cholera and Irish Catholics by expanding the charge of state government beyond quarantine to the exclusion of those foreign-born who menaced the community's health and well-being.

Expanded, too, were the number of states and municipalities with permanent boards of health. While immigrants alone were not responsible for this expansion of governmental power, their sheer numbers hastened such institutional change by highlighting the limitations of more traditional models of public responsibility, ones grounded in notions of moral stewardship and supported by private philanthropy; mass immigration was a catalyst.

Increased public responsibility for the health and well-being of a state or municipality was one thing, adequate funding to meet the health care needs of newcomers, quite another. Prominent Irish physicians such as Dr. William James MacNeven treated their own, thereby shaping the direction and pace of acculturation and enhancing their own prestige within the Irish community. The Catholic Church, with even more at stake, did not hesitate to respond to the medical stigmatization of Irish Catholic immigrants and newcomers' health care needs with an institutional response. The Catholic hospital not only saved souls from Protestant menace but dramatically expanded the availability of hospital care and medical resources, especially to the impoverished urban masses, so many of them immigrants.

The next great spike in immigration beginning in the 1880s would present unprecedented public health problems of such substance and scale that responsibility for the inspection and processing of immigrants would be further formalized and transferred from state to federal jurisdiction. Similarly, newcomers would learn from the experiences of those who had preceded them. Precedents of self-help set by the Catholic Church, by physicians, and by philanthropists who were Irish or Catholic or both, would be expanded upon to meet the needs of millions of immigrants arriving a half-century later.

3

"Proper Precautions": Searching for Illness on Ellis Island

In 1888, a report delivered to the annual conference of superintendents of institutions for the feebleminded clamored for immigration restriction lest the country find itself unable to bear the cost of the institutional facilities it would need to treat, or at least confine the "sewage of vice and crime and physical weakness" washing ashore from Europe and the "nameless abominations" floating in from Asia.[1] The report was not an isolated protest. In the decades following the Civil War, as the nation opened its doors to badly needed laborers from around the globe, many of the native-born yearned to protect the public's health from compromise by foreign menace, whether from Asia or any other corner of the world. Quarantine procedures under state jurisdiction were designed to exclude infectious contagious diseases, to fend off epidemics, if possible. However, that no longer seemed adequate protection. By the last decades of the nineteenth century enough Americans feared that immigration had a detrimental influence upon the public health to move Congress to action.

The controversy that had begun earlier in the nineteenth century over the proper role of the federal government in matters of health and immigration, respectively, continued in the 1870s. Concerns about yellow fever, especially, stirred some support in Congress for a national board of health that would administer an effective national quarantine. The compromise was the short-lived, feeble National Board of Health (1879–1883), an agency intended to support state and local health efforts, but given little power of its own except to take over quarantine responsibilities in cases where the states had proven ineffectual. By the time its funding expired in 1883, the National Board had failed in its efforts to impose a national quarantine, foundering on the shoals of states' rights, localism, and jurisdictional rivalries with the U.S. Marine and Hospital Service.[2]

Less than a decade after the National Board of Health's demise, the federal government acted decisively to exclude diseases borne on the bodies of immigrants with passage of the Act of March 3, 1891 (26 Stat. 1084). Now immigrants would undergo individual health inspections before departure and after arrival in the United States.

The new federal immigration law required that the steamship companies feasting on a growing immigrant trade vaccinate, disinfect, and medically examine emigrants to certify their health prior to departure. Companies were liable for the cost of housing and feeding passengers detained by American authorities for reasons of ill health. Immigrants arriving too sick to admit would be returned to their ports of embarkation at the steamship company's expense. In emigrant villages such as that erected by the Hamburg-Amerika Line to house passengers prior to departure, emigrants had to stand a physician's examination as part of the preboarding routine. After still further congressional action in March 1893, emigrants to the United States had to answer a long list of questions prior to departure, their answers practically constituting a medical history of each individual. Almost always the inspection abroad was cursory. Paying customers were too valuable to turn away, so warring steamship lines gambled that American procedures would be equally lax. Only in Italy did an American official intervene and compel shipping companies to comply with preboarding inspection requirements.

Fiorello H. LaGuardia, who was destined to be the popular mayor of New York City and could debate opponents in any one of eight languages, was the American consul in Fiume from 1903 to 1906. Born to an Italian father and a mother of central European Jewish background, the immigrants' son saw to it that steamship lines took seriously their obligation to certify the health of emigrants boarding their vessels. After 1908, the Italian government, taking its cue from LaGuardia, began to conduct its own exit physical to ensure that those likely to be rejected across the Atlantic not be allowed to depart.[3] Besides sympathy for their own nationals, the Italian government preferred welcoming home migrant workers whose pockets were filled with American dollars rather than those flat broke in need of care and carriage to their home villages.[4] Not until after World War I did European governments of other "donor" nations follow suit and inspect those departing.

Of course, emigrants hardly objected to perfunctory exams; each hurdle vaulted brought them closer to their destination. Most did not realize that they would be better off not departing sick or infirm. Many were unaware that the same law ordering inspection abroad also mandated inspection by American government physicians at the other end of the sea voyage. Those physicians' diagnoses were often the grounds for shipping home those deemed unfit.

Emigrants headed for the United States were part of a tidal wave of migration spreading across the continents greater than at any previous time in recorded history. Not only would 23.5 million emigrants depart for the United States between 1880 and 1924, but many who came were from countries not previously significant in the peopling of America. Population poured westward from countries in southern and eastern Europe, while smaller, but nonetheless significant numbers of arrivals from China, Japan, Mexico, and Canada crossed sea and land to reach the United States, where jobs were more plentiful, salaries higher, and rights guaranteed by laws consistent with the customs and traditions of individual liberty. Not all westward-bound European emigrants or eastward-bound Asians headed for the United States. Canada, Australia, Argentina, and Brazil also attracted newcomers, but the United States drew the greatest number.[5]

Mass immigration posed an administrative challenge unprecedented in peacetime. America's nineteenth-century island communities had barely begun their "search for order" when new arrivals flooded port cities on their own hunt for opportunity.[6] Until the Civil War the federal government had little direct contact with most citizens. Millions of Americans encountered their government in Washington only when they were queried by a federal census taker or picked up mail at the post office.[7] The modern version of the bureaucratic state was born in response to challenges too overwhelming for local government and private enterprise to handle by themselves.[8] Immigration was foremost among these challenges, and health examination of each arrival posed an unprecedented challenge.

On January 1, 1892, Annie Moore, a fifteen-year-old whom the *New York Times* described as a "rosy-cheeked Irish girl" from County Cork, became the first immigrant to enter the United States via Ellis Island, the federal government's new immigration depot in New York harbor. Ferried from her ship, the *Nevada,* to Ellis Island with her two brothers and other immigrants for medical inspection and processing, young Miss Moore had been selected by immigration officials to be first off the boat and into the depot. She was met by a beaming John B. Weber, immigration commissioner for the port of New York. Weber came forward after Moore's name was written in the registry book and presented the girl with a ten-dollar gold piece to commemorate the occasion.[9]

Immigrants traveling steerage or third class after 1891 were required to stand a medical inspection. Those arriving in New York were examined at the Ellis Island immigration depot, opened in 1892. This red brick structure, erected in 1901, replaced the original wooden building destroyed by fire in 1897. (Library of Congress)

After quickly passing before the watchful gaze of the inspecting physicians of the U.S. Marine Hospital Service and answering the questions of Immigration Bureau officials, the three youngsters were admitted. Annie Moore's "rosy-cheeks" had not deceived. According to the Ellis Island physicians, she was physically and morally fit for America.

Annie Moore's arrival received the favorable newspaper coverage it had been designed to attract. However, it was hardly typical of the way most immigrants were welcomed to the United States. For voyagers who could afford the more expensive tickets, those who traveled in first- or second-class accommodations, the medical exam lasted minutes, as it had for Annie Moore and her brothers. It consisted of little more than a few questions and a perfunctory look over by the United States Marine Hospital Service physician who was one of a party that climbed aboard anchored steamships and inspected newcomers in the privacy of their own cabins. Only those who traveled in steerage's cramped confines and the slightly more comfortable third-class bunks were directed aboard small, crowded boats that bounced up and down in New York's choppy waters and ferried immigrants from their tall steamships to Ellis Island, "Isle of Hope" to all who arrived, "Isle of Tears" or "Heartbreak Island" to those whose journey ended in rejection.

Most immigrants spent only a few hours at the island depot, usually fewer then five, but these were often the most traumatic of the entire journey. The medical inspection was the first and usually most vividly recalled episode of the ordeal.[10] As many as five thousand immigrants per day could be processed on Ellis, but even on slower days, the eyes and hands of physicians were rarely idle for long. Each arrival was approached by a uniformed immigration officer who pinned an identity tag on his or her clothing. The tag was inscribed with an identification number, corresponding to that in the ship's manifest. Bearing a label and carrying small articles of hand luggage, each newcomer climbed the stairs to the main hall, or registry room, quite unaware that the medical inspection was already underway.

"Line inspection" took place at the top of the stairs. As each newcomer completed the climb, he or she passed under the gaze of a uniformed U.S. Marine Hospital Service physician who was anxious to scrutinize the newcomer under conditions of physical stress such as that produced by carrying heavy luggage up a flight of stairs. Hands, eyes, and throats were observed. The heart of an immigrant who had toted even light baggage up the stairs could be easily judged strong or weak, and the exertion would reveal deformities and defective posture.

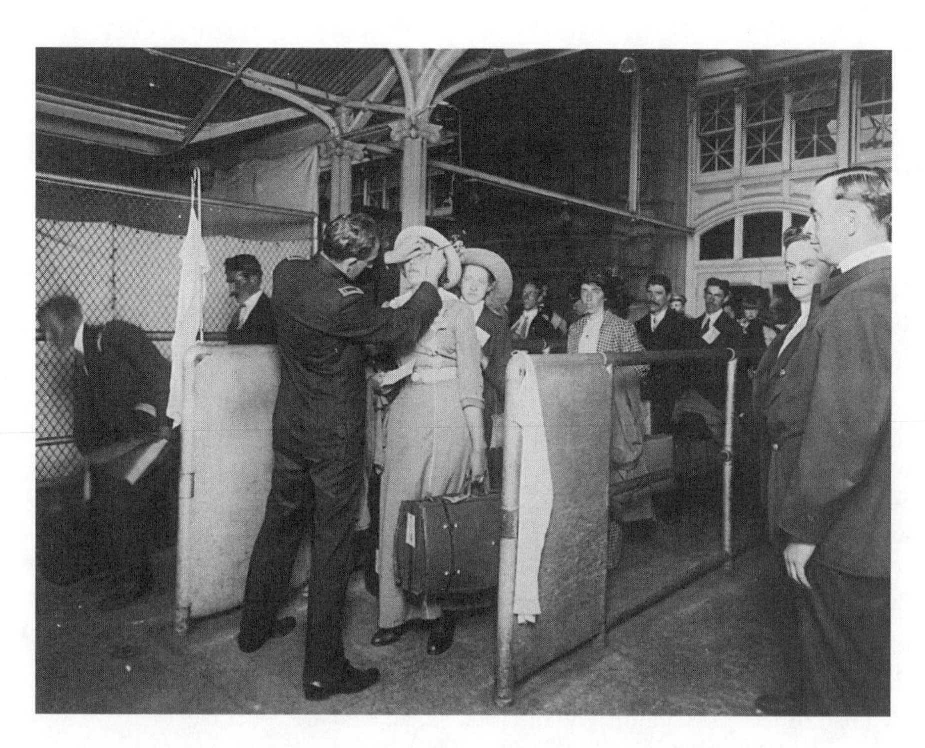

An eye examination on Ellis Island, c. 1907. (U.S. National Archives)

Advances in medical technology changed the inspection and improved physicians' diagnostic abilities between 1892 and the end of the peak period by the mid-1920s. However, immigrant testimony suggests that throughout the era the eye exam remained the most frightening, even painful moment of their trial. Immigrants were given stamped identification cards to hold in their hands. As the newcomer examined the card, he or she was observed by physicians checking for defective eyesight. Eyelids were everted to detect sores, trachoma's telltale sign. Polish immigrant Catherine Bolinski, who had a sty in one eye during passage, was relieved that it had vanished by the time her ship anchored in New York waters, but the eye exam aroused her and her mother's anxieties anyway. She recalled, "They turned your eye over—I had to blink a couple of times that way. I'll never forget it."[11] Fannie Kligerman testified similarly, "This I remember well—the eye exam. It was such a fright, such a fright."[12] Euterpe Dukakis from Greece recalled how unhappy she and her sister were at the "Palace of Tears," as her father referred to Ellis Island, when "the doctor stuck his finger in her [the sister's] eyelid and turned it up for a very bad disease [trachoma] that was very communicable and infectious, contagious."[13]

The immigrant's scalp was probed for lice or scabs, symptoms of favus, a contagious dermatological disorder. Each might also be required to turn the head so that facial expression could be examined. Certain expressions were believed by physicians to indicate mental illness or insufficiency.

As they proceeded past the physicians and into the registry room or "Great Hall," a vast room with its tiled, high vaulted ceiling, the immigrant might have noticed that some in the group had chalk marks on the right shoulders of their garments. Chalked letters meant that a newcomer would be pulled from the line and detained for closer scrutiny. Each letter stood for the particular disability that stalled and might abort the newcomer's rendezvous with America: L for lameness, K for hernia, G for goiter, X for mental illness and so on. Detainees were often separated from others in wire mesh compartments which bore too great a resemblance to an animal pen or a jail cell.

Some of those detained were promptly reexamined in a separate room. Observant of prevailing Victorian codes of public morality, Ellis Island officials were careful to segregate the sexes and to make certain that women physicians and nurses examined female immigrants, who were required to strip to the waist for further medical scrutiny. By 1924 there were four female physicians, two women attendants, and a nurse on duty.[14] Still, it remained a memorable trauma for youthful newcomers raised in modesty and filled with apprehensions. Austrian immigrant Adele Sinko, who grew up just outside Vienna, found it "so embarrassing for me." She "wasn't even twenty-one, very bashful, and there was these big kids running around, but you had to do it. The examination was done by women, but the kids were there. That I resented, when you had to strip to the waistline."[15]

Those detained for several days' observation or recuperation recall their apprehensions but also indignities, such as the daily delousing that Fannie Kligerman remembered as especially humiliating. "The worst thing was, you wouldn't guess! Every morning they came around to delouse us. You know what that means? Our things were taken off, we were naked to be deloused." After repeated fumigation, clothing brought across the Atlantic for the special occasion of arrival in America began to disintegrate. "They took all our clothes, nice dresses, and made rags out of them. As soon as we entered America we had to put them into the garbage."[16]

Asian immigrants who arrived in San Francisco Bay had similar experiences. Some Chinese immigrants cast their feelings in poetry, carving their verses as lasting testimony of their fears and shame into the stone walls of the Angel Island immigration depot opened in 1910. Their honor offended by the cold, bureaucratic manner in which they were processed, Asian arrivals knew all too well that their futures depended upon the gaze and touch of a physician who did not necessarily share their culture's definitions of health, disease, or appropriate interaction between examiner and examinee:

I cannot bear to describe the harsh treatment by the doctors.
Being stabbed for blood samples and examined for hookworm was even more pitiful.
After taking medicine, I also drank liquid,
Like a dumb person eating the *huanglian* [a bitter herb].[17]

It is indeed pitiable the harsh treatment of our fellow countrymen.
The doctor extracting blood caused us the greatest anguish.
Our stomachs are full of grievances, but to whom can we tell them?
We can but pace to and fro, scratch our heads, and question the blue sky.[18]

For most, the indignities of the medical examination were short-lived. The new arrivals departed the immigration depots and plunged into America. However, few forgot their inauspicious welcome on Ellis Island, Angel Island, or other immigration depots. Most of all, they remembered the medical examination when for the first time the United States of America took the measure of those seeking to enter and barred those found physically or mentally unprepared for life in the tough, competitive society that demanded fitness of body as well as of soul. Their testimony reveals that immigrants' ardor for America was cooled, if only briefly, by the sobering realization that its "golden door" opened wide only for those deemed physically and mentally up to the striving and sacrifices required of newcomers.

If newcomers such as Fannie Kligerman and Catherine Bolinski regarded

Ellis Island as an "Isle of Tears," Immigration Bureau inspectors and U.S. Marine Hospital Service physicians saw these federal installations from a different perspective. These proud, uniformed agents of the United States government saw Ellis Island's ornate turrets as towers of vigilance from which they dutifully guarded their country against disease and debility. Those were the enemies, not the immigrants themselves, who were merely unfortunate victims and unwilling bearers of biological menace. Largely unaware that immigrants found the Ellis inspection daunting, officials saw their task not as one of making the process more humane but more effective and evenhanded. U.S. Marine Hospital Service doctors had a special obligation as physicians and members of an increasingly self-conscious profession with rigorous standards of behavior.[19] They were acutely aware of refusing to permit an ever-louder chorus of restrictionists from influencing their medical diagnoses, even if those diagnoses were from time to time unconsciously shaped by their own ethnic biases.

In 1902, Progressive Republican President Theodore Roosevelt acted to bring a fresh honesty and efficiency to Ellis Island, where corruption had mounted during its first decade of operation. He appointed a young Wall Street lawyer who shared his own Progressive belief in scientific expertise, honest, efficient administration, and public service to be the commissioner of immigration of the port of New York. William Williams was a forty-year-old New England patrician, with a bachelor's degree from Yale and a law degree from Harvard. He had briefly served his country in the consular service in the 1890s and as a soldier in the Spanish-American War, until he was taken ill with typhoid fever in October 1898. He was enjoying a lucrative postwar legal career when T.R. recalled him to public service on Ellis Island. Williams served on Ellis from 1902 to 1905, and returned for a second period of service from 1909 to 1914.

The commissioner's first endeavor in 1902 was to rid the island of illegal and inefficient administrative practices, some of which pertained to public health issues. Williams dressed trusted immigration agents and some friends from Wall Street as immigrants and allowed them to undergo a full inspection procedure. These ersatz immigrants helped Williams to root out dishonest inspectors who had been bilking newcomers for money in exchange for fraudulent naturalization papers. Such counterfeit papers allowed immigrants who thought themselves unable to pass the medical examination to pose as American citizens and thus bypass the Ellis Island procedure. Other scams involved immigration officers allowing ship companies to land immigrants whom they knew to have contagious diseases. Williams fired the culprits. More important, he launched a new era in the federal supervision of immigration.[20] A year after Williams took office, Congress transferred immigration responsibilities from the Treasury Department to the newly established Department of

Commerce and Labor. Immigration was henceforth to be considered as an addition to the labor force rather than as a species of import. Central to that change in perspective was Congress's understanding that immigration had an increasingly significant role in many aspects of the public welfare, not the least of which was the public's health.

The same year that Williams marched his friends through the line inspection, United States Commissioner General of Immigration Terence Powderly published a *North American Review* article calling for stricter health controls on arriving aliens. Powderly cautioned that unless "proper precautions" were taken to detect two contagious diseases, trachoma and favus, the future American might be "hairless and sightless." He called upon Americans to refuse to allow their country to become "the hospital of the nations of the earth."[21]

Fiorello LaGuardia, an interpreter on Ellis Island by day between 1907 and 1910 so that he could afford evening law classes at New York University, recalled, "the many heartbreaking scenes on Ellis Island," the "mental anguish" to which he never grew "callous." Others might grow indifferent to the "disappointment and the despair" of those deemed unfit for America, but never the man later known as the "Little Flower." And according to LaGuardia, neither did most of his colleagues, "a hardworking lot, conscientious and loyal."[22]

Whether most federal officials looked upon the newcomers passing before their gaze with the compassion of a LaGuardia or the callousness of a Powderly, federal concern about the immigrant as a potential public health threat was well established by the turn of the century. The germ theory of infectious disease had been embraced by federal authorities, as it was by a slowly increasing number of Americans in the health professions, by the late 1880s.[23] As Robert Koch in Germany had established, bacteria, not the filth that nurtured them, was the cause of disease. Sanitarianism was on the wane, but not the notion that illness often came to America from someplace else.

Cholera epidemics earlier in the nineteenth century, but well within the memory of policymakers, made real the threat of epidemic diseases that knew no national boundaries. Therefore, as early as 1887, the federal government ordered the beginning of cholera studies in a single room on the ground floor of the marine hospital at Stapleton, Staten Island. Dr. Joseph Kinyoun was in charge. The fear that immigrants carried Asiatic cholera had led the federal government to initiate research on bacteria; (a program of research that evolved into the National Institutes of Health).[24]

The federal government could justify establishing Kinyoun's modest laboratory, in part because of a growing tendency in the late nineteenth and early twentieth centuries for Americans to regard health issues as legitimate areas for government involvement and not mere adjuncts to other social concerns.

Earlier, general social policy, largely the responsibility of state and local authorities, subsumed health. To sanitarians who identified the origins of disease in the filthy living conditions of the poor, promoting health and preventing illness seemed to require no more than alleviating poverty and economic deprivation. But now medical science was demonstrating that pathogenic microorganisms did not always recognize class boundaries. Though traces of the older view sometimes surfaced, increasingly officials at all levels of government trusted to physicians and their expertise to identify public health problems and pose solutions. The health and vitality of individual newcomers and their impact upon American society was one such concern. And rather than relying on blanket, antiquated generalizations that suggested that poor and degraded classes were the vessels of disease, the federal government, having assumed responsibility for monitoring immigration, now found itself confronting public health issues that millions of newcomers' presence must raise.[25]

The urgency of Kinyoun's research and the immediacy of the public health's endangerment by diseases from abroad was freshly dramatized by the Asiatic cholera epidemic that reached the United States in the autumn of 1892. Hoping to stall a growing panic, President Benjamin Harrison asked Surgeon General Walter Wyman to issue a circular order declaring that "no vessel from *any foreign port* carrying immigrants shall be admitted to enter any port of the United States until such vessel shall have undergone quarantine detention of twenty days, and such greater number of days as may be fixed in each special case by the State authorities."[26] The cholera epidemic of 1892 hastened an increase in federal public health responsibilities, especially with respect to the potential health menace posed by immigration.[27]

The following February, President Harrison signed an act requiring all ships departing for the United States to carry certificates of health signed by the American consul and a medical officer of the Marine Hospital Service detailed to the port of departure.[28] These certificates had to be submitted by ship captains to the federal quarantine officer in ports of destination before passengers could disembark or cargo could be unloaded. The new law gave the Marine Hospital Service responsibility for approving state and local quarantine facilities and bringing those not up to federal standards into compliance. Nine quarantine stations were already under federal jurisdiction and the 1893 law provided for compensation to other state or municipal governments willing to transfer responsibility and facilities to the federal government. A few states relinquished this critical public health responsibility to federal officers; others eventually followed suit. Although the act also granted the president "by reason of the existence of cholera or other infectious or contagious diseases in a foreign country" the "power to prohibit, in whole or in part, the introduction of persons and property from such countries or places as he shall designate and for such period of time as he may deem necessary," neither Harrison nor any other American president used this power to permanently restrict immigration. Still, more than ever before protecting the nation's public health from diseases borne from foreign ports was evolving into a federal responsibility.[29]

The federal physicians charged with health certification of individual new arrivals at immigration depots belonged to the United States Marine Hospital Service, which was renamed the United States Public Health and Marine Hospital Service after 1902, and finally, the United States Public Health Service in 1912 (hereinafter referred to as the PHS).[30] Their procedures were designed to fulfill their responsibilities as physicians, while respecting sensitive jurisdictional boundaries.

Ships entering New York Harbor after 1892 were first boarded by New York State health officers who conducted cursory inspections for what had become known as the "germ diseases"—typhus, cholera, plague, smallpox, and yellow fever. Only afterward were federal officials permitted to board. Dr. Victor Heiser, a PHS physician assigned to Ellis Island, recalled in his autobiography the confusion that state jurisdiction over enforcing federal quarantine regulations often caused; each state had its own criteria for enforcement. "For reasons not always creditable, the latter [states], pleading their constitutional right to exercise 'police powers,' resented what they considered federal interference with their affairs."[31]

On one particular rainy March morning Heiser almost got caught in jurisdictional crossfire. Federal officers were not permitted to board a ship until New York State officers had certified the ship as having passed quarantine. However, on this foggy morning, with visibility "almost zero," Heiser mistakenly climbed aboard a French ship before the state inspectors had departed. "To my surprise, Dr. Jenkins, the State quarantine officer was still there. The sight of my Federal uniform was as a red rag to a bull," Heiser recalled. Jenkins growled at the young physician, "Doctor, you're under arrest. This ship's in quarantine for smallpox." After quickly calling over the ship's rail to warn off his colleagues, Heiser turned to find a smirking Jenkins who snapped at him, "You'll have to stay in quarantine fourteen days, Doctor." Young Heiser saw his career going up in smoke, as he pondered the negative publicity he could expect, "For a Federal officer to be arrested by a State officer because of assumed ignorance of the law would have made a wonderful newspaper story." However, Heiser quickly recovered his composure and calmly refused to accompany Jenkins to the quarantine station on Hoffman's Island. Jenkins threatened to use force against Heiser until the alert federal physician reminded him that in New York, unlike other states, every ship must fly a yellow flag at her masthead while in quarantine. Glancing upward into the fog with a grin, Heiser said, "If you'll show me any such flag on this boat, . . . I'll go with you. Otherwise, I'm afraid I must say goodby [*sic*]." As Heiser hailed a federal launch, "The little man could do no more than stamp his foot helplessly and utter a fervent damn." When Heiser finally reached his office, he discovered that his colleagues, infuriated by his treatment, had wired Washington, demanding that a writ of habeas corpus be issued to rescue a federal official from state quarantine.[32]

Usually state and federal officers changed places with considerably less tumult. The PHS boarding division inspected passengers fortunate enough to afford first- or second-class passage in their cabins. Every morning Victor Heiser and his colleagues sailed across the bay aboard their federal immigration launch at 5:30 A.M. to board incoming vessels or ships that had arrived the night before. The federal officers were aware that many immigrants who had been rejected on medical grounds when traveling steerage returned in cabin hoping that the higher-priced ticket might save them from all but a perfunctory glance from a physician. Heiser and other PHS physicians were on the lookout, though they often met interference from steamship employees, who were "fearful of offending their best paying clientele." As diplomatic as they tried to be, the federal officers were occasionally overzealous and diagnosed before interrogating. Forgetting to inquire whether the individual before him was an immigrant or a returning citizen, one young federal officer called to the others, "I've a fine case of acne rosacea," a dermatological ailment, only to discover to his acute embarrassment that the man with the rash was an American, indeed, Jay Pierpont Morgan! Fortunately, the renowned financier possessed a sense of humor as well as citizenship.[33]

When William Williams assumed command on Ellis Island, he found that the surgeon general had assigned only eight PHS physicians and one steward to conduct the physical examination of the 497,791 steerage passengers and 68,192 cabin passengers. This was only two more physicians than had been there in 1892, when 445,897 newcomers were inspected. By 1905 still only sixteen officers were assigned to Ellis. This small band conducted the line examinations, operated the immigrant hospital on the island, and conducted the examination of cabin passengers aboard incoming vessels. On the eve of World War I, the number rose to a still modest twenty-five, with four separate line inspections conducted simultaneously.[34]

Still, PHS physicians on Ellis prided themselves on their bureaucratic efficiency and impartiality. Dr. Alfred C. Reed wrote that, "in all the manifold and endless details that make up the immigration plant, there is [a] system, [which is] silent, watchful, swift, efficient."[35] Physicians' approach to sterile technique well illustrates the point. Beginning in 1898, physicians administered the examination for trachoma only to those who exhibited symptoms. However, after 1905, physicians on Ellis Island and at other depots everted all immigrants' eyelids. Because no specialized instrument yet existed to assist the physician in this procedure, the doctors used conveniently shaped buttonhooks, more commonly used to assist ladies in buttoning high-top shoes or long, fashionable gloves. However crude the instrument, physicians realized that it could serve their purpose if used with dexterity and kept sterile so as not to spread infection. They sterilized their instrument with a quick dip of the hook into a disinfecting solution between examinations or a wipe on a Lysol-impregnated towel. The absence of epidemics of trachoma among newly inspected arrivals suggests that the pressure to process the newcomers did not cause PHS physicians to neglect proper medical procedure, using whatever means were available.

Because lengthy, detailed examinations were impossible except for those detained, physicians relied upon clues that consistently proved reliable in weeding out immigrants with physical or psychological problems. Often diagnoses were little more than hunches. PHS physicians came to associate certain facial expressions with particular mental and physical disorders. Assistant PHS Surgeon Howard Knox believed that "facial signs" could reveal "with considerable regularity" such "conditions" as "nationality, temperament, occupation, sexual relations and habits, sensuality, drug addictions . . . and such 'diseases' as brain tumor . . . melancholia, manic depressive insanity, chorea, hydrocephalus, idiocy, imbecility, feeblemindedness, higher feeblemindedness (moron), moral obliquity, local and other paralyses and atrophies . . . renal disease, appendicitis . . . and impending death."[36]

Immigration Bureau officers detained all who failed to pass the medical exam for a more thorough inspection and, if necessary, an appeal to a board of review, prior to exclusion. Sometimes a few days of rest and some nourishing food was sufficient to prepare an immigrant for reexamination and admission. On other occasions, a short stay in the Ellis Island Hospital, opened in 1902, might be sufficient to allow recovery from a minor illness or injury. In the years before completion of the island's hospital, federal officials contracted hospital services from the New York City Health Department and the Long Island College Hospital in Brooklyn, among others.

Though the medical inspection was constantly being modified with procedural changes and improved medical technology, the PHS's examination could provide little more than a snapshot diagnosis of most immigrants. However, in an era when sophisticated diagnostic technology was still limited and personal observation was a highly regarded method of diagnosis, the PHS physicians on Ellis took great pride in their ability to spot illness or disability quickly and with only modest clues. Dr. John C. Thill, recalled W. C. Billings, chief medical officer on the island between 1921 and 1925, was a particularly "clever diagnostician and acute medical observer." Thill remembered one occasion when Billings approached a German woman in the inspection line and said, "*Nehmen Sie die Perücke ab,*" meaning "take off the wig," an article of apparel that the other officers had not noticed. Thill and his colleagues were astounded to see before them "a totally bald lady who had had favus."[37] Dr. Victor Heiser, too, had his favorite diagnostic shortcuts. He observed that "on the left side of any immigrant woman's head was a strand of hair which under normal conditions, was more or less lustrous. If it hung dull and lifeless over her ear, it marked her at once as possibly pregnant." Hardly an illness, pregnancy could be an excludable health condition unless the mother-to-be could prove that she was married and that neither she nor her child was likely to require public assistance. Too intelligent and well trained not to realize that his pet technique might meet with raised eyebrows among medical colleagues, Heiser preempted critics: "Other medical men may ridicule the observation, and I cannot explain it myself scientifically, but time and again suspicion turned out to be fact."[38]

Ellis Island physician Dr. Victor Safford not only defended the instinctual approach used by examining physicians, he saw it as a virtue. After all, physicians were not expected to know what was wrong, merely that the immigrant was defective. More precise diagnosis could occur later under more favorable clinical conditions, such as those at the Ellis Island Hospital. He reasoned, "It is a no more difficult task to detect poorly built, defective or broken down human beings than to recognize a cheap or defective automobile. It requires of course some training and experience to do either, but in any case difficulty usually arises not so much in recognizing that something is wrong as in determining and demonstrating just what is wrong." Safford carried his analogy further, noting that just as an automobile cannot tell a mechanic what has gone awry, neither will a physician get "diagnostic help from the subjective symptoms of a diseased immigrant" because, "The alien who comes before the immigration medical officer is usually interested only in leading the medical officer to believe that nothing is wrong." If immigration officers watched the newcomers closely as they climbed stairs, examined an identity card, or reacted to the discomfort of the eye inspection, there was good reason. Again returning to the automobile analogy, Safford contended that "the wise man who really wants to find out all he can about an automobile or an immigrant, will want to see both in action, performing as well as at rest, and to watch both at a distance as well as to scrutinize them close at hand. Defects, derangements and symptoms of disease which would not be disclosed by a so-called 'careful physical examination,' are often recognizable in watching a person twenty-five feet away."[39] Under time pressure, those physicians facing long lines of immigrants knew that "A man's posture, a movement of his head or the appearance of his ears, requiring only a fraction of a second of the time of an observer to notice, may disclose more than could be detected by puttering around a man's chest with a stethoscope for a week, yet a week might be required to demonstrate what was really wrong with the man."[40] Safford was quick to add that the more precise diagnosis that might cause an immigrant to be rejected was never rushed. "The inspection process might seem to an uninitiated spectator to be a rapid hasty procedure, but no limit was placed by law, by regulation or in actual practice on the time which medical officers might take for the subsequent medical examination of those held for such purposes."[41]

Whatever they saw when they observed the immigrants, PHS physicians did not stop to consider that their professional demeanor could be forbidding or that the garb designed to elicit the respect from their Coast Guard patients could be so misconstrued as to intimidate newcomers passing before them.[42] Many were unaware that some immigrants associated uniforms with the oppressive, authoritarian governments that had caused them to flee their homelands.

Though all were trained to do their job in an unbiased professional manner, some PHS physicians felt special sympathy for the immigrants they examined, much as Fiorello LaGuardia did. Victor Heiser recalled that "heartbreaking incidents were constantly occurring when rejections had to be made." He offered as an apocryphal example a Scandinavian farmer who had slaved and scrimped and saved in Minnesota to bring his family across the Atlantic, only to hear of his daughter the diagnosis from a pained PHS physician, "She has trachoma. She cannot enter."[43] Then as his wife and children waited on Ellis Island for a return ship, choosing to return, too, rather than send the girl home alone, "the father, wretched and unhappy, would haunt the detention quarters, while his family kept up a constant wailing and crying."[44] Was Heiser aware of the discrepancy between his feelings and his duty? Only too aware. "But the law was on the statute books and we had to enforce it, regarding the child as a potential focus of infection rather than as a figure of tragedy."[45]

Others linked the immigrants' physical condition to a supposed inferior character and life-style. Typical of this perspective was the observation of one PHS physician that, "Overcrowding, disregard of privacy, cleanliness and authority, their gregariousness and tendency to congregation along racial lines in cities, are all important factors in the spread of disease among them and by them."[46]

Whether other PHS physicians were as sympathetic to the foreign-born as Victor Heiser or not, there is no evidence that any of them on Ellis Island systematically used their influence to the disadvantage of a particular immigrant group. However, the same may not have been true of other ports less exposed to public scrutiny than the high-profile, busy immigration depot in New York harbor, which federal officials had intended to be a showplace for government efficiency. Certainly Chinese officials in both China and the United States believed that pains were being taken by PHS physicians and Immigration Bureau officers in San Francisco harbor to reject those Chinese not already barred by the 1882 Chinese Exclusion Law.

At first conducted in a small wooden house, the West Coast inspection was moved to a new facility on Angel Island in 1910. Chinese objected to the immigration officers' practice of separating whites from other races prior to the medical inspection, a practice not followed on Ellis Island, although West Indian blacks and some Asians did enter through the New York port. Even more aggravating to Chinese officials was PHS physicians' practice of excluding Chinese newcomers for diseases that were more prevalent among Asian than non-Asian newcomers but treatable and not a serious health threat to the American population. Chinese were often detained on Angel Island for parasite diseases such as uncinariasis, or hookworm, and filariasis. After 1917, clonorchiasis, or liver fluke, was added to the list. In 1910, the Chinese community of San Francisco sent Dr. King H. Kwan, a physician from China, to Washington, D.C., to discuss the exclusions with officials from the Commerce and Labor Department. Kwan succeeded in convincing American officials that filariasis was not a dangerous contagious disease and that symptomatic arrivals should be allowed to remain in the United States for treatment. Not until 1927 did a Chinese physician, Dr. Fred Lam, succeed in persuading Washington that clonorchiasis was not dangerous to the American population.[47]

Eastern European Jews, too, occasionally found themselves the targets of unsympathetic PHS physicians at depots other than Ellis Island. Between 1907 and 1914, approximately ten thousand Jews entered the United States through the port of Galveston, Texas. A project of the Jewish Immigrants' Information Bureau in cooperation with the London-based Jewish Territorial Organization, Jewish philanthropists such as Jacob Schiff supported the Galveston Movement to stop the concentration of Jewish immigrants in the congested industrial cities of the northeastern United States by landing them in Texas instead of New York, and assisting them in finding jobs in the West. Galveston was the Ellis Island of the West. And as on Ellis Island, representatives of Jewish groups monitored medical exclusions carefully to be certain that illness, not anti-Semitism, was the cause. When rejections for "poor physiques" rose dramatically in early 1910, an investigation by Henry Berman, manager of the Jewish Immigrants' Information Bureau, disclosed that Galveston medical examiner Dr. Corput had been making anti-Semitic remarks, including a promise to do everything in his power to exclude Jewish immigrants. Protests to Washington brought some assistance when a new team of Immigration Bureau inspectors arrived and seemed more willing than those they replaced to discount questionable medical diagnoses. However, marginal calls often went against Jewish immigrants. As Jewish groups struggled to get every admission they could, angry immigration officials struggled equally stubbornly against the pressure.[48] In 1913, there was a rash of exclusions for acute inguinal hernia. On this occasion, Jewish representative, Henry Berman, supported Dr. Bahrenberg, who "showed considerable concern that the doctors on the other side do not make a more thorough examination." However, a considerably more skeptical Israel Zangwill wrote from London, "I cannot help feeling . . . that your doctor has developed a morbid flair for hernia. . . . I believe a doctor, like a patient, can find any disease he is looking for. . . . Some cases deported for hernia did not even know they had it, so it could hardly be much of a disability in earning their livelihood."[49] Very likely local pressures to exclude as many eastern European Jews as possible and the absence of the large Jewish organizational presence that existed on Ellis Island created an atmosphere in which some Galveston medical inspectors were willing to cooperate with local immigration opponents by misdiagnosing diseases certifiable for exclusion.

Perennially, the chief ground for rejection at all depots was physical unfitness, trachoma being the most frequent of the loathsome contagious diseases that were grounds for exclusion. In fiscal 1911, a fairly typical year during the peak period of immigration, physicians examined 749,642 aliens upon the newcomers' arrival at Ellis Island; 16,910 aliens, or 2.3 percent, were certified for physical or mental defects. Of these, 1,363 had loathsome or dangerous contagious diseases and 1,167 (85.6 percent) of these had trachoma.[50] Yet throughout the period and especially in the peak period from 1890 to 1924, those inspected at all U.S. immigration depots and returned to their ports of origin because of poor health never exceeded 3 percent of new arrivals in any given year, and the average for the entire period for all depots combined was well below 1 percent. Clearly, the medical inspection at American immigration ports, including Ellis Island, did not markedly reduce immigration to the United States. What did change was the percentage of those debarred for medical reasons. From less than 2 percent in 1898, the percentage increased to 57 percent in 1913 and 69 percent by 1916.[51] Thus, most immigrants who arrived were admitted, but of that small percentage rejected, an increasing proportion were being rejected for reasons of ill health, as documented by medical certification after examination, while a decreasing percentage were being excluded for other causes such as being subversives or anarchists, criminals, contract laborers, or members of the immoral classes (prostitutes).

In part, this increase in the number of immigrants excluded on medical grounds was a direct result of improved diagnostic techniques. In 1891, when PHS physicians first assumed the responsibility of inspecting individual newcomers, tuberculosis was generally diagnosed by auscultation, studying with the aid of a stethoscope the character of the sounds that cavities in diseased lung tissue produced. By 1910, taking X rays of the lungs was common practice, as were laboratory analyses of residue from patients' lungs.[52] Only a slide showing the tubercle bacillus in an immigrant's sputum was sufficient evidence to bar a newcomer for TB. Similarly, the Wassermann test markedly improved rates of syphilis diagnosis over mere observation and examination at immigration depots during the second decade of the twentieth century. Two Wassermann reactions "with an interval of not less than two days" soon became the requirement for certification of syphilis. Gonorrhea, too, could be certified only with a "microscopical examination of the discharge" that would have been of uncertain accuracy in an earlier era.[53]

The changes taking place in medical inspection procedures at all depots reflected increasing bureaucratic efforts to apply advances in medical knowledge in the national interest. Changes in the diseases certifiable for exclusion were added by statute at the recommendation of the surgeon general. However, federal physicians stationed at immigration depots were also sensitive to health information they received from local authorities in cities of departure and arrival concerning the appearance of epidemics. There was often considerable communication between local health officials and U.S. Marine Hospital Service physicians in port cities.

Between 1892 and 1900, the efficiency of the federal physicians varied greatly from port to port. In 1899, 348 newcomers were excluded because they had contagious diseases, over 95 percent of them at Ellis Island. Of these cases, 36 (10.3 percent) had favus and 298 (85.6 percent) had trachoma, "leaving only 14 [4.0 percent] as the total excluded from other ports or from New York with other contagious diseases."[54] According to Immigration Commissioner General Terence Powderly, New York was "where the medical is more thoroughly organized and operated than the funds available under the restrictions imposed by the immigration laws render possible at lesser ports, through which diseased aliens may with less difficulty secure admission."[55]

Whether or not Powderly's analysis was correct, the experiences of the federal physicians and the desire of lawmakers to make the nation safe from foreign germs resulted in the Immigration Act of March 3, 1903 (32 Stat. 1213), a thirty-nine–section law stipulating in detail grounds for exclusion, including medical grounds. A *Book of Instructions for the Medical Inspection of Immigrants* categorized excludable illnesses pursuant to the law. There were two excludable classes. Class A were "dangerous contagious diseases" and Class B were "all diseases and deformities which are likely to render a person unable to earn a living" (see Appendix I). By 1917, the final subcategory under B, "general considerations"—any disease or deformity that did not fit into one of the other categories, but that still "interfere with an immigrant's ability to earn a living"—became a separate category, C (see Appendix II).[56] The 1917 revision of the book also included many more specific conditions in categories

A and B. Improved technology and experience made it practical to specify conditions that would likely have gone unnoticed in 1903, such as nonpulmonary forms of TB.

The increasing number of excludable diseases and the escalation in the proportion of medical exclusions not only suggest improved diagnostic techniques but also that immigration officials had discovered the persuasiveness of "expert testimony" in the form of medical diagnoses as supporting evidence for immigration restriction. In an address to the senior class of 1904 at Princeton University, William Williams called for the PHS physicians, who were answerable to the surgeon general, not to him or even to the commissioner general of immigration, to cooperate in lending their medical diagnosis to the cause of social control. He said,

> I think that in all instances in which the U.S. Marine Hospital Surgeons who conduct the medical examinations at the immigration stations certify in writing that the physical condition of an immigrant, dependent for support upon his own physical exertions, is below a certain standard to be designated by them by some appropriate term, whether this be "low vitality," "poor physique," or some other similar expression, or that he is "senile," such immigrant should be excluded, subject to certain reasonable exceptions in the cases of very close relatives or persons who have resided here a given length of time.[57]

Recognizing the opportunity to shift the responsibility and onus for exclusion from policymakers to physicians who could cite "objective" medical criteria as justification for immigration restriction, Congress passed a law in February 1907 (34 Stat. 898) that gave physicians the option of stating on a medical certificate whether a particular illness or deformity might make a newcomer "likely to become a public charge."[58] William Williams was especially enthusiastic about this new opportunity. He saw in the new law a fresh mechanism for easing the administrative burden on the boards of special inquiry that heard appeals from immigrants facing debarment.

What Williams had not counted on was the reluctance of PHS officials both on Ellis Island and in foreign ports to exceed their function as physicians making medical diagnoses and to involve themselves in the decision to debar particular immigrants. While Williams lamented that more medical certificates were not being issued by PHS physicians, advocates of particular immigrant groups as well as physicians protested the use of these certificates as the sole basis of the case for exclusion. The former criticized physicians' participation in such matters because it called for a judgment that one critic characterized as "well beyond a professional medical opinion."[59] Public Health Service doctors objected because they refused to be agents of exclusion.

For similar reasons, PHS physicians refused to sit on the boards of special inquiry, which made the final decisions on exclusions. No physician ever sat on such a board. Physicians tried to keep their medical assessments separate from final decisions to admit or deport. They would not barter their professional ethics for the sake of bureaucratic harmony. Their adamancy triumphed. The episode suggests how the increasing professionalization of the medical profession and the application of professional ethical standards was replacing the notion of the federal physician as merely another bureaucrat in the growing machine of government. When forced to choose between their professional independence and bureaucratic discipline, physicians chose the former.[60]

President Theodore Roosevelt appointed Philip Cowen, former publisher of the English-language newspaper *American Hebrew* to the Immigration Service in 1905 and assigned him to Ellis Island, where Cowen served on the Board of Special Inquiry. Only successful appeals to the commissioners of immigration in Washington could overturn board decisions and those were rare. Cowen was especially moved by the plight of those given a negative diagnosis by PHS physicians. He recalled a Jewish woman from Russia, brought to spend her last years with her five economically well-off American children. "The doctors certified her as physically defective because of her eyesight, and as of low mentality, because she had failed in the Binet-Simon and other tests." Cowen explained that friends or relatives of an immigrant could bring a medical specialist to Ellis Island at their own expense if they wished to challenge the PHS officer's diagnosis. In this case, they did. The woman stated for the record that she could not see the time on a wall clock, but then added that she had seen well enough to raise her children, doing "all the housework and cooking without medical attention." While she could not assemble the jigsaw puzzles that physicians placed before her to test her mental acuity, she said that if they would allow her to have some meat she would make "a delicious soup for them," and if they gave her flour and other ingredients, she "would bake a loaf of bread finer than they served on Ellis Island." Without further comment, perhaps because of a lump in his throat, Cowen noted that the woman was admitted on appeal.[61]

On another occasion, fate as much as the Board of Inquiry saved from deportation a young Italian man who had a game leg that required use of a crutch and caused him to be certified by a physician as "likely to become a public charge." On that particular morning, the eminent philanthropist and diplomat Henry Morgenthau and his wife were visiting Ellis Island and were being given a tour by Immigration Commissioner Robert Watchorn. The Morgenthaus passed the detention room where the Italian and his young wife were seated and inquired about the case. At Mrs. Morgenthau's request, Watchorn took the case to the Board of Special Inquiry. Mrs. Morgenthau personally pledged sufficient support to ensure that the couple would not become public charges. Eventually the couple was admitted. While rare, such episodes suggest that boards of inquiry viewed newcomers with compassion and were quite willing to admit infirm immigrants as long as board members were confident that their decision would not add to society's economic burdens.[62]

No diagnosis troubled immigration officers and physicians as much as those of mental deficiency and insanity. The Immigration Act of 1882 had excluded "any convict, lunatic, idiot, or any person unable to take care of himself or herself without becoming a public charge" but offered no definitions or means of enforcement.[63] As early as 1890, the *Journal of the American Medical Association* reported that Dr. John Hamilton of the PHS had testified before Congress that if those who arrived insane or became so within their first year in America were returned to their country of origin, "there would be a diminution of one-third of the inmates in the government asylums."[64] Such concerns were embodied in congressional action. The 1903 Immigration Act cited the insane and epileptics among those to be excluded. Four years later, the list included imbeciles, the feebleminded, and the physically or mentally defective, among others. The potential threat seemed enormous, but precisely how to determine either mental deficiency or insanity remained puzzling.

In the 1903 PHS *Book of Instructions for the Medical Inspection of Immigrants,* insanity was defined for inspecting officers:

> Insanity is a deranged and abnormal condition of the mental faculties, accompanied by delusions or hallucinations or illusion, . . . homicidal or suicidal tendencies or persistent mental depression, or inability to distinguish between right and wrong. In the case of immigrants, particularly the ignorant representatives of emotional races, due allowance should be made for temporary demonstrations of excitement, fear or grief.[65]

The manual required that individuals suffering from delirium tremens be certified as insane, but that alcoholics and epileptics not be so diagnosed but be excluded as likely to become public charges. Inspectors were instructed always to include as evidence the physical appearance of those certified as insane. In 1905, Ellis Island Assistant Surgeon Thomas W. Salmon, who later became an internationally recognized authority on mental health problems, noted that during the year fifty-nine immigrants had been certified by medical officers at Ellis as insane and excluded, saving taxpayers $88,500—more than the entire annual cost of inspection at the station. Salmon urged that special forms be created for ships' surgeons, alerting them to physical signs useful in identifying the mentally deficient or insane prior to arrival. Salmon then outlined the signs that PHS physicians watched for to detect those in need of further testing:

> If the manner seems unduly animated, apathetic, supercilious or apprehensive, or if the expression is vacant or abstracted the immigrant is held and carefully examined. A tremor of the lips when the face is contorted during the eversion of the eyelids, a hint of negativism or retardation, an oddity of dress, unequal pupils, or an unusual decoration worn on the clothing—any is sufficient to arouse suspicion. The existence of well-marked stigmata of degeneration always serves to detain the immigrant for further inspection. . . . Old persons are invariably questioned to determine the degree of mental deterioration present, and as a result cases of senile dementia are sometimes found.[66]

The absence of proper facilities and adequate training sometimes resulted in tragedy. In June 1906, a detained alien had become "so obstreperous as to render it necessary to lock him in one of the cells, where he succeeded in grasping an electric light wire, with which he strangled himself." Several months earlier, a woman detained for mental observation had "climbed through a window and jumped to her death." The immigration commissioner at the port of New York, Robert Watchorn, who had succeeded William Williams, blamed the tragedy on the lack of proper facilities for caring for immigrants suspected of being mentally ill.[67]

In spite of Salmon's recommendations and best intentions, eyeball analysis remained the most frequent technique employed in identifying the mentally deficient until the second decade of the twentieth century. Dr. Grover Kempf, an Ellis Island physician from 1912 to 1914, recalled that "The mental examination of immigrants was always haphazard. It couldn't be any other way because of the time given to pass the immigrants along the line. Some questioning—if the immigrant did not respond or looked abnormal he was sent in and given a further examination."[68]

Though many immigrants remarked on the cold, impersonal demeanor of Ellis Island officials, inspectors were often sympathetic and understood that the trauma of the examination itself could cause aberrant behavior and that sensible answers might become muddled in translation. One wrote of the immigrants he observed:

> To say that many are in a state of mental frenzy is well within bounds. Is it then any wonder that many apparently below par on the first examination, give, after a good night's rest combined with quiet and good food, a normal reaction? Even the feeble-minded frequently show some improvement. It must also be remembered that an alien can rarely be addressed in his native tongue by the examiner. . . . Interpreters will at times give voice to their own interpretations of the subject's answer rather than a literal translation.[69]

Interpreter Fiorello H. LaGuardia admitted that his heart went out to newcomers subjected to the psychological examination, describing his feelings in his autobiography:

> I always suffered greatly when I was assigned to interpret for mental cases in the Ellis Island Hospital. I felt then, and I feel the same today, that over fifty percent of the deportations for alleged mental disease were unjustified. Many of those classified as mental cases were so classified because of ignorance on the part of the immigrants or the doctors and the inability of the doctors to understand the particular immigrant's norm, or standard.[70]

LaGuardia was especially haunted by what happened to a young Italian teenage girl from the mountains of northern Italy:

> No one understood her particular dialect very well, and because of her hesitancy in replying to questions she did not understand, she was sent to the hospital for observation. I could imagine the effect in this girl, who had always been carefully sheltered and had never been permitted to be in the company of a man alone, when a doctor suddenly rapped her on the knee, looked into her eyes, turned her on her back and tickled her spine to ascertain her reflexes. The

A mental examination on Ellis Island. (U.S. National Archives)

> child rebelled—and how! It was the cruelest case I ever witnessed on the island. In two weeks' time that child was a raving maniac, although she had been sound and normal when she arrived on Ellis Island.[71]

Had the young woman arrived feebleminded or psychotic or neither? It was this confusion that led to the development of mental tests by an Ellis Island physician and invitations to researchers interested in feeblemindedness to use the depot as a testing laboratory.

Dr. Howard Knox, who served as chief medical officer on Ellis from 1910 to 1916, developed several tests to determine the mental fitness of immigrants. One was a puzzle that consisted of wooden boards from which cookie-cutter blocks of various shapes had been cut. A PHS physician asked each immigrant pulled from the line for further inspection to remove the pieces and replace them in the proper slot. Another was a cube test in which the examiner touched four or five different cubes one after another in a definite order. The subject tried to imitate the examiner, touching the same cubes in the same order. Failure usually resulted in debarment. Knox, who received his bachelor's and medical degrees from Dartmouth College, had a colorful military career as a medical officer with Gen. John J. Pershing in Mexico on the trail of Pancho Villa but no more experience with psychological testing other than most other physicians of the era. Still, Knox's test became a standard diagnostic instrument for feeblemindedness on Ellis.[72]

The efforts of experienced researchers ultimately developed techniques no more certain than Knox's for diagnosing mental deficiencies. The most controversial of these experiments were conducted by one of the most renowned of this country's mental testers. In 1910, the PHS, troubled by journalists' and politicians' charges that it was failing to exclude "undesirables," invited the eminent H. H. Goddard, director of research at the Vineland Institute for Feebleminded Girls and Boys and the originator of the term *moron,* to see if he could identify among arriving immigrants these seemingly normal mental defectives who eluded the untrained eye but whom Goddard and others believed to be responsible for a host of social problems.[73] Goddard, who had introduced the Binet-Simon test to the United States in 1908, believed that feeblemindedness was both detectable and heritable. He was an active member of a study group investigating feeblemindedness that had been organized by the Committee on Eugenics of the American Breeders Association and collaborated with eugenicist Charles Davenport.

Goddard conducted three studies. The first, in 1912, involved inquiring of institution superintendents what percentage of their inmates were foreign-born or the children of foreign-born parents. The data reported by Goddard did not sustain the view that immigration was a major sieve through which the feebleminded poured into American society. Goddard wrote that he had gathered "no adequate statistics to indicate that any one race or nationality is more inclined to mental defectiveness than another" and that the notion of an "undue percentage" of immigrants being mentally defective seemed to him "grossly overestimated." In fact, of the inmates of institutions that had recorded nationality, only 4.5 percent were foreign-born.[74] Well-known families of mental defectives that Goddard had studied, such as the Kallikaks, were all of old native-born stock, not recent immigrants.

The second test was in May 1912, when Goddard returned to Ellis Island with two assistants to try the Binet test on immigrants. One assistant selected newcomers who appeared to be mentally defective. They were taken to another room and tested with the Binet scale. Goddard reported that all nine cases that appeared to be mentally defective were confirmed as such by the test, while a control group of three all tested normal. Later, he tested forty-four immigrants, thirty-three of whom were selected by the Vineland team and eleven of whom were selected by the PHS physicians. Goddard proudly announced that the Vineland workers were corroborated in their choices of mentally defective from the line inspection better than 80 percent of the time. The physicians were accurate less than half the time. Proud but gracious, Goddard made clear that he did not intend to raise doubts about the physicians' expertise, which was in medicine. On the contrary, he wrote, "They do not pretend to be experts on feeble-mindedness. The comparison simply shows what experts can do."[75]

Some PHS physicians were critical. Dr. Henry Knox, who had developed the cube test to detect mental defectives among the newcomers, was skeptical of the results, suggesting that the Vineland workers were culturally biased, noting "to the uninitiated using routine tests for defectives, nearly all peasants from certain European countries appear to be of the moron type; of course this is a fallacy."[76] Four years later, another PHS physician, E. H. Mullan, who like Knox had developed his own variation of a wooden block test for use with immigrants, also demonstrated an awareness of the need to take cultural differences into account when testing immigrants. He said he distrusted the Binet test because "whether or not it is suitable for French or American schoolchildren, [it] is not appropriate to the crude material of the immigrant." Still, Mullan was confident that every precaution was being taken to avoid exclusion through error. He remained certain that "every case certified here as feebleminded is absolutely and without doubt feebleminded."[77]

Goddard's third and final test was his longest and aroused the most controversy. Over seventy-five days in 1913, thirty-five Jews, twenty-two Hungarians, fifty Italians, and forty-five Russians were given the Binet-Simon intelligence test. Of these, Goddard found that 83 percent of the Jews, 87 percent of the Russians, 80 percent of the Hungarians, and 79 percent of the Italians were either feebleminded or below the mental age of twelve. Over three-fourths of the immigrants from these four major groups were morons by Goddard's definition. Somewhat embarrassed by his findings, Goddard "corrected" his data, though inadequately, and still published figures suggesting that over half of the newcomers were morons. At first, Goddard had hesitated, turning down an offer from *American Breeders Magazine,* but in 1916 he reported his results in a paper at the American Psychological Association meeting; in 1917 the article appeared in the *Journal of Delinquency.*[78]

Many Immigration Bureau staff retained a skepticism of psychological testing, often telling apocryphal stories that suggested better ways than the Binet tests to determine whether or not an immigrant had the right stuff to be an American. Immigration inspector Philip Cowen describes such a conversation between a physician and "a heavy-set son of Erin" who seemed qualified:

> "Pat, if I gave you two dogs and my friend here gave you one, how many would you have?" "Four sir," says Pat. "Did you ever go to school, Pat?" "Yes, indeed, sir." "Now Pat, if you had an apple and I gave you one, how many would you have?" "Two, sir." "And if my friend gave you one, how many would you have?" "Three, sir." Then repeating the original dog question, the answer was again "Four, sir." "Why, Pat, how is that?" "Why, sure, I've got a dog home meself."[79]

Despite the quips and barbs of some PHS physicians, there seems little doubt that the attention Goddard's studies called to the issue of feeblemindedness alerted Ellis Island medical officers to an area where inspection procedures were simply inadequate and must be improved. An increase in the number of immigrants excluded for feeblemindedness suggests that PHS physicians were responding to the pressure. Dr. L. L. Williams, chief medical officer on Ellis Island in 1914, observed that deportations for feeblemindedness were on the rise, from 186 in 1908 to 555 in 1913, and perhaps would exceed 900 by the end of 1914.[80] The percentage of newcomers affected was still small, but growing. In the spring of 1914, Dr. Howard Knox estimated the deportation rate for feebleminded newcomers to be 157 per 100,000.[81]

Immigrant advocates argued that newcomers had no greater tendency to be insane or of low mental ability than the native-born. At immigration depots, agents of the Hebrew Immigrant Aid Society (HIAS) and the Italian St. Raphael Society did not hesitate to argue with the PHS if they thought officers were certifying immigrants of their group mentally deficient when the real problem was unfamiliar language and culture. Dr. Antonio Stella, a distinguished Italian physician who consulted for the Manhattan State Hospital for the Insane, denied the validity of intelligence test results on the grounds that such tests were culturally biased. He even more vehemently denied a connection between mental capability and a propensity to mental illness.[82] HIAS often hired private physicians to reexamine rejected immigrants in the hope of winning an appeal. One psychologist, challenging a rejection for feeblemindedness, observed, "there is no specific test which enables one to detect feeblemindedness, for after all we are not dealing with concrete things, the human mind should be looked upon from the point of view of individual make-up, environmental influence and educational advantages."[83] Immigration officials regarded such testimony warily because, "although some of these opinions are undoubtedly honest, unfortunately there are a few physicians who, while declaiming the admission of defective aliens in the abstract, will in individual cases attempt to obstruct the operation of the law by opinions which are plainly disingenuous."[84]

On the one hand, Goddard seemed to stick by his data, arguing that "one can hardly escape the conviction that the intelligence of the average 'third class' [reference to steamship booking] immigrant is low, perhaps of moron grade."[85] Be that as it may, Goddard also observed that "a very large percentage of these immigrants make good after a fashion," their "relatively low mentality" to the contrary notwithstanding. Puzzled by the social implications of all his studies and his own personal observations, Goddard mused over whether or not the mental classification of moron mattered, given "that there is an immense amount of drudgery to be done, an immense amount of work for which we do not wish to pay enough to secure more intelligent workers." Unable to dismiss the usefulness of the term *moron* in describing new arrivals from an array of diverse cultures, or to divorce himself from data that he knew were flawed, a stubborn, bemused Goddard wondered instead, and without irony, "May it be possible the moron has his place?"[86]

Of even greater interest to the government than mental tests were its ongoing studies of germs. During World War I, the slower rate of immigration allowed PHS physicians in the laboratory of the Ellis Island hospital to conduct some medical research, using newcomers as subjects. There was special interest in venereal disease. In June 1915 the surgeon general reported that in addition to testing those actually suspected of having syphilis, Ellis Island physicians had done "about 1,000 Wassermann reactions" on persons in the line inspection. There were also "cultural studies on the gonococcus (bacterium causing gonorrhea) in comparison with M. catarrhalis (bacterium causing catarrh, an inflammation of the mucous membranes, especially of the nose and throat) and the meningococcus (bacterium causing meningitis)." The nation's chief medical officer further reported that studies of trachoma were made with "culture and smears" and that there had been "inoculations into animals."[87] The report concluded with a request for another PHS officer in the laboratory because "there is not another place in the service where there is more opportunity for good research work from the laboratory side; and the expenditures involved would be very modest."[88] Long before the vocabulary of patient rights entered the medical lexicon, immigrants processed on Ellis Island served haplessly as the subject for psychological and physiological research.

There is no epidemiological evidence that the physical condition of immigrants arriving in American ports at the end of the nineteenth century triggered an epidemic or other specific public health crisis in the United States, but policymakers in Washington had no way of knowing this would be the case. No longer content with state-supervised quarantine procedures, the federal government took it upon itself to protect the nation against sickness from abroad.[89] The shift was partly reflective of the wider acceptance of germ theory and a greater understanding of which diseases were highly contagious and which were not. It was also a defensive reaction to the upsurge in immigration, especially from southern and eastern Europe and Asia, which aroused fears of a massive alien public health menace and threat to the genetic vitality of the American population. That periodic cholera epidemics and outbreaks of yellow fever had been blamed on the foreign-born in an earlier era now made linkage of infectious disease and immigration seem more than hypothetical to many. The concern that poor health and innate inferiority would make a substantial portion of the population dependent on the rest had also long been a concern among Americans. There was growing confidence on the part of some Americans that only the federal government had the jurisdiction and the organizational potential to marshal the nation's expertise and resources to cope with large-scale immigration and such attendant problems.

Historian of medicine Charles Rosenberg has reminded us that "medical knowledge is not value-free . . . but, at least in part, a socially constructed and determined belief system, a reflection of arbitrary social arrangements, social need, and the distribution of power."[90] Diseases, too, are negotiated constructs having "biological and social dimensions."[91] On Ellis Island, each diagnosis constituted a negotiation between the biological and the social. Even as most PHS officers consciously refused to allow the agenda of immigration restrictionists to influence their medical assessments, the physicians were unconsciously swayed by their own ethnocentrism. Two such doctors, writing on trachoma, concluded that a perceived increase in the disease on Ellis was a direct result of the "change in the source of arriving immigrants and resulting differences in the character of the people."[92] Dr. Alfred C. Reed, a PHS physician on Ellis, made even more explicit the relationship between the biological and the social: "No alien is desirable as an immigrant," Reed wrote, "if he be mentally or physically unsound, while on the other hand, mental and physical health in the wide sense carries with it moral, social, and economic fitness."[93]

But scrutinizing immigrants closely for specific diseases and debarring them accordingly was no mere matter of ethnocentricity. In a very real sense, PHS physicians navigated between their medical oath to minister unto the individual and their statutory responsibility to guard the health of the public at large. In steering between those two, often conflicting, charges, PHS officers made of Ellis Island an incubator for public health policy and a laboratory for scientific experimentation, a Progressive barricade against the germs and genes that Terence Powderly feared would leave Americans the blind, bald victims of immigration's silent travelers.

4

A Plague of Nativism: The Cases of Chick Gin and "Typhoid Mary"

The medical inspection of individual immigrants on Ellis Island and at other immigration depots did not allay Americans' apprehensions that these newcomers were detrimental to the public health. Officials responsible for protecting the public health, especially United States Public Health Service physicians, state and local health officers, and elected officials continued to find themselves pulled in at least three directions: sound public health policy grounded in state-of-the-art medical understanding; nativists' clamor for excluding immigrants, or at least restricting their behavior as public health menaces; immigrant claims of protection from scurrilous accusations of public health endangerment and unreasonable deprivation of their civil liberties. "America beckons, but Americans repel," ran a familiar immigrant slogan at the end of the nineteenth century. Too often, immigrants were repelled because their very appearance suggested to their hosts' gazes a physical inferiority or vulnerability that the native born feared might be contagious.

Even those sympathetic to immigrants, or at least to particular groups, admitted that newcomers' appearance could evoke repugnance, perhaps persecution. The Chinese, especially, produced such visceral reaction in native-born Caucasian Americans, according to the Stanford University sociologist Mary Coolidge, who had the opportunity to observe it firsthand at the turn of the century: "With physical and social characteristics so different from the rest of the population it was perhaps, inevitable that the Chinaman with his flowing trousers and queue should be a conspicuous mark for race persecution in California at a time when feeling against all foreigners was very strong," the sympathetic Coolidge observed.[1]

Worse yet was the possibility that the newcomers' outward appearance was a clue to even more significant internal distinctions. Some might already be diseased or well on the road to illness, their flawed habits of health and hygiene responsible for their unusual appearance and their vulnerability to contagion. Americans feared that if they risked contact with these aliens, they, too, might find themselves sickly and physically diminished. Whether it was immigrants' faces, their physiques, or their habits of personal hygiene, many native-born Americans feared newcomers as health threats and hoped that the increasingly sophisticated health codes and laws would protect them from the sickness that lurked within and around immigrants. Still, reactions to particular immigrant groups differed markedly.

Chick Gin was Chinese. Mary Mallon was Irish. Both were immigrants to the United States in the late nineteenth century. They might have lived and died in anonymity, as did the vast majority of the "huddled masses" to arrive in the United States between 1880 and 1924. Certainly neither appeared markedly different from other newcomers of their respective groups. However, appearances can be deceiving. Because of bacilli that had invaded their bodies and over which they had no control, Chick Gin's death and Mary Mallon's life sowed consternation and fear among a portion of the citizenry in the land where each had chosen to begin a new life. Their stories are case studies suggesting the diversity and complexity of how Americans perceived and responded to immigrants who were alleged public health threats. A comparison of two very different immigrant experiences demonstrates how preexisting prejudices, immediate political rivalries, jurisdictional disputes among local, state, and federal authorities, and social perceptions of scientific medicine played roles in shaping public reaction to the interconnection of public health and the foreign-born.

Substantial numbers of Chinese immigrants first began arriving on the West Coast in the early 1850s, meeting equally hopeful non-Chinese emigrants from the eastern United States in the gold fields of California. In the decade from 1850 to 1860, 41,400 Chinese arrived, most through West Coast ports. Instant wealth eluded Asians and non-Asians alike in California, low-paying jobs proving far more abundant than gold nuggets. The Chinese, especially, provided a plentiful source of low-cost labor to companies such as the Central Pacific Railroad, laying the track that would bind the nation into an increasingly vast market.

Legally excluded from citizenship, the Chinese emperor's subjects living in the United States served American industry and agriculture in increasing numbers. In shops and factories they rolled cigars, made reins and saddles for the horses that hauled people and commerce, and because the Chinese did not define clean clothes as the proper product of female labor alone, cleaned Caucasians' clothes in laundries. Chinese not in mines or factories or laundries cooked and cleaned for Caucasians, some hoping to save enough to return home wealthy, others planning to go back to China only long enough to find a bride and return and resume their pursuit of prosperity. Few succeeded.[2]

The productivity of Chinese workers did not earn them respect among native-born Caucasians. In fact, undertaking hard labor for low pay contributed to one of the stereotypes that anchored anti-Chinese prejudice, one branding Chinese as all "coolie laborers" who would undermine the dignity and wage scale of American workers.

In the eyes of most nineteenth-century American Caucasians, Chinese workers were stereotypically perceived as little better than the black slaves. Moreover, many Caucasian Americans perceived the Chinese as the product of a heathen tradition, politically bound to a despotic emperor, and like all primitives in the nonwestern world, morally degenerate. The Chinese might view themselves as the subjects of a Heavenly Kingdom and the heirs of an ancient and honorable civilization, but their Caucasian hosts saw Chinese culture as intellectually backward, undemocratic, and technologically primitive.[3]

In his 1855 volume about California, the antebellum Southern critic of black slavery, Hinton Rowan Helper, had called the Chinese he encountered "semibarbarians" as out of place among California's growing population as "flocks of blackbirds . . . in a wheatfield." Helper equated the Chinese in California with black Africans in the South Atlantic states as inevitably destined by their inferiority to become "subordinate to the will of the Anglo-Americans," as had been "the Irish in the North" or "the Indians in New England." A better prophet than he knew, Helper said that he would not be surprised "if the copper in the Pacific yet becomes as great a subject of discord and dissension as the ebony of the Atlantic."[4]

In 1862, Dr. Arthur B. Stout, a California physician and a prominent figure in the American Medical Association, had published a report that carried its conclusion in the title, "Chinese Immigration and the Physiological Causes of the Decay of a Nation." Based upon a collection of dubious data, much of it merely the common wisdom of racial nativism, Stout's report warned that "phthisis or consumption, scrofula, syphilis, mental alienation, and epidemic diseases" were scourges posed by the Chinese, aggravated by their habitual opium smoking. He suggested there were high rates of syphilis among the Chinese, but found their marked fertility inconsistent with the sterility that often resulted from the disease. Stout's report might have been lost in the sea of equally muddled, irresponsible anti-Chinese tracts had the author not been recruited by the secretary of the California Board of Health and a former president of the American Medical Association, Thomas M. Logan, to investigate the harm to San Franciscans that might result from "the combined intermixture of races and the introduction of habits and customs of a sensual and depraved people in our midst . . . with hereditary vices and engrafted peculiarities."

Stout's final report to Logan was couched in terms of a biological struggle that foreshadowed the theories and rhetoric of social Darwinism. What is especially clear is that Stout saw the Chinese as having transformed the level playing field of fair social competition in the United States into a battlefield where Americans faced defeat by "invisible approaches" that could "insidiously poison the well-springs of life, and spreading far and wide, gradually undermine and corrode the vitals of our strength and prosperity." Having made the analogy to a land invasion by enemy forces that could be opposed with "sword and rifled cannon," Stout said that the public health threat posed by the Chinese was greater than if "the hordes of Genghis Khan should overflow the land, and with armed hostility devastate our valleys with the sabre and the firebrand than that these more pernicious hosts in the garb of friends."[5]

American lawmakers in Washington codified their reticence to grant citizenship generously to the foreign-born and their special resistance to the naturalization of Asians. The Fourteenth Amendment stipulated that "all persons born or naturalized in the United States and subject to the jurisdiction thereof" were citizens. Chinese born in the United States were clearly citizens. When Chinese-Americans—mostly males because of the labor migration patterns—went to China to marry and have children, these offspring were citizens and could enter the United States as the children of a citizen. However, the spouse was not eligible for immigration because of the marital tie. Only when the 1906 San Francisco fire destroyed birth records did some spouses manage to elude the law, falsely claiming U.S. nativity. As for citizenship by naturalization rather than birth, the door was slammed shut in 1870. The Naturalization Act passed that year limited citizenship to "white persons and persons of African descent," making Chinese immigrant aliens ineligible for citizenship, a status not changed until 1943. Their race made Chinese arrivals targets of prejudice. Their legal status left them with limited legal or political recourse.

Throughout the 1870s, California Caucasians blamed the Chinese for the presence of various diseases, arguing that the public health menace posed by the Chinese could be curbed only through immigration restriction. California Caucasians blamed an epidemic of smallpox on the Chinese. Some 1,646 cases of smallpox were reported to the San Francisco Health Office between May 19, 1876 and July 1, 1877; 482 died, including 77 Chinese. Dr. J. L. Meares, the health officer of the city and county of San Francisco, observed, "As small-pox [*sic*] is much more fatal among children than adults, and as there are but few children among the Chinese, comparatively, the inference is natural and just that, in proportion to the number of cases, there were fewer deaths than among the whites." Meares estimated that approximately 300 additional cases had been concealed by the Chinese, bringing the total number of cases to 1,946, an increase of 37 in the number of smallpox cases in this epidemic as compared with the previous epidemic in 1868–69.[6] Why blame the Chinese for the smallpox outbreak?

Meares's answer was unequivocal. He attributed responsibility for the epidemic's severity to the increase from approximately 4,000 to 30,000 of "unscrupulous, lying and treacherous Chinamen who have disregarded our sanitary laws, concealed and are concealing their cases of small-pox, which are only known to exist by the certificates of their deaths furnished by the City Physician, unless by accident some living case is discovered."[7] During the earlier epidemic, there had been few Chinese in the city and few cases among them. However, in the late 1870s, the combination of reported and concealed cases estimated by Meares at three hundred, minimum, persuaded him of a linkage between the Chinese and this latest assault on the public health of his city. He complained that not only did the Chinese fail to report smallpox cases, they even disposed of the bodies of the deceased "to some obscure place from the residence in which they died, so that it is impossible to disinfect their houses, for by no ingenuity can it be discovered whence the dead bodies have been removed." And without the corpses, autopsies badly needed by public health authorities to confirm the cause of death and track the epidemic could not be performed. Caucasian San Francisco health officials, then, came to see the Chinese quarter as little more than a "laboratory of infection" situated in the very heart of their city, "distilling its deadly poison by day and by night and sending it forth to contaminate the atmosphere of the streets and houses of a populous, wealthy and intelligent community."[8] To some it seemed no more than the realization of earlier, dire predictions.

Leprosy, too, was identified by social critics of Chinese immigration and concerned physicians as reason to exclude Chinese from the United States. This disease shrouded in myth and mystery has often been depicted as an expression of divine punishment in Western literature, including the Old and New Testaments. Even today, the etiology of leprosy is not thoroughly understood. It is known to be a chronic infectious disease primarily affecting the skin and peripheral nerves, although other tissues may also be affected, including the eye, the mucosa of the upper respiratory tract, muscle, bone, and testes. Because its symptoms are similar to those of other diseases, diagnosis is often uncertain.[9]

By 1870, Caucasian California politicians and journalists were already employing "leprosy" among their litany of metaphors to smear the Chinese on the West Coast of the United States as loathsome. During a federal debate over whether or not individual states would be violating the Burlingame Treaty with China by excluding Chinese from their borders, California Congressman James A. Johnson claimed that the price of allowing residence to "the Hottentot, the cannibal from the jungles of Africa, the West India negro, the wild Indians, and the Chinaman" would be a youth suffering from "rotting bones, decaying and putrid flesh, poisoned blood, leprous bodies and leprous souls."[10]

A 1,200-page report on the state of Chinese immigration presented by a joint congressional committee in 1876 drew upon the testimony of physicians, journalists, labor leaders, and so many others to support the conclusion that the flow of Chinese arrivals must be stemmed.[11] In his testimony before the investigating committee, Dr. C. C. O'Donnell offered the inflammatory and unsubstantiated conclusion that there were 150 Chinese lepers roaming the streets of San Francisco, walking time bombs of infection.[12] News reports that yet more Chinese were in flight from a famine in 1878 only hastened congressional action, as journalists made explicit the comparison with the effects of the potato famine upon Irish emigration to the United States thirty years earlier.[13] By the end of the decade, the leprosy stigma was being slung at the Chinese by Dennis Kearney's nativist clique, who claimed that white workers were losing jobs to low-salaried "coolie labor" from China and other sinophobes in California. In at least one sadistic incident reported even on the opposite coast in the *New York Times*, Kearney's men drove disfigured Chinese males around the city, cruelly putting them on display before crowds to suggest what leprosy spread by the Chinese could do to the human form, though there was no positive evidence that these particular individuals were leprous.[14]

In 1882, the United States Congress passed the Chinese Exclusion Law. The overwhelming congressional support for the Exclusion Law, which transcended all regional divisions, suggests that Americans across the country, not just in California, viewed the Chinese with alarm.[15] But excluding Chinese laborers from emigrating to the United States did not address the alleged public health threat posed by those already arrived. Because the Chinese in San Francisco and other cities rarely consulted physicians who practiced Western medicine, were barred from many health care institutions, and saw no reason to fuel nativist suspicions by complying with health department regulations on reporting contagious diseases, there were no reliable data with which to refute charges that the Chinese were health menaces, but neither was there conclusive proof that they were.[16] Still, when health authorities claimed to have identified a case of bubonic plague in San Francisco's Chinatown, few Caucasians believed that the Black Death would stop at Chinatown's borders.

In 1900, Chick Gin was no longer an immigrant youth who believed that his return to China was imminent. He was forty-one years old and the proprietor of a wood yard, having emigrated at the age of twenty-five in 1884. In January, he fell ill and consulted Dr. Chung Bu Bing. He felt weak and feverish and complained of pain in his head, chest, back, and bladder. His physician diagnosed him as having inflammation of the bladder and prescribed medication. Nine days later Chick Gin's fever was down and his pains were gone, but in their place was a swelling of the glands in his groin and a stiffness on his right side. A urethral discharge caused him to suspect that he had gonorrhea, a not uncommon malady among bachelors in the United States. Chinese bachelors, cut off from family and the possibility of finding a Chinese bride by restrictive immigration laws, were no exception. Perhaps dissuaded by the traditional Chinese code of personal modesty, Chick Gin refused examination. Instead, he accepted more medicine and returned to bed. After two weeks of vomiting and diarrhea, he collapsed and died in the basement of a flophouse, where he had rented a bed.[17]

San Francisco health authorities required examination of all dead Chinese who had not been under the care of a Caucasian physician or been attended by a Caucasian care giver. On the morning of March 6, the police summoned to the dingy hotel where Chick Gin's body had been discovered Dr. F. P. Wilson, whose job it was to conduct such postmortems. The emaciated appearance of the body and the inguinal swelling aroused Wilson's suspicions sufficiently for him to alert A. P. O'Brien, San Francisco's chief health officer. O'Brien notified municipal bacteriologist Dr. Wilfred H. Kellogg, who upon microscopic examination suspected bubonic plague. Kellogg took smears from the enlarged glands to the well-equipped bacteriological laboratory at the Angel Island quarantine station run by the United States Marine Hospital Service and under the direction of Dr. Joseph J. Kinyoun. However, before he did, Kellogg disclosed his suspicion of plague to O'Brien and John Williamson, president of San Francisco's board of health. That evening, San Francisco's panicky health authorities took drastic action even before the lab results arrived.[18]

The chief of police was summoned and ordered to impose a total blockade on the Chinese quarter in San Francisco. The chief sent thirty-two of his men to evacuate all Caucasians whom they could locate and then to cordon off the area. None but Caucasians might enter or leave. On the morning of March 7, 1900, approximately twenty-five thousand Chinese residents awoke to find themselves separated from their neighbors by a rope looped around their fifteen-block neighborhood, subjects of quarantine.

Even the words "bubonic plague" brought images to mind of the Black Death pandemic of the fourteenth century that killed vast numbers of Europeans. Those infected with the bacterium *Yersinia pestis* begin the agonizing cycle of symptoms with fever, chills, weakness, and headache quickly followed by swelling of lymph nodes of the groin, neck, and armpits, all places in the body where the bacteria multiply. The malady is called bubonic plague from the Greek word for groin, *bubo,* because swellings in that region are most dramatic, at times enlarging to the size of a baseball. Within three to five days the untreated condition results in an excruciatingly painful death for approximately 60 to 90 percent of patients.[19]

In the late twentieth century, the rare cases encountered in Western countries are treated successfully with powerful doses of antibiotics. In 1900, antibiotic therapy did not exist and the epidemiology of the disease was not thoroughly understood. Plague is a zoonotic disease, an ailment shared by humans and lesser creatures, especially rodents. Humans receive the infection from the bite of infected rodent fleas. Humans cannot communicate the disease to each other unless the bacterium invades the lung, causing pneumonic plague, which is spread through the droplets produced by coughing. The latter form of the plague is rare, but at the turn of the century, even medical experts were not aware of the distinction between the two kinds of plague or the role of rodent fleas in spreading bubonic plague. At the century's turn, physicians and public health experts believed that plague was always contagious among humans but could be communicated from rats to humans as well. Bacteria were thought to be emitted by victims, whether human or rodent, and thrive in oil, air, or food. Germs were accepted as the cause of the disease but a kind of vestigial sanitarianism caused many to believe that the disease was most effectively assaulted by scouring the environment that nurtured the contagion.

Also, bubonic plague was difficult to diagnose. Still insufficient medical understanding resulted in disputes among medical experts. Few California physicians were sufficiently comfortable with bacteriology—a world of microscopes, slides, lab animals, and petri dishes of germ cultures—to rely on these diagnostic tools. They preferred clinical diagnoses, but early symptoms resembled those of typhoid or typhus fever: high body temperature and exhaustion. Moreover, the buboes, or enlarged and painful lymph nodes in the groin area, were commonly associated with venereal disease, not bubonic plague. Most physicians continued to believe that plague infection was the result of inhaling or ingesting dust particles, a perspective that seemed to justify a public health strategy grounded in the elimination of filth. Some experts thought Chinese vulnerability to plague a product of their rice-based diet, the absence of animal protein leaving their bodies unable to fight off the disease.[20]

The history of plague, then, and the state of medical knowledge about the disease did not make unreasonable the horror that San Francisco public officials felt when they met on March 6 to assess the situation. Quarantine of those suspected of being infected with a contagious disease was a familiar public health preventive. However, the decision to string a rope around Chinatown was an unrealistic response by officials wishing to demonstrate their ability to act quickly and decisively, regardless of the quarantine's meaning for the Chinese on the other side of the rope. Preexisting prejudices and stereotypes drove San Francisco public policy. After all, there had been a long record of public criticism of the unhealthy aspects of Chinatown, and as recently as a decade earlier, a possibly prescient San Francisco health officer had referred to the community as "this plague spot."[21]

Chinese community spokesmen reacted angrily and indignantly to the racism, but not without appreciation for the potential health threat. A newspaper editorial accepted the notion that some houses might have to bear the burden of quarantine, but "never have we heard of blocking the whole town."[22] Chinese merchants publicly complained of the financial losses they anticipated, while crowds gathered outside the offices of the Chinese Consolidated Benevolent Association, more widely known as the Chinese Six Companies, to demand action. The association was cautious and did not issue San Francisco officials a strong rebuke, but the Chinese consul general threatened legal action.

What the restriction meant to Chinese workers unable to get to their jobs if the quarantine continued and fearful that their positions might be forfeited to immigrant competitors of another group can only be imagined. However, a poem that appeared in the *San Francisco Examiner* suggests that employers of Chinese domestics felt as put upon by the quarantine as their servants, though the pervasive racism is not absent even from the protest of the quarantine's foolishness:

Scorn not the humble Chinaman,
Throw not his uses down.
For, as I live, we miss him when
He stays in Chinatown.
When happy Yip and Yellow Sin
Quit the domestic scene,
We have to do the work ourselves
And damn the quarantine.

So ere's to you, yellow Hop Sing Fong,
We're sorry that you're took.
You're a poor benighted 'eathen, but

A first-class fancy cook.
They say your deeds are bloody and
Your morals are unclean.
But goodness, how we miss you
When you're held in quarantine.[23]

On a more sober note, politicians of both parties were quick to accuse their opponents of using the episode for partisan advantage. Republicans charged that appointed health authorities, beneficiaries of Democratic patronage, were manufacturing a public health crisis to justify their requests for higher budgets, some of which might line their own pockets. The ruse was contemptible but also counterproductive according to the Republican press, because the charge of plague, even if untrue, could "terrify the community, paralyze commerce, turn away strangers, and prevent the visits of even neighbors and friends," crippling business and tourism.[24] Plague and the quarantine continued to be the subject of doggerel and disputation, with little attention to the grievances of the Chinese community in the days after authorities discovered Chick Gin's body and until the test results confirmed that he had indeed been a victim of the Black Death.

Even before it was confirmed that the Chinese immigrant had died of bubonic plague, Walter Wyman, the supervising surgeon general of the Marine and Hospital Service had a professional interest in bubonic plague. Now he advised local health authorities via federal officers in San Francisco that several steps be followed including the disinfecting of Chinatown with sulfur, treating the already exposed with a therapeutic serum developed by the French bacteriologist Alexandre Yersin and inoculating previously unexposed Chinatown residents with a prophylactic vaccine developed in 1897 by the Swiss bacteriologist Waldemar Haffkine. As a precaution, he sent supplies of both the serum and the vaccine to San Francisco. While Wyman was growing more concerned about the situation, those in charge on the other side of the continent were beginning to relax and gave in to demands that the quarantine be lifted. The end of the quarantine evoked satires of the "serious" men of medicine:

Have you heard of the deadly bacillus,
Scourge of a populous land,
Bacillus that threatened to kill us
When found in a Chinaman's gland?

Have you heard how the germ incubated
Till a billion bacilli were bred?
How the monkey was then vaccinated,

And the guinea pigs eke, it was said?

Well the monkey is living and thriving,
The guinea pigs seem to be well,
And the Health Board is vainly contriving
Excuses for having raised the deuce.

And the advertised, boasted bacillus
Is a gentle domestic concern,
And the doctors who fill us and pill us
Have libeled it sadly we learn.[25]

The victory verse was premature, as it turned out. On March 11, three laboratory animals inoculated with tissue from Chick Gin's body died.

Even before word got out of the results, some Chinatown residents had begun to flee on the chance that a quarantine might be reimposed. The Board of Health, aware of the magnitude of the crisis, met on the evening of March 11 and called for volunteers to inspect every house in Chinatown to identify additional plague cases and prepare for scouring and disinfecting the quarter. Attorneys hired by the Chinese consulate and the Six Companies were allowed to be present and concurred with the board's decision and pledged their clients' cooperation. Dr. J. Murphy, one of those representing the Six Companies, requested "that you treat those people [the Chinese] kindly and that general instructions to your inspectors will be issued to that effect."[26]

The following day both the Six Companies and the consulate posted notices throughout Chinatown urging public cooperation with authorities. However, residents of Chinatown, frightened by the possibility of another quarantine, secluded the sick and the remains of the dead in an intricate network of tunnels below Chinatown or floated them into the bay on small boats.

In the days and weeks that followed, California's newspapers warred over whether plague really existed and whether the San Francisco Board of Health had taken appropriate action. Not surprisingly merchants both Caucasian and Chinese as well as public officials vociferously denied plague's presence in their city. Meanwhile, the cleansing of Chinatown, house by house, street by street, got underway. City workers fumigated buildings with sulfur dioxide, then scrubbed the interiors with a solution of lye or bichloride of mercury. They also washed down the streets with chloride of lime. Refuse from houses was piled in the streets and burned. San Francisco health authorities based all measures on the presumption that plague germs lurked on objects, as well as in the soil and in the air.[27] As they scrambled to purge the plague with these sanitarian measures, health workers turned up three more possible victims.

Press reports in eastern papers moved San Francisco Mayor James Phelan to draft a letter to fifty other mayors around the country to assure them that his city was not gripped by an epidemic and, between the lines, to beg them not to turn commerce and tourists from his city. Unfortunately, by mid-May another four deaths were made public.

In Washington, Surgeon General Wyman hypothesized that if the trend continued, and San Francisco authorities found more bodies, the crisis might be beyond the capabilities of local and state officials and decided that the Marine Hospital Service must intervene. By telegram Wyman dictated his own battle plan to ranking surgeon Kinyoun, also securing the cooperation of the Chinese minister in Washington. The plan reflected the basic assumption that the plague was primarily a Chinese problem. Kinyoun was to order that the "suspected area" be cordoned and that officers be assigned to ferries and railroad stations to intercept Chinese attempting to depart. Every house in Chinatown was to be inspected and the population inoculated with Haffkine's vaccine. Those found in houses where a plague case was discovered were to be quarantined in a "suspect house in Chinatown" or on Angel Island. All corpses were to be disinfected and all railroad facilities and areas surrounding the city inspected.

Waldemar Haffkine's prophylactic vaccine had been tested with some success in India in 1897, but the side effects were known to be severe. Some patients were merely incapacitated for a day or two after inoculation; many others experienced "localized pain and swelling, erythema, headache and high fever."[28] During a plague outbreak in Honolulu early in 1900, the vaccine had been made available by federal officers on a voluntary basis, mandatory only for those wishing to leave Oahu for other islands. San Francisco's Chinese were nominated as the next guinea pigs by regulatory fiat.

The person in charge of implementation was Joseph Kinyoun, who had been interested in the relationship of immigration and disease for quite a while. Thirteen years before, he had established the federal government's laboratory for the study of cholera. He was a graduate of the New York University's Bellevue Hospital Medical College and, like the cream of American medicine's intellectual crop in the late nineteenth century, had opted for postgraduate work in Germany. After a brief period in Robert Koch's laboratory, Kinyoun had returned and joined the Marine Hospital Service.[29] In the Marine Hospital Service lab at the Angel Island quarantine station, a sample of tissue from Chick Gin's plague-ravaged body had been brought to rest on Kinyoun's microscope slide. Now the federal physician was being asked to deal with the public health consequences of what his tests had revealed.

Although there is sufficient confusion in the language and circumstances to raise some doubts over whether the Chinese spokesmen and their legal counsel willingly and without coercion agreed to allow mandatory inoculation of

Chinese with Haffkine's vaccine, there is little doubt that Kinyoun believed he had achieved such agreement and proceeded under Wyman's direction on that understanding.[30] A May eighteenth telegram from Surgeon General Wyman ordered Kinyoun to contact an official of the Southern Pacific Railroad and "request refusal sale of tickets to Chinese or Japanese without accompanying certificate from Marine Hospital officer."[31]

The anti-Asian racism in the policy, then, was not confined to the Chinese. The Japanese as well as the Chinese were to agree to inoculation with a serum still in its experimental stages of development or they would be refused permission to leave San Francisco. Within days, Kinyoun instructed railroads to refuse transportation to Chinese and Japanese lacking a certificate of inoculation. Federal inspectors were dispatched to stop any undocumented Asian from leaving the state; patrols were stationed at crossing points between California and nearby states, including Arizona, Nevada, and Oregon. Wyman also persuaded the secretary of the treasury to activate an 1890 law giving the secretary authority to take all necessary action to stop the spread of cholera, yellow fever, smallpox, or plague across state and territorial lines. Wyman believed he could legally "forbid the sale or donation of transportation by common carrier to Asiatic or other races particularly liable to the disease."[32]

The Chinese protested. Their leadership did not believe they had ever agreed to such humiliating treatment. *Chung Sai Yat Po,* the Chinese newspaper, denounced the inoculation order. The Six Companies were inundated with demands for legal action. When a newspaper reporter, unable to find the Chinese consul, interviewed his attorney, John Bennett, the latter defended his client's efforts to have the Chinese treated fairly, offering as consolation that it could have been worse and that other measures had been considered and rejected such as a blockade of Chinatown or simply burning it down.[33]

Chinese Consul General Ho Yow and the Six Companies protested to the Chinese minister in Washington, warning that forced inoculation might well lead to bloody violence. Bennett and other lawyers representing the Chinese made a last-ditch attempt on the evening of May 18 to persuade the Board of Health to modify its policy, but the protest fell on deaf ears. The next morning, an army of doctors and city health care workers wielding syringes of Haffkine's serum rushed into Chinatown and began to inoculate all those who would role up their sleeves. Few Chinese did. Instead, they used whatever forms of protest they could. Stores closed in opposition to forced inoculation, as angry Chinese clustered on street corners, their voices and gestures leaving little doubt about the subject of conversation. Others tried to leave the city. Demands for lawsuits among the poor were echoed by the threats of lawsuits from businesspeople.[34]

Undoubtedly, some Chinese were generally suspicious of Western medicine, preferring the herbalists and acupuncturists of their own tradition, but the point that those interviewed made repeatedly was that the vaccine was experimental and posed a danger to them. While Kinyoun and others who had looked through microscopes at infected tissue had little doubt that there was plague in San Francisco, others, especially the Chinese, contended that the absence of numerous deaths suggested that there was no epidemic and possibly no plague in town at all. The Chinese community refused to be pacified by assurances that the vaccine was safe, offers to permit Chinese physicians to vaccinate their own people, or threats that refusal of the vaccine might result in even less acceptable methods of disease control.[35]

Few capitulated and took the shots. Some of those who did fell ill and the Chinese press was quick to report their fever and excruciating discomfort. So the Chinese went to court. Categorical denial of citizenship to Chinese not naturalized prior to 1870 had eliminated any political threat of ethnic bloc voting in state or municipal elections. However, the Six Companies and the Chinese consul, as well as many merchants in the Chinese community, had not completely lost confidence in American institutions to treat Asians fairly. They saw in the judicial system some hope of redress from what appeared to them a nightmare of nativism cloaked in the garb of public health policy.

The Six Companies hired the prominent law firm of Reddy, Campbell and Metson to represent Chinese merchant Wong Wai in a suit against Joseph Kinyoun and the members of the San Francisco Board of Health. The suit charged that compulsory inoculation with an experimental drug under threat of being forbidden to leave the city constituted a "purely arbitrary, unreasonable, unwarranted, wrongful, and oppressive interference" with Wong Wai's personal liberty. Though not a class-action suit per se, Wong Wai's counsel asked the court to view their client as representing a class of complainants whose right to pursue "a lawful business" was being curtailed. Because the Chinese were being singled out, the brief argued, the Chinese of San Francisco were being denied "equal protection of the laws."[36] The state argued that it had the right to compel behavior in the interest of the public's health, even if it meant regulating against just the Chinese. The state justified its position by observing that thus far all the victims had been Chinese. U.S. Attorney Frank L. Coombs, arguing for the state, also asserted without a shred of evidence that the state's actions were justified because the Chinese were more susceptible to plague than other groups.

Judge William Morrow, who had been a three-term Republican congressman with a reputation for being unsympathetic to the Chinese, validated the Chinese claims. Morrow noted that the measures adopted were "not based upon any established distinction in the conditions that are supposed to attend this plague, or the persons exposed to its contagion." Rather, the measures taken, including compulsory inoculation, were "boldly directed against the Asiatic or Mongolian race as a class, without regard to the previous condition, habits, exposure or disease, or residence of the individual" on the unproven assumption that this "race" was more liable to the plague than any other. Defense counsel was unable to offer evidence to demonstrate differences between the Chinese and other groups when so directed by Morrow. The judge found that the racial provision of the order to inoculate clearly violated the equal protection clause of the Fourteenth Amendment. Morrow's decision not only saved the Chinese from compulsory inoculation, but it also set legal precedent that would limit government's ability to violate the right of individuals in the name of public health.[37]

With the federal government having had its wrist slapped by the court, the California State Board of Health was back in control. At the suggestion of Dr. W. F. Blunt, a health officer from Texas who had been invited to observe the board's meetings, the State Board decided to again quarantine Chinatown. The San Francisco Board of Health concurred as did the city's Board of Supervisors, fearing that if Chinatown was not sealed off, the whole city might find itself quarantined. Joseph Kinyoun, who no longer had authority to speak on the subject following Judge Morrow's decision, unofficially advised that Chinatown be quarantined and that after an inspection, plague cases be moved to hospitals and suspicious cases be detained. He also advised that a plan be devised for the hunting and extermination of rats as disease vectors. Though Kinyoun was as unaware as others of the rat's central role in spreading plague contagion in Chinatown, he seems to have been more aware than most that simply quarantining Asiatics was insufficient precaution. The meeting passed a quarantine resolution confining Chinatown's Chinese residents to their neighborhood but leaving to officials' discretion the fate of any Caucasians who lived within Chinatown's limits.[38]

This second quarantine of Chinatown again raised the issue of whether or not plague really existed and the extent to which it posed a real danger. The Chinese community and its lawyers did not rush to litigation. Recognizing the volatility of the situation, the Chinese pledged their support to wiping out plague in San Francisco, if it in fact existed. However, spokesmen did ask why, if Chinatown was the nexus of infection, there seemed to be no plan to protect the Chinese living under quarantine, most of whom showed no signs of infection. Should not those suspected of having plague be isolated from the other residents of Chinatown? Chinese leaders also wanted to know what plans were being drafted to feed and provide care for the thousands of Chinese under quarantine who were now unable to go to their jobs and earn their livings outside of the community. The *San Francisco Examiner* reported that Consul General Ho Yow insisted that municipal officials had an obligation to provide care for the Chinese from public monies, a view echoed by others who cited previous court decisions in which persons in quarantine were held to be as "public prisoners" and therefore "properly a charge on the public

Treasury." However, the city had neither a plan nor a sympathetic view of the Chinese plight.[39]

Conditions in Chinatown were deteriorating rapidly. A newspaper reporter who had toured the quarter the day after the imposition of the quarantine described the siege-look of the area as armed guards patrolled its perimeter. Business was not being conducted and "stores are closed, doors barred." On street corners he saw groups of "Asiatic," who were "excitedly gesticulating" as they talked. The reporter was reminded of "inhabitants of a beleaguered town." Soon the food supply began to dwindle and the prices of what remained on store shelves skyrocketed. The Board of Health took no action on food, choosing to confine itself to strictly medical decisions such as ordering an autopsy for all who died in Chinatown, conceding only so far as to allow a physician designated by the Six Companies to attend to ensure that false claims of plague were not recorded. The board also mandated that Chinese laundries throughout the city be inspected and fumigated in case the owners were hiding escapees from Chinatown. Finally, the board resolved to request that the mayor obtain federal assistance in identifying sites outside the city to be used as detention centers for those suspected of exposure to plague, an idea that had originated with Joseph Kinyoun.[40]

In late May, there was some discussion in the press of burning Chinatown to the ground and starting over. At least one newspaper demanded, "Clear the foul spot from San Francisco and give the debris to the flames."[41] The Chinese press reported the call for the burning of Chinatown adjacent to stories about possible detention sites. Though the two actions were not directly related, at least some Chinese found it quite conceivable that their neighbors intended to relocate them, then burn Chinese homes and businesses to the ground.[42]

Early in June, the Board of Health tightened the quarantine, forbidding streetcars from even passing through Chinatown, which they had been allowed to do as long as they did not pick up passengers. City officials increased the size of the police guard patrolling the borders of Chinatown, and the police strung barbed wire where it appeared to be useful in sealing off the Chinese still further. Actually, officials exceeded the order for fencing and began the construction of a high wooden wall around Chinatown somewhat reminiscent of the ghetto walls used to enclose the Jewish quarter of some European cities centuries earlier. Most ominous, though, were plans to evacuate approximately 1,500 Chinese that officials suspected of being exposed to infection. The Chinese assumed that this was the first step of a broader policy to evict them from San Francisco altogether. Consul Ho was quoted in the press as saying that any attempt to forcibly remove the Chinese from their homes would be physically resisted.[43] With violence appearing increasingly likely, both sides ended up back in federal court.

A Caucasian depiction of congested and unclean conditions in San Francisco's Chinatown during the bubonic plague scare, *Harper's Weekly,* June 2, 1900. (National Library of Medicine)

The litigant was Jew Ho, a Chinatown grocer. He filed his complaint with the U.S. Circuit Court for the Northern District of California and, like Wong Wai, wished his suit to be considered on behalf of other residents of the quarantined district as well.[44] Jew Ho was challenging the arbitrariness and discriminatory character of the quarantine. During oral argument, counsel demonstrated that Caucasian residences and businesses on all sides of Chinatown were not being included in the quarantine. Moreover, though the Board of Health claimed to be fighting a threatened plague epidemic, it had made no provision to feed or care for members of the Chinese community, even barring physicians hired by the Chinese from the quarantine area. Threatened evacuations and the physical barriers being erected were also raised in argument.[45]

Judge Morrow and another of the three judges who had heard the Wong Wai complaint heard this one, too. Morrow, choosing to avoid delay and to deliver his opinion in open court, claimed jurisdiction on grounds of the Fourteenth Amendment and diversity of citizenship. This allowed him to rule on all complaints as if his court were a California state court. The judges found quarantine a reasonable approach to the control of infectious disease, but not when established around an entire section of a city, regardless of whether the residents had shown signs of illness or had positively been exposed to the disease. Likewise, no specific action had been taken to confine non-Asians within Chinatown who had been so exposed. The arbitrary nature of the quarantine and its racially discriminatory character thus warranted an injunction to terminate. However, Morrow did allow imposition of a quarantine around specific buildings that the Board of Health believed to be contaminated with plague. He also insisted that a physician employed by the Chinese had the right to care for those suspected of having plague and to witness autopsies of alleged plague victims.[46]

The Board of Health complied with Morrow's order, though the board ordered the cleaning and fumigation of Chinatown's sewers as a precaution. Reporters who witnessed the end of the quarantine described it as the end of a siege as "a horde of Chinese poured through the lines like the advance guard of a relief column."[47] Within hours, business as usual was resumed.

No other major court battles were necessary. However, a stubborn Joseph Kinyoun ended up being ordered to desist by his superiors after the governor of California complained to President William McKinley that Kinyoun had refused successful litigant Wong Wai a certificate to leave San Francisco on the grounds that he had been resident within Chinatown when it was under quarantine. Wong Wai asked that Kinyoun be cited for contempt of court. Though this never occurred, the incident ended Kinyoun's efforts to confine the Chinese and left him an angry and bitter individual, given to racist invective. He was later transferred from his Angel Island post and reassigned to Detroit. Although free with his denunciation of the Chinese as crafty, deceitful, and hopelessly contemptuous of science, he saved his sharpest knives for the San Francisco business community, which he accused of sacrificing the public's health on the altar of profit.[48]

Bubonic plague was present in San Francisco. On August 11, 1900, it took its first Caucasian victim. The disease continued to claim lives, most of them Chinese, at the rate of a victim every ten days to two weeks. By the end of 1900, the federal government again intervened, but in a very different manner than it had months earlier. The Secretary of the Treasury, acting at the behest of the chief of the Division of Domestic Quarantine of the Marine Hospital Service, appointed a distinguished panel of three biologists to investigate and determine once and for all whether plague existed in San Francisco. The commission, comprised of professors Simon Flexner of the University of Pennsylvania, F. G. Novy of the University of Michigan, and L. F. Barker of the University of Chicago, was not universally welcomed. State officials, including the governor, regarded the body as an unwelcome federal intrusion and continued to fear the negative publicity that would cost the state commerce and revenues.[49]

The commission established its own bacteriological laboratory and examined the tissue of thirteen deceased Chinese. Six had, indeed, died of plague. The commission did not recommend use of any serum or vaccine, so in the ensuing months, cleaning streets, cleaning and fumigating dwellings, and disinfecting personal property were the only plague preventives employed. By the following year, attention had shifted primarily to catching rats and ratproofing buildings. Although health officials did not yet have evidence that fleas on rats were plague's vectors, public health physicians observed the association of rats with plague by hoping to catch and kill as many of the rodents as possible. Not until early 1904 did victims cease appearing on a regular basis. Since the discovery of Chick Gin's remains, there had been 120 additional cases and 112 more deaths from bubonic plague; patients were overwhelmingly Chinese.[50]

Several years later, the bubonic plague struck again, but few blamed the Chinese. At 5:14 A.M. on Wednesday, April 18, 1906, San Francisco was devastated by a massive earthquake. Recovery was slow and the city suffered a public health crisis with repercussions that lasted for years. The bubonic plague outbreak beginning in May 1907 lasted eighteen months, consisted of 160 reported cases, and claimed 78 lives, almost all non-Chinese. This time, there was no accompanying plague of prejudice. The fleas on rats, not the Chinese, were held responsible for the epidemic.[51]

In the case of typhoid, it was well known by the early twentieth century that the carriers were human. What was not well known, however, was that a perfectly healthy individual could infect others. That lesson public health officers would learn from their encounter with Irish immigrant Mary Mallon. In American legend and lore, Mary Mallon has become synonymous with the health menace posed by the foreign-born. During the 1980s, newspaper coverage of a case of typhoid fever at a McDonald's restaurant in suburban Maryland traced to an immigrant worker who had removed plastic gloves when mixing vegetables for the salad bar resulted in popular quips about the existence of a "Typhoid McMary," but it was only one in a long history of such associations in the public mind.[52] However, in her own time, the public reaction to the Irish immigrant woman and her condition was far more ambivalent.

Mallon, or Typhoid Mary as she came to be known, arrived in the United States as a young girl in her early teens from County Tyrone, Ireland, where she was born on Sept. 23, 1869. Although Robert Koch and some of his colleagues had presented papers on the probable role of healthy carriers in producing typhoid, the phenomenon was not familiar to many American epidemiologists and was certainly not comprehended by the general public.[53] Some, such as Walter Reed, had merely hypothesized the possibility of a healthy carrier. In 1900, Reed, discussing the communicability of typhoid, wrote, "Apparently trustworthy bacteriologists have reported the finding of [the typhoid] microorganism in the most unexpected places," including "in the stools of healthy persons."[54] Mallon was the first healthy typhoid carrier to be positively identified by health authorities in the United States.

Much of what is known about Mallon has been gleaned from the epidemiologist who tracked her down and identified her as a danger to the public's health. Dr. George A. Soper, a thirty-six-year-old epidemiologist and sanitary engineer with a Ph.D. from Columbia University, investigated several cases of typhoid that had occurred at Oyster Bay, Long Island, in August 1906, and the trail he followed led him to Mary Mallon.[55]

Typhoid fever is the result of infection by the typhoid bacillus, which exists in over fifty known strains. Transmission is by food or water that has been contaminated by the feces or urine of a victim or a carrier, an individual who does not suffer from the fever but is capable of infecting others.[56] One to three weeks is the usual incubation period. Symptoms are fever, headache, malaise, absence of appetite, spots on the trunk of the body, and constipation. Left untreated, approximately 10 percent of those infected die. Modern antibiotic treatment has slashed mortality rates to less than 1 percent. Two to 5 percent of all cases become carriers.[57]

While hardly limited to those at the bottom of society's ladder, typhoid fever has long found a substantial portion of its victims among the poor and uneducated, those who live under conditions that promote infection and lack knowledge of risk-reducing precautions. In 1906, typhoid was not at all an uncommon disease: 3,467 cases were reported in New York City and 639 New Yorkers died of typhoid that year.

The outbreak of 1906 might have attracted less public attention, then, had it not affected a wealthy and well-known New Yorker. Gen. Henry Warren, who was a prominent New York banker, his family, and servants were spending a restful summer at a rented summer retreat in Oyster Bay until six of the eleven persons in the household were stricken by typhoid. George Soper, a meticulous investigator already known for tracing the source of an Ithaca, New York, epidemic, was hired the following winter by George Thompson, who owned the country place at Oyster Bay. Thompson feared that the publicity from the preceding summer's tragedy would make his property worthless as a rental unless the health menace was identified and eliminated.

Soper started with the usual suspects and tests. Water from the cesspool, privy, overhead tank, and bathing facilities tested negative for the bacillus. So did the facilities of properties in the vicinity. As he followed every available lead, the tenacious medical detective found his way to the Irish immigrant cook who had worked for the Warrens and left soon after the epidemic started. Mary Mallon was hired on August 4, 1906. The fever struck its first victim on August 27 and the other five took to their beds within a week.

Backtracking through employment agency records, Soper was able to locate some of the positions Mary had held, except for those she had obtained answering newspaper advertisements. Seven households had been struck by typhoid. The profile was similar in all cases: a wealthy family that had fled New York's summer heat and had hired servants, including a cook. Because some of the typhoid cases occurred among the other servants, Soper speculated that the families ate cooked meals in which the typhoid bacilli were killed; uncooked desserts would have been served by a butler or maid. However, Mary may well have served her peers directly and in one particular case, a family member recalled Mary cutting fresh peaches directly onto some ice cream.[58]

When Soper finally found Mallon she had already infected yet another wealthy family, this one living at a fashionable address at 60th Street and Park Avenue. A laundress had been removed to a hospital and the family's only daughter was near death. Soper found Mallon in the kitchen. She was about forty years of age, "five feet six inches tall, a blond with clear blue eyes, a healthy color and a somewhat determined mouth and jaw." She "had a good figure and might have been called athletic had she not been a little too heavy," a woman who "prided herself on her strength and endurance," according to Soper.[59]

Fortunately for Soper, Mallon was not even more of an athlete than she appeared. In a manner that Soper recalled as low-keyed and nonaccusatory, but may have been otherwise, he explained to the cook his mission and asked that she cooperate by giving him samples of her blood, feces, and urine. However, Mallon did not respond in kind. Soper testified, "She seized a carving fork and advanced in my direction."[60] A later encounter at the apartment of Mallon's male friend proved no more productive. This time Soper was accompanied by a former assistant and old friend, Dr. B. Raymond Hoobler, who later became head of the Detroit Children's Hospital. Mallon flew into a rage, vehemently denying the possibility that she was infecting others with typhoid, pointing to her own excellent health as evidence. The two men beat a swift retreat.

Unsuccessful on his own in getting Mallon's cooperation, Soper reported his findings to City Health Commissioner Thomas Darlington and to Dr. Hermann M. Biggs, chief medical officer at the New York City Department of Health. Soper recommended that Mallon be arrested so that her excreta could be submitted for laboratory analysis. A female physician and health department inspector, Dr. S. Josephine Baker, was given the unenviable task of obtaining samples for laboratory analysis. In her autobiography, Baker, who later became a prominent figure in public health, founding the pioneering New York Bureau of Child Hygiene, describes her ordeal with Mallon: "When I first interviewed her, Mary was busy at her job in the kitchen of a prosperous Park Avenue household, . . . the traditional brownstone-front house in the Sixties. Mary was a clean, neat obviously self-respecting Irish-woman with a firm mouth and her hair done in a tight knot at the back of her head."[61] Baker's first visit to Mary ended when Mallon, "jaw set" and eyes glinting, angrily denied Baker's request. The rebuff aside, Baker recalled that her first impression of Mary was that this dangerous woman bore no outward sign of the evil that lurked within her intestines. Mallon appeared a well-groomed Irish woman, but firm in her resolve to remain unmolested.[62] Baker was equally determined to do her job and returned the following day with several burly New York City policemen. Mallon answered the doorbell armed with a "long kitchen fork in her hand." She lunged at her guests and fled, eventually finding herself cornered in an outdoor shed where she had hidden, very likely with the assistance of her protective fellow servants, according to Baker. Captured, a screeching, clawing Mallon was taken to the hospital in an ambulance with Dr. Baker sitting on her chest to restrain her during the drive.[63]

For the next several months, Mallon was kept in isolation at Willard Parker Hospital, her excreta tested three times weekly and almost always testing positive for *Bacillus typhosus*.[64] Various drug therapies were tried unsuccessfully to eliminate her infectiveness. She also refused to allow removal of her gallbladder, a procedure that sometimes stopped the production of typhus bacilli, but often did not.[65] Eventually Mallon was sent to Riverside Hospital on North Brother Island. On the island she was permitted to live by herself in a small bungalow. She was allowed to cook for herself the food that health authorities supplied her.

Mary Mallon did not come to popular attention until June 20, 1909, when

William Randolph Hearst's sensationalistic *New York American* ran a story in its Sunday magazine section with the dramatic title, "'TYPHOID MARY'—MOST HARMLESS AND YET THE MOST DANGEROUS WOMAN IN AMERICA." The nickname was the invention of health department officials and had been picked up by two newspaper reporters.[66] The popularity of Mary's cause and her public image was enthusiastically fueled by Hearst's yellow press. Letters poured in and probably offers of support for legal action as well, because Mary eventually had a lawyer.

After two years on the island, Mallon, represented by George Francis O'Neill, sued the city for her release. The presiding judge dismissed the case on July 20, 1909, accepting physicians' testimony that Mallon was a public health menace. During the case, Mary Mallon actually took the stand. Her argument was that she "never had typhoid in my life and have always been healthy." She blamed "the drinking water" for the cases of typhoid that had occurred in her places of employment and, hoping to turn medical evidence to her own advantage, noted that "My own doctors say I have no typhoid germs." Undoubtedly hoping to arouse the sympathy of the court, Mary described herself as an "innocent human being" who had "committed no crime," but nevertheless was being "treated like an outcast—a criminal." She characterized her treatment as "unjust, outrageous, uncivilized," invoking the conscience of the court "that in a Christian community a defenseless woman can be treated in this manner."[67]

The court sustained the right of the New York Board of Health to isolate an infected individual, a decision that in the court's view hardly broke new legal ground in light of quarantine procedures. However, Mary Mallon herself was not sick. The turn in scientific understanding cried out for fresh legal and administrative interpretation. What Judge Mitchell L. Erlanger would not do, New York Health Commissioner Dr. Ernest J. Lederle did. Mary Mallon had aroused his compassion. He decided to trust Mallon and was persuaded that she understood the danger within her. After receiving her promise not to cook or handle others' food, the commissioner released Mallon on her own recognizance on February 19, 1910, requiring that she report to the Health Department every three months. Lederle reassured the public by explaining to an *American* reporter that while Mary Mallon was not cured, she had been "taught how to take care of herself." Lederle observed that Mallon would have posed no danger to anyone except for "the fact that she was a cook."[68] In direct response to reporters' questions, Lederle explained that he could not assure the public that Mallon generated fewer typhoid germs as a result of the treatment she had received at Riverside Hospital. However, he did provide reassurance that "Her safety to the public . . . lies in the fact that she now knows how to prevent their [the typhoid germs'] spread, and that she will change her employment."[69] Lederle then made a plea for Mary Mallon almost as passionate as the one she had made in her own behalf in court: "For Heaven's sake—can't the poor creature be given a chance in life? An opportunity to make her living, and have her past forgotten? She is to blame for nothing—and look at the life she has led!"[70] Lederle offered Mary that chance.

A better physician than a judge of character, Lederle's trust was abused when the freed Mallon went underground, living under several aliases including Breshoff and Brown and cooking for guests in hotels, restaurants, and even sanatoria. She also ran her own boarding house briefly. Details of her years on the run are sketchy, but Mallon was definitely linked to at least one typhoid outbreak in 1914 at Newfoundland, New Jersey, and probably another in Marblehead, Massachusetts. A member of the Bryant family in Marblehead that had employed Mallon described her as an Irish nationalist committed to raising money in support of her homeland's struggle against the British. The same source described Mallon as convinced that her plight was the fabrication of English conspirators to punish her for her political commitment to Irish liberation.[71] However, there is no hard evidence that Mary Mallon was in any way connected to the Irish nationalist cause. The silence of the New York Irish newspapers in covering her case suggests that the Irish political community may have feared being stigmatized as endangering native-born Americans, a burden it hardly needed as it sought to rally support and funds to oppose British colonialism. More likely a tortured, deluded soul in search of some explanation for her tragedy rather than a political activist, Mallon's attribution of her suffering to ethnic antagonisms possibly suggests a desire to blame her misfortune on forces outside of herself, diverting attention from the violation of her person by bacilli and answering the question that every victim asks, why me?

Mary Mallon's rampage ended in 1915, when twenty cases of typhoid at New York's Sloane Hospital for Women was traced to the kitchen. Perhaps a subconscious cry for help and a desire to be subdued, perhaps an act of sheer defiance and bravado, Mary had taken a job working under the very noses of New York health authorities under the alias Mary Brown. Workers at Sloane had nicknamed their colleague Mary Brown "Typhoid Mary" as a joke when the hospital epidemic occurred because a few had recalled the earlier accusations lodged against Mary Mallon in the press. None had considered the possibility that their Mary Brown was the notorious Mary Mallon. George Soper was contacted by hospital officials and arrived after a suspicious Mallon had fled. Presented with a sample of the cook's handwriting, Soper identified the absent hospital employee's handwriting as Mary Mallon's.

The forty-eight-year-old woman was captured and remained in detention on North Brother's Island for the rest of her life. At first she had periodic rages described by one journalist as like "a moody, caged, jungle cat."[72] However, with age came resignation and calm. She worked in a laboratory at the hospital and led a quiet life, eventually even receiving permission to visit friends in Manhattan and Corona, Queens. She always returned. On December 25, 1932, "Typhoid Mary" was found on the floor of her small house, paralyzed by a stroke. She spent the last six years of her life a paralytic in Riverside Hospital, dying on November 11, 1938. Her Catholic funeral was conducted at St. Luke's Roman Catholic Church at 138th Street in the Bronx, with interment not far away at St. Raymond's cemetery.[73] Dead of bronchopneumonia, Mallon left behind a death certificate that lists typhoid carrier for twenty-four years as a contributory cause, though there is no medical evidence that Mallon was ever affected, much less felled by, the contagion inside her.[74] Even at the very end, not all physicians could accept that Mary Mallon was a perfectly healthy carrier of typhoid.

Mary Mallon is known to have infected a minimum of fifty-three individuals with typhoid fever, three of whom died, probably a conservative count. However, her notoriety was and remains out of proportion to the sickness or death, or even popular fear, that she caused. Perhaps it was simply Mary Mallon's bad fortune to be a healthy carrier at a time when the phenomenon was still on the periphery of medical understanding.[75] There is little doubt that headline-hungry journalists capitalized extensively on her misfortune.

Sympathy for Mallon from those who wrote to her and some who wrote about her appears to be grounded in an image of a woman alone being victimized by both her body and a cold, scientific, somewhat insensitive medical/state nexus. Some of the articles that appeared in the yellow press and even a lengthy profile of Mallon, "Typhoid Carrier No. 36," that appeared in *The New Yorker* three years before Mallon's death saw the rapid growth of medical understanding and the bulging corpus of public health

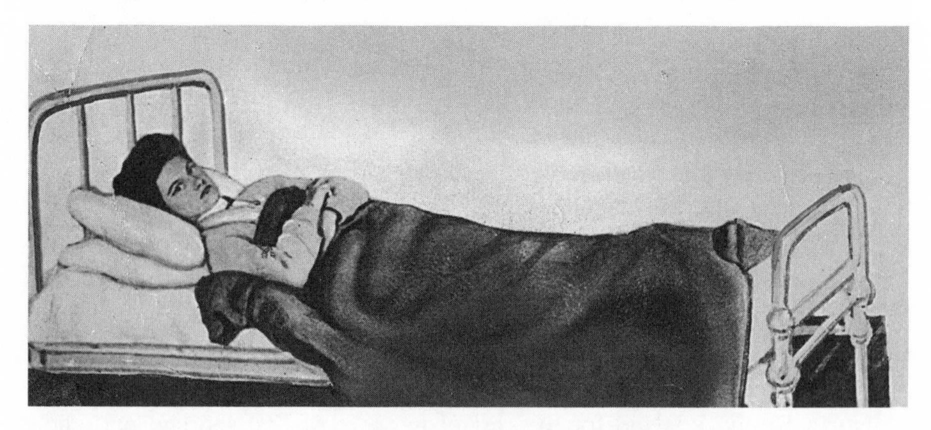

Mary Mallon ("Typhoid Mary") confined by public health authorities to North Brother's Island, c. 1915. (Brown Brothers)

rules and administrative apparatus as increasingly restrictive of individual rights, though the legal tide was already flowing in the opposite direction.[76] In 1905 the Supreme Court ruled in *Jacobson v. Massachusetts* that a statute empowering local boards of health to require compulsory smallpox vaccination was a reasonable exercise of police power and not a derogation of any individual liberties guaranteed in the Constitution.

On the other hand, many of those critical of Mallon often raised her gender as an issue, too. From a conventional, stereotypical, early-twentieth-century masculine perspective, the classically domestic image of a nurturing woman making food for others to eat was violently defiled by Mallon's behavior. It should not be surprising, then, that some men of the era who encountered Mallon, a strong and bright woman, perceived a nonfeminine quality in her visage, carriage, and general demeanor. George Soper recalled, "Nothing was so distinctive about her as her walk, unless it was her mind. The two had a peculiarity in common. Those who knew her best in the long years of her custody said Mary walked more like a man than a woman and that her mind had a distinctly masculine character, also."[77] Mary seemed a perversion of true womanhood as invented by males in the late nineteenth and early twentieth centuries.[78]

Mallon was often identified in articles as an Irish immigrant. The homes of the wealthy were the workplaces of many Irish immigrant girls and women who had found that America's door to opportunity often opened onto the kitchen.[79] Why, then, did Mallon's adventures not fuel a wave of anti-Irish sentiment similar to the anti-Chinese sentiment on the West Coast? Why were Irish cooks and maids not fired en masse?

Ethnicity and race played very different roles in each episode. Mary Mallon was a Caucasian Irish Catholic, a member of a group increasingly assimilated socially and dispersed geographically among the American population. An increasing number of prominent Americans claimed Irish ancestry openly and with pride, including the California oil magnate Edward L. Doheny and the shipping millionaire William R. Grace. In cities such as New York and Boston, the Irish and local politics were all but synonymous.[80] Mallon's role as a healthy typhoid carrier brought no harm to the Irish. Contrast this with the medical scapegoating of Chinese immigrants at the turn of the century. Preexisting patterns of racial nativism in American society and an especially combustible anti-Asian atmosphere on the Pacific Coast of the United States set the stage for ethnic confrontation touched off by several cases of bubonic plague in the Chinese quarter of San Francisco. Chinese visibility, the result of racial distinctiveness, and their segregated residential pattern, sometimes voluntary, at other times imposed, only increased their vulnerability to stigma and smear.

Sinophobia, unlike anti-Irish prejudice, was not on the wane at the turn of the century—it was cresting. The Chinese were different to the gaze, different in life-style, different in culture, different in religion, and yet, competitive with American workers. Anti-Asian racism was the political coin of the realm in California, and the Chinese, legally excluded from citizenship, could never hope to pose a political counterweight to their adversaries. Only the courts and Chinese economic resources to mount a first-rate legal assault prevented even worse abuses than occurred.

West Coast economic vulnerability and inexperience in urban administration also appears to have played a role. San Francisco, unlike New York, was a young city and slightly behind East Coast cities in developing the necessary administrative apparatus to handle rapid growth and modernization. New York's Board of Health, made permanent in 1866, had existed for a decade longer than San Francisco's in 1900. It enjoyed the leadership of outstanding public health professionals such as Drs. Hermann M. Biggs, T. Michael Prudden, and William H. Park. Major private medical institutions existed in the East and Midwest, but not yet on the West Coast. Community leaders, realizing that their city was dependent upon distant eastern capital and tourism, perceived little margin for error in maintaining a highly positive public image.

Finally, the state of medical knowledge and the tack taken by public health officials differed markedly in the two episodes. New York public health authorities seemed to act decisively to curb the threat. They knew the cause of typhoid and its epidemiology, and what needed to be done to control its spread. Less was known about the origins and control of bubonic plague. Moreover, the terror in the public mind stirred by the Black Death well exceeded public dread of typhoid, an all-too-familiar companion of urban life. This left officials of the relatively young San Francisco public health establishment and, later, the U.S. Marine Hospital Service with an administrative and public health challenge of unprecedented proportion. Joseph Kinyoun was a good laboratory scientist who can hardly be faulted for not fully understanding the epidemiology of bubonic plague. However, he was culpable for failing to comprehend the racial hysteria that was distorting public health policy. Kinyoun and his colleagues were undermined not by obsolete medicine but by their failure to progress beyond the racial prejudices and stereotypes pervasive in American society.

5

"That Is the American Way. And in America You Should Do as Americans Do": Italian Customs, American Standards

During the sweltering summer of 1915, Philadelphia was stunned by an outbreak of typhoid. Such seasonal eruptions were not uncommon, but there was good reason for health officials to be especially chagrined and frustrated. The city had undergone a wrenching political conflict over the purity and healthfulness of its water and had constructed an expensive waterworks, completed in the late fall of 1911. A year after citywide water filtration began, the typhoid death rate, which had been 75 in 1899, dropped to 13 per 100,000, the lowest since the city began record keeping in 1861; by 1915, the death rate had descended still further to 7 per 100,000. City fathers beamed with

Title taken from John Foster Carr, *Guide for the Immigrant Italian in the United States of America*, published under the auspices of the Connecticut Daughters of the American Revolution (New York: Arno Press, 1975; orig. 1911), 48.

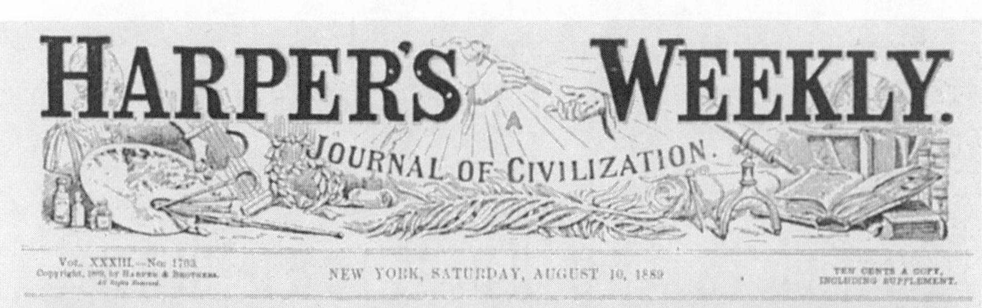

A Board of Health doctor in New York tenement, *Harper's Weekly*, August 10, 1889. (National Library of Medicine)

pride, but no longer. Why, now that the waterworks were completed, should new cases of typhoid suddenly menace the city?

Health officials traced the source of typhoid to Italian immigrants working as pickers on suburban farms. There, they drank unfiltered water and suffered infection. Returning home to Philadelphia, sick laborers infected their families, who unknowingly infected others by touching produce on streetside pushcarts where they shopped. Municipal officials announced that the Italian preference for eating fresh salads and raw vegetables was the source of the problem, not the water, which they boastfully and with great relief announced was "free from contamination."[1]

Faced with an immigration that seemed to grow larger with each passing year and more varied in its origins than any other in American history, turn-of-the-century public officials and physicians charged with protecting the public health could hardly ignore the fact that newcomers' customs and traditions concerning health and hygiene were competing with their policies and recommendations for immigrants' attention. When immigrants from one group or another became ill or physically disabled after arrival, it was not uncommon for cultural negotiations over health and hygiene to degenerate into an exchange of accusations. American public health officials not uncommonly blamed immigrants for their own poor health, charging that traditional practices of health and hygiene were deficient and often little more than magic and superstition. Immigrant spokespersons fired back that poverty, congestion, and generally unhealthy conditions immigrants encountered in the United States after arrival compromised the well-being of bodies that had arrived here healthy and robust.

As health officials in city, state, and federal governments carefully compiled public health data, it did indeed reveal differences in the morbidity and mortality between immigrant groups and native-born as well as among various immigrant groups. But why? Were the Italians, for example, the single largest group arriving between 1890 and 1920, inherently unhealthy or did the health of Italians deteriorate in the United States, and if so, what caused this declension in the health and vitality of those who arrived well and became sick?

One Italian physician boldly asserted that whatever health problems Italians in the United States might suffer, they were acquired on this side of the Atlantic. Speaking of the Italians in their homeland, he described them as "one of the healthiest in the world," instead blaming the congested housing and dark, unventilated sweatshops for weakening the bodies of "agriculturalists, fresh yet from the sunny hills and green valleys of Tuscany and Sicily."[2] Such advocates resisted social constructions that portrayed southern Italian arrivals as sickly and inherently enfeebled. They argued that health in the immigrant community was a negotiation between culture and environment. On the one hand, they encouraged newcomers to relinquish sole reliance upon traditional remedies and accept state-of-the-art Western medicine and hygiene to improve their health. On the other, they were vocal and active adversaries of the poverty and congestion that allowed disease to breed and spread in the immigrant neighborhoods of American cities.

The typhoid scare in Philidelphia was not an isolated episode. A year later, the Italians suffered identification with a polio epidemic. Polio had been rare in western Europe and the United States until 1907, when a major epidemic hit New York City and killed an estimated twenty-five hundred persons. After 1907, polio deaths rose considerably; between 1910 and 1914 health officials across the country reported approximately thirty thousand cases, including five thousand deaths from the disease. Observed in children as early as 1774 as a sudden paralysis subsequent to a fleeting illness, polio was identified as a filterable virus by Simon Flexner and Hideyo Noguchi in 1914. However, even after they cultivated and isolated the virus, researchers still had no idea how to prevent, control, or cure polio.[3]

Ironically, the very improvements in personal hygiene and public sanitation that were eliminating epidemics of typhoid and cholera by the early twentieth century contributed to the fresh virulence of polio. As improved sanitation killed off the milder, immunity-giving viruses, the more potent, crippling viral strains thrived. In the summers of 1907 and 1910, New Yorkers from affluent and poor families alike were felled in large numbers, most of them children.[4] In the summer of 1916, even as the world's attention was turned to the war devastating Europe, polio attacked children in cities all along the East Coast.[5] Physicians reported 8,927 polio cases in New York City alone.

New York City health officials received news of the first reported case June 6, 1916. The victim lived in a heavily populated section of Brooklyn, near the waterfront. The health department ordered visiting nurses to make a search of nearby tenements. Apartment by apartment, they pinpointed another twenty-two victims. Some had been stricken with illness for several weeks, but their families had not called in a physician. Among those that nurses' reports described as especially reticent were frightened and confused Italian immigrants.

By 1916, millions of Italian immigrants, most from the south of Italy, had arrived in the United States. Approximately 4.5 million came between 1880 and 1921, more than any other group. Many settled in the New York area, at least initially. In New York, the polio death rate per 1,000 estimated population of children under ten years of age was 1.63 for Italian children, well below the 3.42 for the native-born or the 3.27 for German youngsters. The reasons remain unknown, but as Health Commissioner Haven Emerson observed, "Certainly the social and economic conditions under which these people live are no more favorable than those under which the Americans,

Germans and Irish live, among whom the mortality of the disease is the highest."[6] However, while the Italian mortality rate for polio was low, the 1,348 polio cases contracted by those of Italian nativity in New York City was the highest for any immigrant group, second only to the 3,825 cases among the native-born.[7]

By their very nature, epidemics are intense. Polio epidemics struck suddenly, in summer, when New York youngsters of modest circumstances were enjoying the freedoms of childhood, frolicking in the city streets.[8] Because there were so many Italian immigrants, living in tightly concentrated neighborhoods, and because immigrants were viewed by many as a marginal and potentially subversive influence upon society, the incidence of Italian polio made a dramatic impact upon the imagination of a public already shaken by the virulence of the epidemic and the youth of its many victims.[9]

Rumors spread that the epidemic had been brought by immigrants from Italy to the United States rather than contracted here by the newcomers.[10] Although that rumor was quickly dispelled by the U.S. Public Health Service, some were still prepared to connect this particular immigrant group to the epidemic, arguing that Italians' poor hygiene and unhealthy life-styles made them vulnerable to polio, which they then spread to natives and other newcomers alike.[11]

The charge that Italians were especially unclean and unhealthy was not new. The Italians in New York City, mostly impoverished workers from the southern provinces, were frequently denigrated by critics. In 1890, Richmond Mayo-Smith, a Columbia University professor of political economy, described Italian immigrants in New York's tenements: "Huddled together in miserable apartments in filth and rags, without the slightest regard to decency or health, they present a picture of squalid existence degrading to any civilization and a menace to the health of the whole community."[12] In 1914, two years before the polio epidemic, an equally uncharitable E. A. Ross wrote, "Steerage passengers from a Naples boat show a distressing frequency of low foreheads, open mouths, weak chins, poor features, skew faces, small or knobby crania, and backless heads. Such people lack the power to take rational care of themselves; hence their death-rate in New York is twice the general death-rate and thrice that of the Germans."[13]

Ross was far better at polemics than statistics. A careful statistical analysis of mortality data by Louis Dublin of the Metropolitan Life Insurance Company using sex and age data from the 1910 census for New York State and Pennsylvania confirms that in New York the Italian death rate was generally lower, not higher, than that for the native-born of the same age and gender and considerably lower than that for the comparable age and sex category of the German group. (See table 5.1.) [14]

TABLE 5.1
Mortality Rates for Native-Born (N.-B.) of Native Parents, Italian Immigrants, and German Immigrants, All Ages, in New York State, 1910[15]

	N.-B.	Italian	German
Male	15.9	9.2	27.5
Female	13.9	9.7	22.6

What, then, was the link between Italian immigrants and polio in some minds? The answer appears to be filth, especially filth resulting from the unsanitary habits and personal hygiene of newcomers in the eyes of their hosts. Even as they searched for the living organisms responsible for infantile paralysis, public health professionals could not completely wean themselves from sanitarian patterns of thought. Reports suggested that the high number of polio cases in the Italian population might have something to do with their habits, including cultural customs and traditions.

In the summer of 1916, during the height of the epidemic, New York public health officials labored mightily to curb polio where many thought it was originating—among the immigrants, but especially among the Italians, because they had the highest number of cases. Italian neighborhoods were deluged with health department pamphlets and signs in the Italian language, warning immigrant mothers about polio and urging them to be hygienic in their personal habits and child-rearing practices. The Department of Social Betterment of the Brooklyn Bureau of Charities issued one hundred thousand leaflets printed in Italian, Yiddish, and English.[16] All public gatherings of the urban poor were monitored by health officials, including block parties and public playgrounds. However, a large gathering of Italians must have seemed a particular health threat because in New York City, the three-day *festa* of Our Lady of Mount Carmel was cancelled by order of the health department.[17] Although eastern European Jews and Polish immigrants were sometimes mentioned as carriers of polio, Italians—the immigrant group most victimized by polio—were generally held in greatest suspicion of implication in spreading the disease.

When education and bans on public assembly proved insufficient, quarantine was tried. Decisions to quarantine a household were made upon the recommendation of visiting nurses. Public health nurses who climbed tall tenement staircases to report on polio cases as part of New York City's Special Investigation of Infantile Paralysis under the Rockefeller Institute's Dr. Simon Flexner were often angry and impatient with immigrant families. Italian families more than others irritated nurses and aroused their suspicions. Italians brought with them to America a basic distrust of people in positions of authority; this encompassed nurses and doctors. Not surprisingly, because they possessed the power to recommend that a family be confined to its home, nurses were often feared and resented as intruders by Italian immigrants, especially the parents of children stricken with the disease.[18] Mutual distrust, then, prompted an anti-Italian sentiment in many of the nurses' reports. After visiting one-year-old Petro Pollizzi, one nurse wrote, "These people neither understand nor speak Eng. [*sic*] so I could get very little accurate information. . . . These people are very ignorant and suspicious."[19] Often nurses noted whether families with stricken children were being treated by an Italian physician, with the implication that the care of such doctors was suspect. Similar comments on the physicians treating other groups of newcomers are absent.

Insensitivity to cultural differences between the native-born and the Italians nurtured a disdain among the nurses for those they were treating. Nurse Ida May Shlevin heaped scorn on the practice of kissing the dead, an Italian custom that was part of the ritualized expression of grief and respect for the departed. Shlevin saw, instead, disrespect for both the living and medical science in the custom and indicted this ethnic tradition as one cause of the epidemic ravaging the city.[20]

Some members of the Italian community vigorously sought to sever the link between ethnicity and polio in the public forum, pointing instead to anti-Italian nativism as the source of the stigma suffered by the group. One writer who refused to give his last name wrote to Mayor John P. Mitchel, denouncing the prejudice. He fumed, "I wish to say emphatically that the American Italian is not to be singled out and charged with anything." Their only culpability was "nationality," he wrote. Instead, the protester urged the mayor to consider poor sewage systems and inadequate garbage disposal facilities as the source of the scourge.[21]

More often, though, impoverished and fearful Italian immigrants simply resisted what they perceived to be the intrusion of health officials, who might even be spreading the disease themselves simply by going from house to house. Apprehensive immigrant parents barred their doors to visiting nurses. On at least one occasion, an irate Italian parent tried an Old World technique of intimidation. A pediatric clinic nurse, who often reported cases of polio and violations of the sanitary code in a Brooklyn neighborhood known as Pigtown, had her life threatened in a letter sent by the "black hand," the traditional name of the Mafia. Whether or not the nurse was the target of the feared criminal organization, after the threat she was escorted to and from the clinic by a policeman.[22]

The behavior of individual Italian families, who dealt with being stigmatized by firing off indignant letters and slamming their doors in the face of municipal health workers in 1916, suggests that they perceived public health measures as unwelcome and unwarranted intrusions. They resented the pressure to submit to institutionalized forms of social control at the price of social ostracism. They perceived such intrusions as almost always motivated by a lack of understanding of their ways and American ethnocentric prejudice.

Beginning in the latter part of the nineteenth century, most Italian immigrants to the United States were townspeople, or *contadini,* from the Mezzogiorno, the impoverished southern provinces. For *contadini,* how one lived was determined not by the government but by *la famiglia* and *l'ordine della famiglia* (the rules of family behavior and responsibility). Long and bitter experience with public officials—and the large landowners they shielded—had left the *contadini* with a cynical attitude toward all forms of authority other than the family. Such cynicism, embodied in the adage *la legge va contrai cristiani* (the law works against people), was exacerbated by unification in the 1870s. The lot of the *contadini* deteriorated; reform never filtered south.

The northern Italians who controlled the new government were largely unconcerned about the south. Those who thought of the region at all considered it populated by lazy, impoverished, irresponsible, crude, and often violent primitives, rather than shrewd, hardworking, independence-loving, family-oriented *contadini.* The results were legal and economic burdens that crushed the Mezzogiorno. Taxes strangled the south and the region was trapped in a maze of increasing debt. While cows and horses, most often the property of the wealthier *latifondisti,* were exempted from taxation, mules and donkeys, crucial to the *contadini,* were taxed heavily. Church lands, confiscated by the new government, were almost always transferred to the *latifondisti,* almost none being distributed to the *contadini.*[23]

Increasingly heavy tax burdens compelled *contadini* to mortgage their lands and take loans, often from agents of the landowners. Interest rates ranged from 400 percent to 1,000 percent. Inexorably, land was captured through default; a report to the Italian parliament in 1910 stated that several hundred *latifondisti* owned or controlled over 50 percent of the land in the Mezzogiorno.[24]

Not all southern Italians farmed. Some were artisans, who worked in factories manufacturing silks and other textiles. To protect northern industry, high tariffs were passed by the central government. These crippled the unprotected industries and agricultural markets of Mezzogiorno merchants; unemployment rose markedly by the end of the nineteenth century.

The southern provinces were anything but an integrated ethnic whole; people did not even speak the same dialect from one province to the next. Still, much in the life-style and patterns of belief of the *contadini* concerning health and disease was similar, though hardly identical, from province to province.

As they had been for centuries, illness and remedies were bound closely to a blend of religion and folk beliefs that varied somewhat from place to place.[25] Christianity had come to the cities long before the countryside, where pagan deities continued to hold the attention of peasants.[26] From Rome northward, Italian Roman Catholicism resembled that practiced in the rest of northern Europe, but south of Rome, it was blended with traditions and customs of pagan origin. Even in the south, there was great diversity. Campania and Sicily retained much from the Greco-Roman tradition. Coastal areas colonized by the Greeks became part of Magna Graecia and were influenced by the Hellenistic medical tradition through the migration of Greek physicians. Other regions in the south, such as Abruzzi-Molise, remained isolated from Greco-Roman influence; there reliance upon superstition and magic remained especially strong.[27]

Transcending regional differences, certain patterns were generally common in the south. Preservation of health was tied closely to worship of objects, especially statues and sacred relics, and the attachment of particular powers and qualities to individual saints. "Saint Rocco protected devotees against illness, Saint Lucy guarded their eyesight, and Saint Anna helped during the pangs and dangers of childbirth."[28] Such saints found in southern Italian worship appear in neither the Bible nor the writings of the early Christian fathers, but were folk substitutes for old Greek and Roman gods and spirits of the forests, rivers, and mountains. When peasants prayed to a local Madonna for good health or the cure for a particular affliction, they were engaged in a pre-Christan ritual. The Catholic Church had assimilated such folk customs across Europe, having found they could not be successfully excluded from popular worship.[29]

At the annual celebration of a town's patron saint, worshippers took the saint's image from the church and carried it in a procession through the streets. Offerings, especially money and jewelry, would be cast at the saint's feet. On such festive occasions, the personal clothing of a sick person might be placed at the statue's foot. Believers contended that, when the garments were again worn, the goodness of the saint imbued the body of the worshipper and restored health.

Many southern Italian beliefs about illness are common in most peasant cultures, such as attributing illness to the influence of one who practiced *jettatura* (sorcery), through use of the *mal'occhio* (evil eye), a belief that had no basis in Roman Catholic theology and that the church never succeeded in supplanting.[30] According to the contemporary sociologist Phyllis Williams, southern Italian peasants blamed illness or misfortune on the influence of "an ever-present menace, the power of envy."[31] Those jealous men or women who possessed the evil eye could, with a glance, cause physical injury, sickness, or even death. Amulets could ward off the *mal'occhio.* Popular amulets whether of a precious metal, coral, or lava represented animals' horns, claws, or teeth. Fish, scissors, knives, and a male—but not a female—hunchback were common symbols for amulets. They could be placed over a doorpost, on a bedroom wall, on a chain around the neck, in pockets, or in the lining of clothing. When someone suspected of having the evil eye approached, the amulet was to be grasped and pointed unobtrusively in the offending individual's direction. In the absence of an amulet, one might protect health and well-being by extending the index and pinkie fingers from an extended fist to represent a set of horns.[32]

Wizards and witches also survived into the twentieth century in the minds of southern Italian villagers. Nor were all embodiments of evil. Witches, *maghi* and *maghe* (male and female, respectively), often worked as herb doctors practicing folk medicine, curing diseases, and healing broken hearts with their potions. Individuals who wished to safeguard their homes from evil witches put salt or garlic in keyholes, on windowsills, or on the doorstep. A broom placed behind a door was also regarded as a preventive. An American physician, unfamiliar with local beliefs and fears, was puzzled by his failure to persuade his neighbors that sleeping with an open window at night was healthy. Their conviction that evil spirits traveled through the night air made it "a task for a modern Hercules," he later wrote.[33]

Southern Italians coped with illness by applying folk remedies derived from a reservoir of folk traditions and customs, and by consulting specialists such as witches, barbers, midwives, and herbalists. There were few physicians; small towns usually had one who was paid by the state out of community coffers. His salary was fixed, though he often received free use of a house in exchange for treating the most impoverished gratis. Wealthier patients paid for special attention, which allowed the physician to supplement his income. These town physicians were not highly respected. In part, townspeople were suspicious of scientific medicine. But equally important, folk wisdom taught them "when it don't cost anything you might know it is no good."[34] The level of suspicion in which town doctors and scientific medicine were held is suggested by an outsider's account of how *contadini* reacted during the 1892 cholera epidemic:

> They [*contadini*] resist and resent every effort to purify and ventilate their houses, and the most natural and simple precautions are neglected. As for the physicians, provided for them at the public expense, they look upon them with horror, and it is dangerous for them even to walk the streets. About a week since Dr. C—, on his way to visit a patient, excited a veritable tumult. "Give it to him! give it to him," was cried out from the infuriated crowd. "He is one of the doctors paid by the municipality to poison the poor people," and had not the carabineers interfered for his protection, he would have been torn to pieces.[35]

Such fear and loathing of the physician as intruder, common in rural southern towns, was not the case in larger cities such as Naples and Palermo, where a more cosmopolitan view of medicine and its practitioners prevailed.

Contadini in southern Italy found many virtues in folk medicine. It cost little because most medicine was practiced by experienced housewives or neighbors. Should an herbalist be called, he or she could make diagnoses and concoct required medicine from materials readily available in nearby fields and forests.

The diagnoses of folk healers were grounded in adages associating physiological characteristics with particular patterns of personality or disease, such as "The face without color is false" or "He who has long ears will live long."[36] People with long necks were thought to be susceptible to tuberculosis, while those with hooked noses and livid faces were likely to possess the evil eye. Those with thick lips were unusually sensual; large mouths were a certain indicator of greed.[37] Even as American nativists such as E. A. Ross had come to associate certain facial features and body types with inferiority, so, too, did southern Italian peasants have their schema. Given that most Italian peasants were short as compared with northern Europeans, it is not surprising to find "Tall men have little enterprise" among the generalizations upon which equally ethnocentric southern Italian folk diagnosticians relied.[38]

The ailments that both folk healers and physicians confronted in southern Italy at the end of the nineteenth century were endemic to poor rural populations. Respiratory diseases such as tuberculosis, bronchitis, and pneumonia were relatively rare. Indeed, Dr. Antonio Stella, an Italian physician in the United States, described Italy as a country that "yields less victims annually to consumption than any other nation on the continent under similar demographic conditions."[39] Nonetheless, TB was especially feared because accurate diagnosis was rarely made until the disease was in its last stages. Sicilians called it the *male sottile* (insidious sickness) because it seemed to sneak up on its victims and then quickly overcome them. Peasants so feared the disease and its stigma as an almost certain death sentence that those infected would refuse to use a receptacle for sputum, preferring to spit on the dirt floor of the house as did healthy people.

Most common were diseases accompanied by high temperatures such as malaria, typhoid, and rheumatic fever. Always, the disease was defined by its symptoms and it became the focus of folk-healers and physicians alike. Also quite prevalent were childhood diseases such as measles, chicken pox, and scarlet fever. Peasants understood that if they exposed their children to these ailments at a young age, mild cases ensued and recovery was virtually certain. Smallpox was so common that a Sicilian proverb cautioned, "a girl cannot be termed beautiful until she has had smallpox," so it could be determined whether scarring had destroyed her beauty.[40]

Among the very young, cholera infantum and intestinal infections were common. A lengthy period of nursing, as long as two years, made for healthy infants, free from the ravages of contaminated milk, but the quick transition to full adult fare at an early age may account for the complaints of intestinal irritation. Cures for children varied from sitting the child in ritualized body positions—right leg to left arm and vice versa—or dressing the child in a miniature copy of the black habit worn by St. Anthony for a part of each day for several weeks.

Cures varied from prayers and rituals separate from the body to efforts at restoring a chemical balance to the body. In the former category was the custom of humiliation and sacrifice. Women with chronic diseases sometimes begged for money door-to-door. The funds were then taken to a priest as a contribution for a mass to be said for her recovery. A family surrogate begged if the patient was too ill to move about. In the latter case, excesses of acid or salt in the body or the accumulation of too much blood in one location were often blamed for inflammation. Aside from allowing diseases to run their course, a first principle of folk medicine, cures were administered until one worked. As one observer noted, "It never seemed to enter anyone's mind that death might be due to the conglomeration of treatments or that recovery took place in spite of them."[41] The therapeutic transformation away from traditional remedies was an uneven process in modernizing societies. In rural communities such as those in Southern Italy, the move to modern therapeutics was especially slow. A physician, if and when he was summoned at all, frequently had to content himself with allowing his therapies to compete with those offered by family members. His medical degree did not earn him a priority in the eyes of those at the bedside of a sick relative or neighbor.[42]

Folk medicines could be animal, vegetable, or mineral in substance. Common vegetable cures were olive oil, lemon juice, wine, vinegar, garlic, onion, lettuce, wild mallow, flour baked into bread, rue, and tobacco, known as the *erba santa* (sacred plant). As for animals, those used in whole or part were wolf, chicken, viper, lizard, frog, pig, dog, mouse, and sea horse. Of necessity, all were available to southern Italian peasants from a nearby forest or purchasable from merchants. Similarly, minerals most frequently used were those available locally, such as rock salt and sulfur, especially popular with Sicilians who could find sulfur in mines along Sicily's southern coast.

Bodily secretions, especially saliva, urine, mother's milk, blood, and ear wax, were all valued curatives. Saliva was thought to be especially valuable as "fasting spittle," sputum taken from the mouth early in the morning before the ingestion of food. Mothers used it to bathe the eyes of children suffering from conjunctivitis. When taken from a seventh male child, fasting spittle was used to treat impetigo. Spittle was a critical antidote to the evil eye, as well. In some parts of southern Italy, it was customary to spit three times behind the back of a woman suspected of being a whore who had kissed a newly born infant. Obviously, several episodes of infant death had led to a diagnosis grounded in moral values, and spittle was the purgative. Because spittle could break curses and spells, those paying sick calls on neighbors fended off contagion's threat by spitting hard at the house door in self-defense. Women in extended labor moved the process along by asking a neighbor to spit out the window of the birthing room, banishing any curse or spell that might be stalling a healthy, normal delivery.[43]

Contadini blended various substances to produce medicines. Sulfur and lemon juice were mixed as an ointment for scabies, for example. Others were simply fastened to the body. A live frog fastened to the temple was thought to be a cure for some eye ailments. Slices of potato or lemon were bound to the wrists to reduce fever. Sometimes cures were affected through an incantation or by wearing a particular garment. Sicilians found the wearing of a red scarf a cure for erysipelas, a dermatological disease characterized by red lesions. Black silk scarves were tied around the neck to cure sore throats.

As they had since the Middle Ages, barbers cured through bleeding, cupping, and scarifying. They set fractures, cauterized wounds, and opened abscesses. External growths could be removed, but internal growths, such as cancers, were regarded as incurable. Barbers were of little help with venereal disease. Known as the "French sickness" or the "woman's sickness," gonorrhea was held to be cured only by contact with a virgin. In Sicily, young women regarded as feebleminded became the cure in some cases.

The feebleminded as well as those diagnosed as suffering from nervousness, hysteria, or other mental disorders were generally thought to be possessed by evil spirits. On the theory that such spirits could be purged by making their environment inhospitable, patients were shaken, beaten, or subjected to other physical abuses. In single women, some of these conditions were attributed to the need for a husband, a diagnosis that reflected the social pattern of early marriage.

Witches could use charms to banish the influence of the evil eye. One such incantation suggests the synthesis of non-Christan and Christian traditions present in much of southern Italian folk medicine. Treating a person struck mad by the evil eye, a witch might chant:

> Two eyes have harmed you,
> May three persons relieve you:
> Father, Son, and Holy Spirit.
> Away with envy and iniquity
> May they scorch and burn in the flaming fire.
> Drive away all evil.
> In this house there are four evangelists:
> Luke, John, Mark, and Matthew.[44]

Sociologist Williams notes as significant that "Luke the Physician" is mentioned first, intentionally jumbling the usual order in Christian texts: Matthew, Mark, Luke, and John. Christianity was thus not abandoned but rather adapted to complement non-Christian belief, perhaps lending a legitimacy to pagan practices and thereby increasing the sick's confidence in their efficaciousness.

In the 1930s, Dr. Carlo Levi, a Jewish physician, writer, and critic of fascism, was forced into internal exile in Lucania at the start of the Abyssinian War. There, he encountered the ubiquitous belief in magical cures. In the small town of Gagliano, Levi became familiar with his hosts' beliefs that "Magic can cure almost any ill, and usually by the mere pronouncement of a spell or incantation." In his moving memoir of his time in exile, Levi described these cures, some of which were "of local origin; others belonged to the *corpus* of classical lore which came to these parts who knows when and how."[45] He was especially impressed by the abracadabra, a tiny roll of paper or a metal plate bearing a triangle constructed of the letters spelling abracadabra. Although the peasants at first hid their amulets in embarrassment in front of the physician, Levi wisely showed respect for the amulets and their curative powers. Knowing that the townspeople would not soon have the opportunity to replace superstition with scientific medicine, he preferred not to disparage what worked for his neighbors. "I respected the amulets, paying tribute to their ancient origin and mysterious simplicity, and preferring to be their ally rather than their enemy."[46]

Levi's neighbors called jaundice, *male dell'arco*, rainbow sickness, because it made the victim's complexion change color to the strongest in the sun's spectrum, yellow. One might contract the disease in several ways. Rainbows walk across the sky with their feet on the ground. Therefore clothes hung out to dry became impregnated with the rainbow's colors. Those donning the garments fell ill and turned yellow. Others fell ill because they urinated in the direction of the rainbow. Because the curved arc of the urine from a male resembles the rainbow arc, he might be transformed into the image of the rainbow and turn yellow. In either case, the cure was the same. The patient must go to the top of a nearby hill, where the black handle of a knife was applied to his forehead to make a cross. The same gesture was made upon the body's various joints with the knife handle. The ceremony was performed three times on three consecutive mornings to restore the skin's white color and the victim's health.[47]

Contadini at the turn of the century saw the hospital as primarily a place to die and the institution was shrouded in superstition and mythology. Some thought that those who died in hospitals did not go immediately to purgatory because they had not received appropriate funeral rites. Therefore, hospitals were haunted by the ghosts of the recently dead who terrorized those about to die. Patients in delirium were said to be conversing with the ghosts of former patients. The hot climate requiring rapid burial after death merely fueled tales of ghosts.

Because *ospedai* (hospitals) were largely charitable institutions, many *contadini* despised them. In their world, nothing of worth was received in acts of altruism. Even as they believed that doctors offered free care only when they expected to get a body for experimentation, many *contadini* devalued hospital care because they felt that it was merely another institution designed by intruders to interfere in their lives, separating them from family, friends, and therapies that they trusted at the very time when all three were most needed. In larger cities such as Naples, *cliniche private* (private clinics) and *case di salute* (sanatoria) served the wealthy.[48]

The closed world of the Mezzogiorno at the end of the nineteenth and the early years of the twentieth century and its complex patterns of folk ritual and Christianity was not shaken by the arrival of nonbelievers but by the departure of natives for other countries in search of jobs. The already crippled economy of the region was further worsened at the turn of the twentieth century by a series of natural disasters such as blights, earthquakes, and volcanic eruptions. Natural disasters such as these were rarely the root causes of emigration, but often they were the catalysts. The peak exodus year was 1907, with 285,731 departures. The scramble also altered the gender composition of Italian immigration. Prior to the turn of the century, most emigrants were young males in their teens or early adulthood, who left their parents or young wives and children behind as they pursued opportunity abroad. Immigration officials estimated that 78 percent of the Italian immigrants were men. After 1910, the character of migration changed. By 1920, 48 percent of southern Italian arrivals were female.[49]

As did members of other immigrant groups, millions of Italians found jobs in urban areas of the United States and lived in tenement hovels and congested neighborhoods near the factories or construction sites, where they found employment and no shortage of advice on improving their health and hygiene. Some of it came from assimilated Italians such as Dr. Antonio Stella and other physicians. In an essay on "The Effects of Urban Congestion on Italian Women and Children," Stella concluded by urging policymakers to engage in social planning and institute a program of education aimed at the individual immigrant. He called "for a better distribution of the immigrants *not after* they have reached Ellis Island, but *before* they decide to leave their motherland, by informing them of the wonderful resources of this vast continent, outside and beyond the large cities; let us educate them to the principles of hygiene and life, when they are settled here, and above all, let us distribute the work in appropriate areas outside of the city limits, so that proximity of the factory should not be . . . the chief reason for their congesting the city." Stella also urged the building of model tenements at a rent low enough so that it would not absorb more than a third of a worker's salary. The physician quite explicitly asserted the relationship between improving the physical and moral condition of newcomers, because "when we shall have given the people clean, healthy homes, full of light and sunshine, we shall have accomplished the physical and moral regeneration of the masses; we shall have given them that to which every human being is entitled, health and happiness."[50]

Perhaps no native-born reformer could have more clearly stated the need for the newcomers to change and for American society to provide the opportunity. But who should make the first move? Stella thought that society had the obligation to offer newcomers better housing, education, and a healthy environment, leaving implicit the obligation of the newcomers to avail themselves of the opportunities. More typical, though, were the native-born reformers who saw the burden falling primarily upon newcomers to educate themselves and change habits.

The Daughters of the American Revolution (DAR) in Connecticut commissioned John Foster Carr to write a *Guide For The Immigrant Italian In The United States Of America* (1911). The ladies of the DAR made their aims quite clear. Because so many Italians were arriving in their state, the DAR wished to "help them become Americans."[51]

The *Guide* offered information on a wide range of topics, from a thumbnail sketch of American history to how to send a telegram. However, health matters were a priority and even the section advising immigrants where to settle suggested its importance. "There is Prosperity and Health on the Farm," the section is titled. Pitching country life to many whose families had been on the land for generations, Carr advised, "Country life is healthier for yourself and your family. You are protected from diseases common in the city, and, more important still, the moral health of your boys and girls will be better protected."[52]

Realizing that most Italians would settle in the city whatever they were advised, the *Guide* preached "public hygiene" as a disease preventive in an urban environment. In urging newcomers not to spit in public places such as sidewalks, the floors of public buildings, or the floors of elevated train stations and cars, John Foster Carr explained, "Spitting is not only a disgusting habit; it is the cause of tuberculosis and other diseases."[53] Similarly, immigrants were warned that it was against the law to "beat or shake a mat, carpet, rug, or garment out of a window, in a street, or in such a manner that the dust therefrom passes into the street or into occupied premises" or to "keep a live chicken within the built-up sections of New York without a permit from the Board of Health, or to kill such chickens within the city limits."[54] Clearly, some habits of rural living would have to change.

The lengthiest section of the *Guide* was "The Importance of Caring for the Health," because "A working man's capital is a strong, well body." However, threats to the worker's wealth in himself were everywhere. "When men live crowded together, as they do in our tenement houses and in the shanties of a camp, their vitality is lowered, and they become ready subjects to such diseases as pneumonia, and, what is far worse, consumption."[55] And so, "To avoid disease and lowered vitality you should keep very clean, eat well, sleep in well-ventilated rooms, and live much in the open air."[56] Knowing, perhaps, that Italians slept with closed windows to exclude witches and other malevolent spirits, Carr made it a point to advise Italian immigrants, "It is never dangerous in America to sleep with your windows open. If there are mosquitoes put nets on the windows. *Prevention is the best cure for disease. Avoid bad air, bad food, bad water, bad habits*."[57]

After many paragraphs urging bathing, care in purchase of food, sweeping and dusting, and many other good habits, Carr encouraged his readers to cooperate with government officials and to trust them. They must not hesitate to report unhygienic or dangerous conditions to the board of health, because "That is the American way. And in America you should do as Americans do."[58] He carefully explained that boards of health in American cities had power to coerce those who disobeyed standards of hygiene and endangered the public welfare. "It can oblige people to keep their houses and living-rooms in a sanitary condition. It has power to force employers to keep their shops and factories in a sanitary condition." He especially mentioned the boards' role in sending doctors to schools to ensure the children's welfare, because "The sickly child is always behind in his studies. Only well children make progress."

Perhaps comprehending the hesitation of southern Italians to spend large sums on health care, Carr was careful to emphasize that many aspects of health care in America were "paid for by taxes." And to banish the gloom and doom often linked to health, he made a point of mentioning "large public baths," "playgrounds for children," and "open air gymnasiums for men and boys." New Yorkers were advised to avail themselves of "recreation piers built out onto the river, where mothers can take their small children during the hot weather, and where it is pleasant to promenade in the evening, and often upon these piers excellent concerts are given," and to frequent parks, which offer the city dweller "opportunities for rest and for breathing the fresh country air."

And should, despite all these precautions, an Italian immigrant or a family member fall ill? First, the foreign-born ill must "Beware of the Medical Institutes that advertise in the Italian papers, that pretend to cure every kind of disease, even those that are incurable." Immigrants were instructed to shun quacks and patent medicines and go to a hospital or dispensary. Again, Carr sought to allay fears of cost, emphasizing that "American hospitals are supported by the taxes and by the gifts of the wealthy. They are entirely free to the poor. They are splendidly equipped, and in them rich and poor are treated with equal skill and tenderness."[59] Aware that Italians' fear of hospitals was heightened by their suspicion of Americans, the author mentioned two Italian hospitals in New York City, the Italian Hospital on West Houston Street and further uptown, Columbus Hospital at East Twentieth Street. The latter was a Roman Catholic hospital founded in 1891 and operated by the Missionary Sisters of the Sacred Heart.[60]

No discussion of urban health would be complete without special mention of tuberculosis. The *Guide* started with the reassurance that consumption was "not a fatal disease," especially if detected and treated early. Offering the popular wisdom of the day, Carr, medically naive by even 1911 standards, described "pure air and sunshine, outdoor life, and nourishing food" as the "only cures" rather than as treatments that might bring the condition under control. Those who suspected themselves infected were urged to head for Clinica Morgagni in New York, "an Italian institution that takes special care of the tuberculosis poor, and provides in case of need for sending them to sanatoria."[61] Carr concluded, however, with the stern warning that "One sick with contagious disease is liable to be punished if he *exposes himself or another similarly* sick in any public place."[62]

There is no evidence of how many Italian immigrants read Carr's book or the others written in this era. However, the DAR was sufficiently well financed that it had the capability of broad distribution. Advice books were fine for those who could read, and took the time. But what happened to illiterate Italians—or those too intransigent—to heed the warnings of the DAR?

Most Italian immigrants found themselves in poor housing in congested neighborhoods such as those of New York, Philadelphia, and Chicago. That they were willing to consign themselves to such conditions, resisting the blandishments of reformers, can be attributed to the Italian belief that their stay in America would be temporary. Described by American immigration officers as "birds of passage," Italian males, and their Greek and Slavic counterparts, ventured forth in the spring, remained abroad until late fall, and returned home for the winter months. Data from both Italian and American sources suggest that of the 2,200,000 emigrants to depart for the United States between 1902 and 1910, 1,058,000 returned, or 48 returns for every 100 departures.[63] Often they left carrying enough cash to marry or set their families up on a plot of land or in a small business. Just as often, they left in poor health, victims of economic ephemera that had drawn them to the United States.

Southern Italian immigrants in the United States had a medical spokesman who mediated between them and the larger, American community. Dr. Antonio Stella compiled data, constructed arguments, and published articles defending southern Italian immigrants against those who saw the newcomers as sub par.

Antonio Stella's roots were not humble, nor was his life an epic in overcoming obstacles.[64] He was born in Muro, Lucania, a southern province, in 1868, but his father was a lawyer and a noted numismatist, not a peasant. Stella was educated at Naples, where he attended the Royal Lyceum. He received his M.D. from the Royal University in 1893. After graduation, he emigrated to the United States and was naturalized in 1909.

Specializing in internal medicine, Stella, a private practitioner, developed an international reputation for his publications on tuberculosis and its relationship to urban poverty. In New York City, he became a prominent figure in the medical community, treating such important figures as opera star Enrico Caruso. He was a consulting physician to Manhattan Hospital and the Italian Hospital, served as a visiting physician at Columbus Hospital, and was appointed an examiner in lunacy for the state of New York. The year he was naturalized, but prior to his oath, he served as a delegate of the Italian government to the Sixth International Tuberculosis Congress, held in Washington, D.C. He was the author of many medical monographs and articles on immigration, a fellow of the New York Academy of Medicine, a member of the advisory council of the New York Board of Health, and an active participant in many national and local professional societies.

Stella wrote a book designed to shed the most favorable light upon the Italian contribution to the United States. The volume's subtitle, "Statistical Data and General Considerations Based Chiefly Upon the United States Census and Other Official Publications," was clearly an effort to establish the credibility of the arguments within by grounding them in data generated by Americans themselves. Among the many topics explored through the use of his data were the health and vitality of Italian arrivals.

Stella boldly asserted that whatever health problems Italians in the United States might suffer, they were acquired on this side of the Atlantic. Speaking of the Italians in their homeland, he described them as "one of the healthiest in the world on account of [their] proverbial sobriety and frugality, also perhaps on account of the fact, that natural selection has had there [in Italy] freer play than elsewhere."[65] He vehemently denied the "wild accusations" that syphilis was arriving with the newcomers from Italy, citing inspection data from Ellis Island: "In the month of July 1921, 11,000 immigrants were given intensive examination at Ellis Island, performed with removal of clothing, and of all this number only 43 were found to be afflicted with venereal diseases."[66] Stella appealed to empirical evidence, noting "The freedom of italian [*sic*] immigrants from syphilis is not only a matter of common experience, but is strikingly evinced by the sturdy and vigorous physique of these peoples and by their extraordinary fertility." He proudly boasted, "The very increase in population of this country is almost wholly due to the high nativity [*sic*] [fertility] of the immigrants."[67]

Absolving American Italians of any charges of slovenliness or weak moral fiber, Stella reminded readers that "the Italians, together with the Slav, more than any other foreign group, engage in dangerous and hazardous occupations, in mines, steel mills, blasting, excavations, besides all sorts of dusty and unhealthy trades—thus many times they pay with their health and life, while adding to the prosperity of the United States."[68] Citing data from the 1900 census, Stella observed that Italians had the second-lowest rate of mortality from tuberculosis, 113.6 per 100,000; only native-born Americans with a mortality rate from TB of 112.8 were better off. This low rate Stella attributed to Italians not remaining in America long enough to be counted: "any medical man who has been brought into close contact with them, will bear witness to the fact, that only a few of the italian [*sic*] tuberculous die in the district in which they have contracted the disease. Their fear of consumption is much greater than among any other nationality, and the belief in climate as the only cure for pulmonary disease is so firmly rooted, that the first suggestion of anything abnormal with their lungs leads them to make immediate preparation for a trip to the home country."[69] Always careful to support his assertions with statistics, "when such statistics were kept," Stella offered data from 1903–04, when only two cases of tuberculosis among Italians coming to America were treated in ships' hospitals, "a rate of 0.006 per cent." However, among the homeward bound to Italy, during the same two years, "there was [*sic*] 457 in the ship's hospital, besides 17 who died at sea, without counting those who traveled as first and second cabin passengers and thus escaped enumeration."[70]

In 1904, Stella published an article on tuberculosis among Italian immigrants in *Charities*, the publication of the Charity Organization Society of New York. He told readers that "exact information" was often missing because of Italians' mobility. He advised, "one must follow the Italian population as it moves in the tenement districts; study them closely in their daily struggle for air and space; see them in the daytime crowded in sweat-shops and factories; at night heaped together in dark windowless rooms; then visit the hospitals' dispensaries; and finally watch the out-going steamships, and count the wan emaciated forms, with glistening eyes and racking cough that return to their native land with a hope of recuperating health, but often times only to find a quicker death."[71]

Stella wandered the streets of the Italian neighborhood on New York's Lower East Side, so he knew firsthand about the discrepancy between the number of cases of tuberculosis reported and the number that really existed. In tenements on Elizabeth and Mulberry streets there were barely "twelve and fifteen cases of consumption reported to the Board of Health since

1894," but from his "personal experiences with some of the houses in that particular neighborhood," Stella reported that "the average has been not less than thirty or forty cases of infection for each tenement yearly, the element of house-infection being so great." Frustrated, the physician rhetorically asked, "And how could it be otherwise?"[72]

Stella blamed the congested housing and dark, unventilated sweatshops for weakening the bodies of robust Italian agrarians, making them fertile soil for "the Koch bacillus." The knowledgeable doctor educated his audience as to how important that process of physical degeneration was, because "We know now-a-days that the penetration of a pathogenic germ into our system is not sufficient to cause a disease. It must find our body in a state of temporary paralysis of all its natural defenses, to be able to give rise to certain morbid processes, the evolution of which constitutes a disease."[73] Stella acknowledged the lower mortality rates of eastern European Jewish immigrants living near the Italians, but he rejected behavioral explanations suggesting that Jewish religious laws and customs mandated a healthier lifestyle. Jews' resistance to disease, Stella wrote, was "due, less to religious restrictions, than to a certain immunity against the ills of urban life, acquired through long residence in cities."[74] Dirty linen and impure air did not cause disease, Stella wanted his readers to know, but "Six months of life in the tenements are sufficient to turn the sturdy youth from Calabria, the brawny fisherman of Sicily, the robust women from Abruzzi and Basilicata, into the pale, flabby, undersized creatures we see, dragging along the streets of New York and Chicago, such a painful contrast to the native population! Six months more of this gradual deterioration, and the soil for the bacillus tuberculosis is amply prepared."[75]

Stella's observations were more than amply confirmed by Italian immigration officials, who watched as their "birds of passage" flocked home wilted and sick. Ship physicians reported that the rate of illness was higher among those returning to Italy than among the westward bound and that tuberculosis was the most frequently diagnosed ailment. In 1904, a list of the most common diseases treated on ocean voyages back to Italy showed that the 278 cases of tuberculosis detected were greater than all the others combined; malaria was a distant second place with 49 cases, contracted by Italians engaged in rural migrant labor.[76]

Contagious disease among returning immigrants was not the only health problem, however. Some appeared to be victims of homesickness or nostalgia. Others were diagnosed as mentally ill. In 1906, only one case of mental illness was recorded among the thousands traveling third class to America, but of those returning, 72 sought treatment for mental disorder, including 55 men and 15 women.[77] There were also a greater number of suicides on the return voyage. Dr. T. Rosati observed in 1910 that there had been two suicides on a recent return voyage from America, "something to think about—that suicide is more frequent among the repatriates." Council of Emigration member Luigi Rossi recorded what he had heard about returnees in the port of Genoa. There he heard that returnees from different countries could be distinguished by the condition of their bodies and their wallets, because "those returning from the United States come with sufficient health and money, those from Argentine return with their health but no money, and those from Brazil bring neither health nor money." Bad as conditions in America might be for the "bird of passage," conditions elsewhere were worse.[78]

Italians who remained in the United States fared little better than trans-Atlantic migrants. Children suffered greatly. (See table 5.2.) Dr. John Shaw Billings's data for New York City for a six-year period ending in May 1890 (prior to the creation of Greater New York in 1898) suggest that, while Italians had a mortality rate of 35.29 per 1,000 (based on mother's nativity), exceeded only by the Bohemians (43.57), much of the problem lay in the mortality of those under the age of fifteen, with high levels of infant and childhood death. Those under the age of fifteen with native-born white mothers had an average death rate per 1,000 of 32.43, while those with mothers born in Italy had a death rate of 76.41. Among those fifteen years and older, the Italian mortality rate dropped to 12.27, the third-lowest, bettered only by those from Hungary, 8.45, and Russia/Poland (largely Jewish), 6.21.[79] Billings explained that the excessive death rate among those having Italian mothers during the six-year period was due to excessively high rates of measles, diarrheal diseases, stillbirths, pneumonia, and homicides.

In 1906, Department of Health Records Registrar Dr. William H. Guilfoy

TABLE 5.2

Comparison of Average Annual Death Rates per 100,000 of Mean Population Among Those Having Native-Born (N.-B.) White Mothers and Those with Mothers Born in Italy and Residing in New York City and Brooklyn, 1884–1890[80]

	NEW YORK		BROOKLYN	
	N.-B.	**Italian**	**N.-B.**	**Italian**
Measles	46.21	198.75	17.75	91.54
Diarrheal disease	318.14	425.58	NA	NA
Stillbirths	224.09	383.15	180.48	273.18
Pneumonia	287.25	455.89	229.27	280.33
Homicides	3.75	9.89	1.12	11.44

presented data demonstrating that, when the nativity of parents was used as a variable rather than simply nativity of the individual, the mortality rate increased for groups with high birthrates, such as the Italians.[81] Italians and Austro-Hungarians had the highest rates for mortality from typhoid fever (where number of cases exceeded twenty). As was Billings, Guilfoy was impressed by the figures on mortality from measles, a childhood disease: "Italian children have a mortality of almost five times that of the entire city." They also showed the highest mortality from scarlet fever and whooping cough. As for diphtheria, Guilfoy blamed "neglect of the use of diphtheria antitoxin," which he claimed produced a mortality rate among Italian children "of three times that of all other nationalities combined." Although Italians did not show especially high rates of tuberculosis, "the most surprising rate of the whole table" is the rate of bronchopneumonia among Italian children, which reached "the enormous height of 710 per 100,000, more than 7 1/2 times higher than that of American children."[82]

Guilfoy, as did Stella, blamed the Italians' predisposition to respiratory disease on the congested conditions they encountered in American cities. "When we consider under what conditions the Italian families herd together and the opportunities this affords for the spread of contagious and infectious diseases, the difficulties encountered by the sanitary officers can be imagined but not described." To illustrate his point, Guilfoy asked his readers to imagine "seven tenement houses with a population of 1,500 [people], of a square block that by actual count harbors 5,100 souls, surrounded by other blocks harboring an average of 2,000, and you obtain an idea of the necessity for eternal vigilance on the part of health officials."[83] The guardian of New York's public health data, Guilfoy saw clear distinctions among nationalities as evidenced by the diseases to which they were the most susceptible. To Guilfoy, "The lobar pneumonia column reiterates the story as to the Italian predisposition to disease of the respiratory tract; here the Italian leads, with none of the others [nationality groups] a close second." In 1906, 513 Italians died of lobar pneumonia (infection of an entire lobe of the lung), a rate of 333.1 per 100,000, far and away the highest in the city.[84]

The picture was not completely bleak. In 1906, rates of mortality from diarrheal disease, always a clue to infant mortality, were higher for individuals of Italian parentage living in New York City than for any other group, 483.7 per 100,000; Austro-Hungarians were second, at 259.7.[85] The data confirm the generally poor living conditions of Italian families described by Stella. However, data gathered for eight smaller industrial cities with substantial immigrant populations between 1911 and 1915 suggest that infant mortality in the Italian immigrant community was considerably lower than for other groups (see table 5.3).[86]

TABLE 5.3

Infant Mortality Rates by Nationality of Mother in Eight U.S. Cities, 1911–1915

White native-born	93.8
Italian	103.8
Jewish	53.5
French-Canadian	171.3
German	103.1
Polish	157.2
Portuguese	200.3
Other foreign-born	129.6

The precise cause of these lower rates is difficult to determine. However, the markedly lower rate of gastrointestinal diseases as the cause of death has suggested to some scholars that the low economic status of most Italian immigrants as compared with white native-born Americans may well have been offset by the extended period of breast-feeding long traditional to the group, a pattern that the Italians shared with eastern European Jews and Poles (see table 5.4).[87]

TABLE 5.4

Rates of Infant Mortality from Gastrointestinal Disorders by Mother's Nationality in Eight U.S. Cities, 1911–1915

Native-born white	25.2
Italian	21.7
Jewish	10.5
French-Canadian	64.2
German	27.1
Polish	64.0
Portuguese	101.6
Other foreign-born	38.5

American reformers, even those willing to concede the influence of unfavorable socioeconomic conditions, laid much of the blame for Italian ill health at the feet of the immigrants themselves. First-generation approaches to health and disease did not differ markedly from those in southern Italy. The home or domus, the term preferred by some social scientists to describe the collectivity of family and possessions, remained the main focus of life. Matters of health and disease were treated there, if at all possible.[88] Obviously, remedies used in folk cures required some modification. Phyllis Williams, a sociologist whose handbook was designed for social workers, physicians, and others serving the Italian immigrant community, observed changes occurring almost immediately after arrival because of "inability to procure materials, such as wolf bones, from which to compound accustomed remedies." However, immigrants found an array of substitutes for the "medicine cupboard," including "a bewildering array of mushrooms and other foods as well as . . . plants, berries, and barks."[89] Plants that did not grow wild, such as basil and rue, were cultivated in gardens or in window boxes that sat on the sills of tenement apartments high above city sidewalks.

Illness was still often defined as an enemy's curse, a work of human jealousy or spite administered through the gesture of the *mal'occhio.* Restored health was a divine blessing, often a sign of the Madonna's indirect intervention. Such a blessing was meant to be shared with others. In a ritual carried from villages in the Mezzogiorno to New York's East Harlem, where the Madonna of Mount Carmel presided, the clothing of recently healed children was donated by their parents to the church for distribution to the community's poor, a transaction that expressed appreciation of "the intimate connection between private grief and joy and the claims and contributions of the community."[90] The transaction was conducted on the day of the *festa* in the saint's honor and suggested that "the Madonna's healing action created the requirement of social reciprocity and entailed a moral response."[91]

With the exception of prominent public figures such as Antonio Stella, physicians remained unpopular objects of suspicion and distrust, often consulted only to mollify authorities. At times, Italian immigrant parents preferred the assistance of witches, *maghi* [singular *maga*] (*iannare* to Neapolitans), to battle polio-induced paralysis. One mother claimed that a witch had cured her child of polio by rubbing her with some salve and mumbling an incantation. Although the child was also under treatment by an Italian physician, the mother gave all the credit for the cure to the *maga,* explaining, "They had only gone to the physician because the nurses at the hospital where the child had been treated at the time of the paralysis insisted upon it."[92] When a physician was summoned at all, he had to be an Italian physician; one such from New York described his group's preferences:

> Italians almost always call an Italian doctor because of the mutual sympathy and common language. The Italians are very fond of their families and will spend every cent to care for a member if ill. They are not satisfied with the American doctors because they make a short visit, prescribe and leave. This leaves the family in much doubt and accounts somewhat for their calling in another doctor if there isn't a marked improvement in a few hours. The Italian doctors tell the

> family what the malady is, and explain to them all about it, and this is what they expect. . . . They always pay cash and as a consequence they are inclined to call various doctors at different illnesses, just as they patronize different stores.[93]

This behavior echoes Old World healing patterns. There, too, many choices were available, depending upon the illness, from the local healer to the barber, from a witch to a physician.

Still torn between two worlds, Italian immigrants clung to tradition with a tenacity that complicated the task of physicians. Individuals who trusted a barber with bloodletting vehemently opposed blood tests. The logic was that blood drawn off by the barber was unhealthy, while that taken for a blood test was healthy blood unnecessarily removed from a healthy arm. Many immigrants echoed the traditional argument so well known among southern Italians that, because diseases often run their course and go away, it is preferable to wait for testing and treatment until it is apparent that the virtue of *pazienza* (patience) will not do. This approach to health is consistent with the broader fatalism that southern Italians imported with them, a view that assumes one should not expect a great deal from life and disappointment will thereby be minimized.

Psychiatrists, especially, had difficulty with Italian immigrant patients. To those who believed in witches and evil eyes, the probing stares and questions of a psychiatrist appeared threatening but in a way that could be diffused, much as evil spells had been combated on the other side of the ocean. In her handbook, Phyllis Williams cites a typical episode:

> A woman consented to a psychiatric interview, even though she had no faith in the physician. She did so purely because she was fond of the visiting nurse and wished to please her. She arrived at the clinic with a large handbag clutched in her arms with which she was greatly preoccupied. Nothing could persuade her to part with it. "I tell you after," she confided mysteriously and went into the psychiatrist's office. When she came out, she opened the bag and exhibited a quantity of amulets against the Evil Eye. She had added all she could borrow from her neighbors to those possessed by her family. "The doctor," she triumphantly asserted, "he no hurt me."[94]

Williams sensitively advised psychiatrists who would be seeing Italian immigrant patients to inform themselves of the "vagaries of Italian folklore," including the influence of witches and witchcraft, as an aid to understanding their patients.

At the turn of the twentieth century, Italian immigrants felt no more warmly toward hospitals in the United States than they did toward American physicians. Hospitals here seemed as cold and distant institutions as they did in Italy. Moreover, Americans hospitals charged a great deal for care and offered little hope of cure. Italians in the United States often spent a great deal of money to house relatives in private rooms, avoiding the "disgrace" of care on the ward.[95] Pregnant immigrant women usually chose home confinement with a midwife or Italian physician in attendance. However, some preferred hospital to home delivery because it allowed them to elude the traditional sexual intercourse rites at the onset of labor. Many women found this a painful practice and younger immigrant women and American-born Italian women increasingly shunned it. Once they gave birth, though, the same women who had sought hospital care now wanted to leave, even when their physicians recommended a longer stay.[96] In general, Italian immigrant hostility to hospital care was most frequently an expression of the centrality of the home, bed, family—the domus—to the lives of the newly arrived. For them family and the life cycle of birth, marriage, and death was the center of human existence, to be celebrated in the intimacy and privacy of the home among blood relations and only the most trusted friends here, as they had been in home villages.

The relationship of Italian immigrants and their children to modern medicine did not remain stagnant. Customs, traditions, and beliefs altered in the pressure cooker of assimilation with each succeeding generation raised in the United States. A study of southern Italian women in the North End of Boston suggests what might be described as a one-and-half generational depth to traditional beliefs after immigration. Women of the immigrant generation, largely from the regions of Abruzzi-Molise, Campania, and Sicily, and their older American-born daughters raised prior to World War II tended to cling to traditional perceptions of illness and cures. "The older women . . . had been recipients or observers of the traditional folk cures in childhood or early adulthood more often than the younger second generation women."[97] By contrast, those younger second-generation women had begun to rear their children in the 1940s "at a time when effective therapy in the form of vitamins, immunization and antimicrobial drugs had become generally available."[98] These women and third-generation women largely avoided the religious aspects of *festas*. Unlike the elderly, they did not "view the roles of the saints and the physician as complementary," nor did they perceive recovery from illness as "a miracle wrought by the timely intervention of both the saint and the doctor."[99] Younger second-generation women tended to attach more importance than their mothers or older sisters to "the skills and technology of modern medicine," although they, too, sometimes resorted to "prayer and petition" in an extreme health crisis.[100]

Traditional beliefs associated with the evil eye as the cause of illness still found expression at times in all generations, but the belief was hardly pervasive and appears to be importantly linked to the class position and individual personality of believers. Regional differences among southern Italians were evident, but it is difficult to determine the role that contact in the United States among the different regional groups may have exerted. Patron saint observance was among the more notable differences. Abruzzese immigrants had not organized patron saint societies as did newcomers from Campania and Sicily. However, in the streets of Boston's North End, the former often attended the *festas* of other groups.[101]

Though it may be difficult to unravel the influence that Italians from different regions had upon one another, it is easier to trace the impact of native-born American reformers, who often operated from urban settlement houses.[102] The motives of settlement house reformers has been a battleground for social historians. Some have pointed to the good intentions and strenuous efforts of settlement workers to meet newcomers' physical needs, even as they helped them to retain a strong sense of their own culture. They cite the *Settlement Cook Book*, which emerged from a Milwaukee settlement, and Jane Addams's programs to conserve and celebrate holidays, customs, folksongs, and needlework techniques as examples of such cultural conservation.[103] Others have denounced the settlement as a cornerstone of American cultural imperialism, a plant for transforming the foreign-born into Americans with lures of assistance.[104] However, there were a variety of settlements and considerable complexity in the interaction between immigrants and settlement workers, subtleties often trampled by both sides in the historical discourse.[105]

In Philadelphia, the St. Mary Street Library Association (1884) became the College Settlement in 1892, then the Starr Center in 1900. In 1907, the Starr Center opened the Casa Ravello Branch to serve the needs of the huge southern Italian community in South Philadelphia. Casa Ravello was a milk distribution center and medical office. Reformers assigned to it felt the need to "understand the language . . . of the daily toil, the daily need, the daily worry, the daily interests of those we would help. An acquaintance with the Italians will do us good." With no little admiration, reformers said of their clients, "They love beautiful music, pictures, flowers—the things that add grace to life—the grace which eludes definition, but which in its highest interpretation is indeed past understanding."[106]

Much as settlement house workers at Casa Ravello might revere the ethnic characteristics of their clients, there remained no doubt in their minds that the services they rendered to the community, improving health and hygiene, were a means to an even broader end, assimilation. In the first annual report after the Casa's opening, settlement workers candidly discussed the tactic:

> The huge problem of the Americanization of the immigrant is approached from an infinite variety of sides, and the providing of pure honest milk to a helpless mother in the hour of her baby's greatest need affords an avenue of approach not likely to be ignored, and that road leads one straight to the immigrant's

> heart. . . . We welcome every token that we are winning their trust, their esteem, their confidence.[107]

Four years after Casa Ravello opened its doors, its staff was confident that their efforts would help the Italian community to raise children who would be "a physically robust contribution to American life," even though all the reformers could claim credit for was advice: "the material commodities, which they need, they pray for."[108]

Italian immigrants in Philadelphia, New York, and other cities generally felt a mixture of emotions toward the visiting nurses sometimes headquartered in settlement houses. The perspiration-soaked nurses who climbed tenement stairs were often treated with gratitude for the care they brought to poor and perplexed newcomers, but nurses sometimes reported cases of disease that required quarantine, as during the 1916 polio epidemic. When nurses did this part of their job, Italian immigrant clients reviled them as intruders who invaded people's homes, trampled upon their privacy, and even curbed their coming and going in the name of a cold abstraction—public health policy. How much cooperation could nurses expect from a people whose historical experience told them that laws were instruments outsiders used to undermine family, community, and ancient patterns of loyalty? With little prior experience or sensitivity to Italians' customs and mores, it is small wonder that the well-meaning nurses were at times menaced by those whose children they sought to save. In contrast, settlement workers not involved in nursing invited the newcomers into *their* home, the domus of the settlement house. Their intrusion was more subtle courses in cooking or infant care—tendering newcomers the promise of a healthy, prosperous future in exchange for relinquishing Old World ways. Still, settlement house workers touched the lives and altered the behavior of only a fraction of Italian immigrants.

The loudest calls for change in Italian habits of health and hygiene came from within the community itself. Often the voice was that of an Italian physician, bitter over the sickness and suffering he could not spare his patients. In 1904, Dr. Rocco Brindisi observed with some sympathy, "The Italians, like all peoples with ancient habits and traditions, cling to many prejudices and superstitions, which often hamper those who work with them." Brindisi was confident that his compatriots were on the road to "regeneration" and that he himself was an instrument of change: "It is education through the public institutions and the missionary work of the physicians, that will bring the principles of hygiene and their practical benefits into the Italian homes, while waiting for the more substantial fruits of the schools." For Rocco Brindisi "there was not the slightest doubt in my mind that the rising generation of our Italians will be, in regard to sanitary conditions, on the same level with the American people."[109]

Dr. Brindisi was correct. Eventually, Italian immigrants, even "birds of passage," conformed to American standards of public health and hygiene. However, the path to conformity was neither short nor straight, sometimes taking more than a generation or two. Strong cultural ties bound Italians to their unique definitions of health and the etiology of disease and the relationship of both to a belief system that was a synthesis of Christian and pagan traditions varying from region to region within the provinces south of Rome. Italian immigrants' resistance to alien ways and intrusions from those unrelated to them by blood or Old World regional bonds proved a powerful obstacle even to well-meaning public health officers and reformers.

American public health officials were not generally virulent nativists. However, ethnocentrism, whether conscious or not, frequently caused them to discuss the linkages they perceived between disease and Italian immigrants, casting the newcomers as public health villains whose habits and customs could endanger others. Public health officials' construction of the relationship of alien ways to disease did not go unchallenged. Physicians such as Dr. Antonio Stella raised their voices in defense of their group's health and vitality, charging that the circumstances of life for impoverished immigrant workers, not the workers' habits, were responsible for high rates of morbidity and mortality among *paesani* in the United States. Stella was articulate, well educated, and financially well off. Most of his fellow immigrants were not. The majority were poor, unskilled laborers. The seasonal migratory patterns of many and their regional differences—Sicilian, Neapolitan, and so on—dividing them by custom and dialect, limited the quantity and quality of self-help available to Italian newcomers at the turn of the century, leaving them especially vulnerable to their surroundings.

Often public health officials agreed, blaming urban squalor, not southern Italian life-styles, for the sickness they saw. Touring the Italian neighborhood in Brooklyn during the 1916 polio epidemic, General Secretary of the Brooklyn Bureau of Charities Thomas J. Riley placed a red dot on his map to designate "every ill-fated house or tenement" where the disease had taken a victim. The sympathetic Riley did not blame the foreign-born for their own suffering. Instead, he identified filthy conditions as the breeding ground for the guilty pathogens, placing responsibility on the city government for its failure to achieve "good municipal housekeeping." Echoing Dr. Antonio Stella, Riley hoped that a cleaner, healthier city would eventually help to "bring the humblest family in our midst a fair chance for a wholesome living."[110]

In tracing the cause of Italian illness to the city itself, Thomas Riley stood on much firmer ground than nativists who pointed to what they saw as the innate biological and intellectual inferiority of such newcomers. Most Italian immigrants came from a rural culture with habits and customs suited to agricultural conditions, relishing fresh fruits and vegetables that became deadly when washed in the contaminated water of the Philadelphia suburbs. These *contadini* brought with them a long-standing and well-founded distrust of government institutions and persons in authority, including physicians. Their suspicion of outsiders turned them inward to their families, cutting them off from whatever benefits turn-of-the-century medicine offered. Italian immigrants from the southern provinces perceived American doctors and nurses as ethnocentric and responded by resisting the intrusion. Only time and the exhortations of such as Dr. Antonio Stella eventually persuaded these newcomers and their children to seek out modern medical care. Even then it was most readily accepted when offered by health care providers of their own group. Italian immigrants responded logically to the experience of illness in America, drawing upon time-proven customs and traditions for remedies. That their responses seemed unequal to the task of keeping them healthy on this side of the Atlantic can be attributed to the alien nature of the American urban environment in the early twentieth century.

6

Gezunthayt iz besser vi Krankhayt: Fighting the Stigma of the "Jewish Disease"

Just as Italians came to America from an agricultural setting, many eastern European Jews came from the industrial cities of Europe, already once removed from rural villages. In the cities of Russia and Poland they had accommodated their lives to urban crowding, urban squalor, urban illness. This is not to say that all arrived healthy or remained healthy in America. But their accommodation differs markedly from that of the Italians.

Of the 23.5 million immigrants who arrived in the United States between 1880 and 1921, over 2.25 million were eastern European Jews, making them

Title taken from the Yiddish "Health is better than illness," a common expression often uttered after someone has sneezed.

second only to Italians in numbers of new arrivals. One of them was Bertha Pearl, a Jewish immigrant from Russia.[1] In 1916, Bertha Pearl lived in Ward East II, at the Montefiore Home and Hospital in the still sparsely populated northwest Bronx. With her body confined to a sanatorium, her mind escaped through the narrow passageway of a pen point. Pearl passed the weeks and months at Montefiore writing numerous brief essays that she published in the *Montefiore Echo.* Patient publications are rare and valuable finds for historians because those ill enough to require hospitalization rarely had sufficient energy or patience to keep diaries or maintain extensive correspondence, much less author articles. However, TB patients "chasing the cure," as immigrant Jews described the process, often spent long months resting, taking clean, fresh air into their diseased lungs, and eating large quantities of nutritious food in special hospital facilities. Boredom and homesickness were not uncommon. Some reminisced and reflected upon life before Montefiore, when, having traversed continents and oceans in flight from poverty and oppression, they fell victim to fresh hardship in tenements on the Lower East Side of Manhattan.

In "Those Days," Bertha Pearl described experiences that had brought her and many like her to tuberculosis sanatoria and other health care facilities:

> When you came from the other side you lived on Pike Street, didn't you? Well, it was somewhere on the East Side, that's sure; everybody I ever knew did, and they are real nice people, too. . . . And isn't it funny the way we all have to do the same thing when we first come over—take bitter salt, give bananas a first try-out, and sleep on the floor at our poor relations's—
>
> Then, finally, your people find rooms—somewhere in a basement, most likely—and your father refusing to work on *Shabas,* you go into business. Maybe a soda water stand, maybe a grocery store; but it isn't long any way, is it, before you hear that all the money is gone? And the way your father finally goes to work at buttonholes, and your mother gets to be the "apple woman." It's great sport those days—Then your father dies, and your neighbors report you to the United Hebrew Charities, and you get less and less to eat till you get bananas and bread for a steady diet. You remember, don't you, those days? [2]

Pearl's bittersweet memoir recalled for her fellow tuberculosis patients—Jewish and non-Jewish alike—the world they knew at the conclusion of their journey to America. Unrelenting poverty, long hours of work in unhealthy environments, poor hygiene, and inadequate health care weakened them, making their bodies vulnerable to disease. Some may have carried contagion, undetected, past the Public Health Service physicians at immigration depots. However, many others arrived robust and healthy, only to fall victim to disease in their new home, sickened by America emotionally or physically.

The persecution of European Jewry that drove Bertha Pearl and millions of other Jews to seek their freedom as well as their fortunes in the United States was founded upon ancient hatreds and superstitions derived from biblical interpretations blaming Jews for Jesus' crucifixion and subsequent anti-Semitic folklore, often including blood libel legends.[3] However, by the eighteenth century anti-Semites increasingly turned to science to rationalize their bigotry. Scientists throughout Europe subscribed to a polygenetic vision of distinct and unequal races. In the nineteenth century, it seemed clear to some that particular races were innately degenerate and declining while others were on the ascent. As the central European scholar Sander Gilman has observed, "In the world of nineteenth-century medicine, this difference [between Jews and non-Jews] was thought to be exemplified by the 'pathological' or 'pathogenic' qualities of the Jewish body."[4]

Some western European Jews, such as Martin Englander and Cesare Lombroso, became adept at collecting and analyzing data useful in refuting the racial arguments of anti-Semites.[5] From the other side of the Atlantic, in the United States, Dr. Maurice Fishberg vehemently denied the biological construction that defined Jews as racially inferior to non-Jews and in a state of physiological and moral decline. Born in 1872, Fishberg was himself a Jewish immigrant from Russia, emigrating in 1889. A physician, Fishberg became clinical professor of medicine at the New York University and Bellevue Medical College.[6]

Unlike Dr. Antonio Stella, who described the villages and towns of southern Italy as warm, sunny, healthful places that yielded a robust, healthy populace, Fishberg described the villages he knew as a frigid breeding ground for disease. In rural areas of Russia, where so many Jews lived, conditions contributed to the nurture of pathogens. Towns there were "in a pitiful condition from the standpoint of health and sanitation." He observed, "There is no water supply, no system of public sewage; domestic refuse is hardly removed from proximity to dwellings, and the emanations from fermenting and decomposing organic matter, particularly during the summer months, are simply unbearable. The streets are usually unpaved, and in moderately large towns, only the principal streets are improved, and the quarters inhabited by the Jews—which are generally the poorest sections—are usually neglected."[7] Water sources were wells or rivers. However, as Fishberg cautioned, "This water is usually polluted by percolation below the surface from cesspools and decaying animal and vegetable matter in the neighborhood. In towns where the number of wells is insufficient, water must be purchased by every family from water-carriers by pail or barrel or kept for days."[8] One shtetl dweller recalled, "There was a stable with the urine from the cows. Nearby was a well, and I'm sure it was seeping in. There was a lot of typhoid as long as I can remember."[9] Noting that public health experts generally cited "twenty-five gallons of water per head per day" as required to preserve the cleanliness and health of the body, Fishberg estimated the amount available to Russian Jews as well below that figure.

Houses were badly overcrowded in the towns with substantial Jewish settlement. In Kovno, the Jews constituted 70 percent of the total population but occupied only 52 percent of the total houses in the city; in Grodno, 80 percent of the population and 65 percent of the houses; in Vilna, 60 percent of the population and 40 percent of the houses.[10] Moreover, many houses were in deplorable condition, with dirt floors and small windows that permitted minimal air and sunshine to enter.

Fishberg characterized Jewish existence in Russia as "a complete picture of medieval life."[11] And yet, in spite of abysmal poverty, Russian Jews and Jews throughout eastern Europe generally enjoyed better health and longevity than their non-Jewish neighbors. In 1907, death rates in Russia per 1,000 were 18.08 for Jews and 32.51 for Gentiles.[12] During typhoid, typhus, diphtheria, smallpox, and cholera epidemics, "the Jews are stated to have had lesser mortality and morbidity from the scourges," according to Fishberg's sources, though he was quick to point out that there were also times in Russia when "Jews are stated to have been preferred by the infection, and their mortality is said to have been higher than that of the Gentiles."[13]

Why did Jews generally have lower rates of mortality than their non-Jewish neighbors in Europe? If their environment was so unhealthy, was culture or some evolved biological advantage the answer? Would the pattern prevail in the United States, where eastern European Jews could expect to encounter the same unhealthy conditions that Antonio Stella blamed for the decline in the health of Italian newcomers? Also, how did Jewish immigrants in the United States and their advocates respond to the American counterparts of European race thinkers, especially those who spread anxieties that sickly enfeebled Jews would constitute a dependent class and a public burden to their hosts?

In the swirl of speculation and allegation, Jewish newcomers to the United States, fearful of fueling anti-Semitism, found themselves in a defensive posture, attempting to demonstrate that they were neither biologically inferior to their neighbors nor a threat to the public health or welfare. The Jewish community in the United States mounted a dual response to smears aimed at the health and vitality of the group. Health experts collected, analyzed, and published data that not only refuted the notion that Jews were especially prone to diseases but demonstrated with data that rates of some illnesses such as consumption were lower among Jews in the United States than rates for non-Jews, echoing the European pattern. They often argued, as did Antonio Stella on the Italians' behalf, that the struggle of life in America, not biological inferiority, was the origin of immigrants' unclean and unhealthy ways, so disturbing to native-born observers. Second, American Jewish philanthropists financed health care for the indigent, building hospitals and sanatoria in communities with high concentrations of eastern European Jews. The goal was to defuse charges that Jewish immigrants were becoming public charges in their new country. Historically linked to a medical tradition that extended back through Maimonides and into antiquity, Jews, especially the more assimilated German Jews who had arrived in an earlier era, established and supported institutions to care for themselves and their newly arrived coreligionists, the Russian/Polish newcomers linked to them by blood and history.

To Russian Jews and those throughout eastern Europe, nothing, not even the loss of income, equaled the calamity occasioned by loss of health (*gezunt* in Yiddish, the group's mother tongue). Yiddish is filled with idiomatic expressions testifying to the importance of health and dread of disease. Friends departing from one another say *Abi gezunt* (As long as you're healthy), or sharing a drink they say, *Tsu gezunt!* (To health!). However, one swears at an enemy, *Chapn a chalerye* or in shorthand, *A chalerye* (May a cholera catch you).[14] The entire household was involved in caring for the sick family member and no expense was spared. Traditionally, Jews did not distrust the physician, as many southern Italian peasants and townsfolk did, but their poverty made a physician's services prohibitively expensive except in dire emergencies. Fully half of the Jewish population in many communities depended on charity, and up to 40 percent were the families of *luftmenschen,* those without specific skills, investment capital, or even regular jobs. So these impoverished shtetlach dwellers would consult folk healers before calling a doctor.[15]

When a child fell ill, the first response was usually an enema administered by a member of the family or by the professional "enema woman," whose primitive tool was often a calf's bladder attached to a goose quill. Next came folk remedies.[16] Clare Kaplan recalls that during a girlhood illness in Russia, her mother asked her to urinate and then rubbed the urine on Clare's forehead to bring down her fever. On another occasion, her mother brought down Clare's fever by ripping the top of her ear and letting the blood flow.[17] Another woman recalls her grandmother pulverizing cooked lima beans to rub on a severe burn.[18] If home remedies, including compresses or tea with raspberry syrup, failed to ameliorate the condition, the ministrations of a *feldsher* (male) or *feldsherkeh* (female) were obtained. This salaried, licensed physician's assistant (paramedic) might employ castor oil, gargle, cupping, or, in the case of severe illness, leeches.[19]

Should familiar remedies fail, the eastern European Jew, as did the southern Italian Catholic, turned to faith. They sought the spiritual intervention of great rebbes, renowned scholars and teachers, some of whom were also said to possess mystical powers. In sickness, as in all other matters, eastern European Jews petitioned God not as a cold, distant abstraction but as a living force, a spiritual presence with whom they had an intimate, even a speaking relationship. They prayed to Him in Hebrew but spoke to him in Yiddish, the mixture of Middle High German, Hebrew, and Russian that was the colloquial language of the Pale. To sustain themselves culturally as well as physically, Jews disciplined their lives with religious ritual. There were ritual commandments to cover almost every aspect of life, including health and hygiene. Judaism taught the value of human life and that all measures must be taken to cure disease, heal the sick, and sustain life. The Talmud (Sanhedrin 37a) made clear that he who saves one life saves a whole world.

Rabbis taught that the physician was God's instrument in healing the sick and recommended summoning a physician whenever the severity of the illness and economic circumstances permitted. The issue of whether or not Jewish law permits one to study medicine, become a physician, and heal the sick was a matter of no little Talmudic dispute. Some argued that only God could heal, but the weight of commentary throughout the ages was with those who found biblical sanction for the physician.

Similarly, opinion in Talmudic debate weighed heavily on the side of those who argued that even as the physician must heal, the patient must seek his cures rather than rely upon faith alone. Indeed, the Talmud (Sanhedrin 17b) warned wise persons not to live in a city that lacked a physician because to be able to devote oneself fully to the Law and its study one must be healthy and "avoid things which are detrimental to the body and to acclimate himself to things which heal and fortify it."[20]

Those challenging the hegemony of orthodox Jewish teachings were no less committed to modern medicine. By the last decades of the nineteenth century, orthodoxy was already under siege by liberal ideas imported from western Europe. This *Haskala,* or Enlightenment, captured the hearts of many Russian Jews, especially the young. Fresh ideas such as individualism, liberty, socialism, and Zionism were joined to a belief in science and scientific medicine and became the intellectual baggage of those who craved change and despised the rabbinate only a little less than the tsar.[21]

The resonance between Judaism and medicine was evident in the respectful admiration with which a doctor was received in a shtetl home:

> He arrives by carriage or sleigh, a figure of awe, waited on hand and foot and deferred to by everyone. He sits in state while the family stand about him, the women craning and staring. His instruments, his learning, his foreign appearance and manner—he is usually a Gentile or a highly assimilated Jew—inspire wonder and uneasiness. Everyone competes to bring him whatever he asks for—a spoon, some water, a towel—and to fetch his coat when he is ready to leave.[22]

The physician was, after all, the court of last resort on earth. However, even Jews living in larger cities found medical service scarce.

Although an oppressive regime had reduced the likelihood that many of the brightest Jewish students would have access to medical educations, some continued to become physicians. Doctors of medicine were among those permitted to live outside of the Pale, but not until the midnineteenth century. Before that, medical students could live in Moscow but had to leave once they received their degrees. However, the Russian medical corps and civil service were in short supply of medical personnel, so after 1865 Jewish physicians serving the empire could live in cities, except St. Petersburg and Moscow.

The assassination of the relatively liberal Alexander II in 1881 by a clique of nihilists who had been living in the home of a Jewish woman precipitated a ferocious anti-Semitic backlash. The severity of the new tsar, Alexander III, and patterns of escalating violence kindled a fresh enthusiasm for migration. Some Jews headed for western Europe and the United States. Others chose to go only as far as the larger cities of the Pale, where jobs were more plentiful and there was less exposure to random terror from roving bands of cossacks, whose atrocities enjoyed official acquiescence if not encouragement.[23]

Younger Jews, frequently single males, departed shtetls only to find themselves in the slums of Minsk, Vilna, Warsaw, Lodz, and Bialystok in search of jobs. What they found was a nascent industrial revolution. There, at the mercy of employers, Jews often worked fourteen to sixteen hours per day for wages as low as two or three rubles per week.

Weakened bodies are easily compromised by disease. In Russia, typhus and smallpox were endemic. Cholera epidemics occurred periodically.[24] The Asiatic cholera epidemic of the early 1890s spread even faster than earlier epidemics, following the path of newly constructed railroad trains that barreled westward across Russia from the Caucasus Mountains to Congress Poland. An outbreak in 1892 spread quickly this way from Afghanistan to Russia. Muscovites were falling ill by July; by mid-August cholera was in Kiev. Under advice from Robert Koch, Prussian authorities sealed the border. After July 18, emigrants headed for the United States and departing from Hamburg and Bremen chugged across Germany in sealed trains. Stops were made only if equipment malfunctions required and then passengers were not allowed to leave the train. On the rare occasions when some had to be escorted off, stations were cleared of other passengers and later were disinfected.[25] Newspapers in the United States, including the *New York Times,* indignantly condemned steamship lines for not voluntarily curbing the transportation of immigrants and continuing to "rush its filthy Russians and Poles across the ocean and into the United States after the cholera had become epidemic in Hamburg."[26] The *Times* was owned and operated by the German Jewish Ochs/Sulzberger family, financially charitable but equally unwilling to mute its distaste for the unclean and unhealthy appearance of new arrivals from eastern Europe, most of whom were eastern European Jews. What the *Times* refrained from saying explicitly in its editorials, but Jewish health experts feared, was that the newcomers would need medical care on a scale that would make many of them public charges, especially now that emigration was rapidly escalating.

By the turn of the century, eastern European emigration to the United States was soaring in response to a sharp rise in religious persecution. Eastern European Jewish immigrants presented a host of problems and challenges to members of their own community in the United States and to public officials in the cities and towns where they sought to begin life anew. Matters of health and disease were not least among them. After arriving in the United States, the great majority of eastern European Jewish immigrants flocked to New York, Chicago, Philadelphia, and other cities where they found jobs and community with other Jews. However, they also found congested streets and crowded tenements.

New York City became the "promised city" to the eastern European masses, as it was to millions of other foreign-born.[27] Three-fourths of all immigrants arriving in the port of New York began their new lives in the city, though many later moved on. Concentrations of immigrant groups in particular neighborhoods and streets made them especially visible within a burgeoning urban population. In 1870, there were an estimated 80,000 Jews in New York City, only 9 percent of the population. However, by 1915 they made up almost 28 percent of the population, with 1,400,000 persons.[28] Though there were Jewish families scattered all over the city, the majority congregated on Manhattan's Lower East Side, south of Fourteenth Street and east of the Bowery because of its proximity to the city's main factory area. The seventh, tenth, eleventh, and thirteenth wards choked with newly arrived Jews and were demonstrably the most densely packed corners of the United States. The tenth ward had over seven hundred persons to the acre, the thirteenth about six hundred.

In some cities, immigrants lived in shanties near railroad tracks or mines; in others such as New York, the tenement house became almost synonymous with the foreign-born, especially with eastern European Jews. Apartment houses known as "double-deckers," or dumbbell-style tenements, were buildings six to seven stories high and approximately twenty-five feet wide, constructed upon a lot of the same width and one hundred feet deep.[29] Each floor was usually divided into four sets of apartments with seven rooms on each side. The front apartments generally consisted of four rooms each, the rear of three rooms each, making altogether fourteen rooms per floor, only four receiving direct light and air. Ventilation of the inner rooms and some minimal light was via air shafts. The air provided was often foul because it contained odors coming from the windows of the other apartments. Also, some tenants, unused to urban living, used air shafts as convenient dumps for garbage, which often remained rotting at the bottom for weeks. Others spanned the shaft with clotheslines; drying laundry blocked air and light.

Toilets were communal, four per floor, located off the air shafts. Residents who could not wait their turn were compelled to use vacant lots or, at night, the sides of wagons. Streets emitted a putrid odor magnified considerably by the summer heat. When the temperature and the stench was unbearable, tenants fled with mattresses to the roofs and fire escapes in search of sleep.

Fire escapes, when not used for hot weather slumber or outdoor playpens for toddlers, provided storage space for tenement dwellers, who piled them high with furniture and other belongings, rendering them useless as escape routes. In fires, tenements became death traps. Of the 250 recorded deaths in Manhattan fires between 1902 and 1909, one-third were Lower East Side tenement fire victims.[30]

By blocking the very fire escapes that might save their lives, impoverished immigrants undermined the efforts of reformers such as Jacob Riis, who tirelessly lobbied for tenement reform and sometimes won. The New York State legislature passed a Tenement House Law of 1901 establishing new guidelines for builders, designed to ensure healthier living conditions for tenants. New structures had to have windows opening a minimum of twelve feet from the building opposite it. Each unit had to have its own toilet and running water, clear access to fire escapes, and sturdy staircases. Older buildings had to be renovated, at least to the extent of installing modern toilet facilities, to improve the health and hygiene of tenement life. The absence of light was also remedied. Dr. Maurice Fishberg observed that "old model tenements are now generally being fitted up with gas fixtures. But it is peculiar that only the front room and kitchen are thus provided and the bedroom almost invariably neglected." Perhaps from his own experience as a physician to the poor, Fishberg complained that "bedrooms are also sickrooms, and the absence of proper light works hard upon the physicians who have to attend patients in their homes."[31] A Tenement House Department was organized in New York to enforce the new legislation, but inefficiency and corruption plagued its operation. Inspectors were too few; graft and bribery, ubiquitous.[32]

Tenement buildings were incubators of disease, and the likelihood that germs would thrive in such environs caused critics to refer to the tenement air shafts as "culture tubes on a gigantic scale."[33] Novelist and social commentator William Dean Howells described the unhealthy filth and squalor of the tenement in less clinical terms after a visit to the Lower East Side. All his senses were aroused. Visually and at a safe distance, the tall buildings seemed almost stately to his midwestern gaze. However, "to be in it [a tenement], and not to have the distance, is to inhale the stenches of the neglected street, and to catch the yet fouler and dreadfuller poverty-smell which breathes from the open doorways."[34]

The physical hardships of life on the Lower East Side were written on the faces of those who lived it. For novelist William Dean Howells touring the neighborhood, "It is to see the workworn look of mothers, the squalor of the babies, the haggish ugliness of the old women, and the slovenly frowsiness of the young girls."[35] Others, looking more closely than Howells, observed that poverty had different faces and that separate ethnic groups differed markedly in practices promoting health and hygiene. The nature of those perceived differences depended, in no little measure, upon the cultural biases of the observer.

In their writings, American nativists and anti-Semites sought to sketch the Jew as a public health menace, one who might end up on the public relief roles in droves, deficient in the physical vitality to stand the test of the rugged American environment, as did the pioneer forebears of the native-born and the sturdier stock that had emigrated to the United States from northern and western Europe. Even those sympathetic to newcomers, such as journalist and social critic Jacob Riis, believed the habits of eastern European Jewish immigrants were likely to nurture disease among them. According to Riis, hygiene was all but absent from the "Jewtown" he patrolled with his notepad and camera.[36]

In 1908, Dr. Manly H. Simons, medical director in the U.S. Navy, complained, "The poorer classes of Jews are very unsanitary; they work and live in dirty and badly ventilated quarters. Though special virtue is claimed for the Jewish method of killing the animals they use for food this is offset by the dirtiness of the shops in which the meat is sold."[37]

Though Simons granted that the "upper class Jew is healthier," nevertheless, "as a type Jews are beginning to show mental and physical degradation, as evidenced by the great variability of development, great brilliancy, idiocy, moral perversity, epilepsy, physical deformity, anarchistic and lawless tendencies." Never hesitating to substitute assertion for evidence, Simons expressed firm conviction that from the densely packed neighborhoods where Jewish immigrants lived, "come the greatest proportion of the distorted forms and minds, the beggars, tramps, burglars and other perverts who make life burdensome and fill our prisons with criminals, our asylums with insane."[38]

To the eye of nativist scholar E. A. Ross at the University of Wisconsin, Jews appeared wasted and lacking in the physical vitality that he attributed to America's sturdier Anglo-Saxon stock, deficiencies that were inborn and then exacerbated by the vicissitudes of life in America. In 1914, Ross wrote, "On the physical side the Hebrews are the polar opposite of our pioneer breed. Not only are they undersized and weak muscled, but they shun bodily activity and are exceedingly sensitive to pain." As evidence, Ross quoted an unnamed settlement house worker who testified, "You can't make boy scouts out of the Jews. There's not a troop of them in all New York." Another told Ross, "They are absolute babies about pain. Their young fellows will scream with a hard lick." Students, perhaps in Ross's classes at Wisconsin, confirmed that "husky young Hebrews on the foot-ball team lack grit, and will 'take on' if they are bumped into hard."[39]

Two years after Ross published his book, Madison Grant cast aspersions on the slight appearance of Jewish immigrants and raised doubts as to the possibility of their assimilating. In *The Passing of the Great Race,* Grant commented that just as a Syrian or Egyptian freedman could not be transformed into a Roman by "wearing a toga and applauding his favorite gladiator in the amphitheater," so it would be impossible to transform into an American "the Polish Jew, whose dwarf stature, peculiar mentality, and ruthless concentration on self-interest are being engrafted upon the stock of the nation."[40]

Ross, Grant, and other nativists associated slight build with innate inferiority of type, including a vulnerability to disease. They did not allow themselves to be persuaded otherwise by data such as the municipal health records from New York City, where many eastern European Jews resided after emigration. In 1908 those data indicated that Russian/Polish Jews had the second-lowest death rate of thirteen ethnic groups on which information was available—1,354 per 100,000 (only the Swedes were lower)—while the native-born ranked sixth, with a mortality rate of 1,849. The mortality rate for all groups combined was 1,811, well above that for the Russian/Polish Jews.[41]

Nor were these nativist "scholars" dissuaded from their beliefs by Columbia University anthropologist Franz Boas's studies demonstrating that the American environment was modifying the very "racial characteristics" that nativists so detested in newcomers. Because the slope of the cranium was often regarded as a reliable index of race, Boas, himself a German-Jewish immigrant, measured the skulls of second-generation immigrants as well as other physical dimensions and discovered that many no longer physically resembled their parents' generation. Long-headed types grew shorter and round-headed types often developed elongated heads. In a 1911 report presented to the congressional commission investigating immigration, Boas concluded that nutrition and other aspects of living conditions determined these "racial characteristics" more than heredity.[42]

Ironically, even as anti-Semitic critics in the United States portrayed the Jew as unclean, unhealthy, and disease-prone, physicians and public health officials, both Jewish and non-Jewish, were compiling statistics that refuted the notion that Jews were physically degenerate. Data seemed to suggest that Jews were healthier and less susceptible to certain diseases than their Gentile neighbors in the United States as they had been in eastern Europe. The 1908 New York City data, compiled and analyzed by William H. Guilfoy, registrar of records in the New York City Department of Health, noted that those born in Russia/Poland were less likely to die of typhoid fever, pulmonary tuberculosis, bronchopneumonia, lobar pneumonia, or chronic Bright's disease than most other nationalities, but more likely to die from cancer and heart disease.[43]

The reasons for such surprising differences remained unclear at the turn of the century. Many physicians and public health officials argued that cultural differences between Jew and Gentile protected Jews from contracting some diseases and improved their overall health and vitality. Others suggested that nature, not nurture, was at work. Jews had acquired a biological advantage over others. Jews, long exposed to some diseases in the confined quarters of European ghettos, had developed herd immunities through a long process of natural selection. Thus, when Jews in European or American cities did suffer ill health or decreased vitality, such observers contended, it resulted from environmental conditions beyond their control. Such conditions originated in the poverty that strangled Jews after they arrived in the United States, the very conditions that had left Bertha Pearl confined to a tuberculosis sanatarium.

Because of his own immigrant background, Dr. Maurice Fishberg was deeply sympathetic to the Jewish newcomers he treated. He was quick to defend them when necessary, alternating between natural selection and culture to explain the Jewish immigrant's condition. Social reformers' and journalists' descriptions of Jewish immigrants' unclean and sickly appearance following a visit to the Lower East Side of New York or some other similar neighborhood stirred Fishberg to a cultural explanation of Jewish health and hygiene.

Fishberg attributed charges of Jewish uncleanliness to the perceptual disadvantage of investigators, the limited circumstances of observation. "Philanthropists and reformers . . . are apt to regard the immigrant Jewish population of New York as a 'slum' population; as degraded residents of narrow, filthy streets, whose foul general habits are beyond their power to describe. They reach such conclusions after paying a passing visit to Hester, Essex, Ludlow, Suffolk, Norfolk, and other streets, say on Thursday afternoon, or Friday morning. But these ladies and gentlemen have never visited the homes of these apparently unclean people who sell and buy from push-cart, stands and small stores in these streets, which are the only places for marketing the city fathers have thought it advisable to give them."[44] Fishberg contended that "of the homes of the poor population of the city, the Jewish home is the cleanest." The sink, where germs could breed, "is in the majority of cases kept as clean as in any home of the American family, and much cleaner than by people of other nationalities (for instance, Poles, Bohemians, Italians, etc.) of the same social status." Moreover, "personal cleanliness of the Russian Jew is far above that of the average slum population."[45]

In arguing that cleanliness was central to Jewish culture, sacred and secular, Fishberg not only cited the ritual bath, or *mikva,* essential to Orthodox

Jews, but the Russian steam baths so popular with recent immigrants. During the summer, the public baths on the East River were also "crowded with Jewish humanity from daybreak till late in the evening."[46] He noted that religious law required the observant Jew to clip finger- and toenails weekly, wash hands before and after eating, and wash face and hands each morning.[47] Hoping to shatter the impression that Jews were a primitive people living in their own squalor, Fishberg cited the words of one European commentator on Jewish habits who characterized Judaism as having "made religion the handmaid of science; it has utilized piety for the preservation of health."[48] The discrepancy between his own perception of Jewish cleanliness and that of "charity and settlement workers" Fishberg attributed to the socioeconomic class of newcomers under observation. He contended that social work professionals had the opportunity to observe only "those who have recently arrived to the United States and are having a hard struggle to get along under altered conditions of life," and "those unfortunate people who, although they have been longer here in New York City, but for reasons of illness, incapacity, or even degeneration, have not been able to accommodate themselves to conditions around them."[49] Unlike Fishberg, these native-born outsiders never saw the working poor, many of whom needed economic assistance from the United Hebrew Charities but, nevertheless, were clean and given to healthy habits.

Especially important among these healthy habits was diet. The food consumed by Russian Jews, according to Maurice Fishberg, was "considered to be above reproach, even by those who are prejudiced." The physician was especially referring to the dietary laws of *Kashrut,* which, he correctly observed, demanded a "thorough inspection as to the health of the animal killed."[50] The word *Kosher* means, literally, ritually permitted or fit for consumption. Judaism's laws defining what is and what is not "fit" to eat are extensive and complex, even detailing the preparation of acceptable food.[51] Many of the laws of *Kashrut* codify the slaughter and preparation of meat. Orthodox Jews believe that, while God prefers that meat not be eaten, *Kashrut* is a compromise or a concession by the divine to human weakness and need.[52] Only tame, domestic animals that are herbivorous may be eaten, and these must be slaughtered in the most humane fashion. The meat is then *kashered,* or cleansed of blood, because blood is the symbol of life. *Kashering* keeps Jews aware that they are eating flesh and makes them reverent toward life.[53] Certain foods and certain combinations of foods are taboo. Pork is forbidden because the pig is not exclusively herbivorous and, therefore, is considered unclean.[54] Mixing milk with meat is also prohibited because it would be considered ungrateful and callous to take the child of an animal such as the cow, to which humans are indebted for so many necessities of life, and cook it in the very milk that nourishes humans and is given them so freely by the animal's mother.[55] The rules concerning this prohibition are so strict that separate kitchen utensils and dinnerware must be used for making milk and meat meals. Although few eastern European Jewish immigrants could have quoted the laws and interpretations from the Talmud, most fully or partially observed *Kashrut* after arrival.

Not all those who ventured into the Lower East Side were as impressed as Fishberg with the healthfulness of the fare. Some were repulsed by the wares and ways of curbside merchants, such as the *New York Times* reporter who stood windward of a cheese vendor and later wrote, "Phew, how that cheese did smell! Yet in spite of the fact that the cheese was a reeking mass of rottenness and alive with worms, the long-whiskered descendants of Abraham, Isaac, Jacob and Judah on the East Side would put their fingers in it and then suck them with great and evident relish." Especially offensive to the reporter were the purveyors of food who seemed "filthy persons," whose clothing was often infested with vermin and who, themselves, were often afflicted with hideous sores.[56] Such a profile would hardly inspire readers' confidence in the nutritiousness of the Lower East Sider's fare.

Discussions of what food and drink Jewish immigrants consumed and their willingness to consult physicians for health care spoke, if somewhat obliquely at times, to the larger issue of whether eastern European Jewish immigrants were as healthy and robust as the native-born and whether they were healthier or more enfeebled than other immigrant groups arriving in the United States during the same era. Jews, responding to their Gentile hosts' anxieties about their physical ability to contribute to American productivity, sometimes cited their dietary preferences and taboos as cultural imperatives contributing to their health and vitality. Soda water, or seltzer as it was more commonly known, was a popular beverage among eastern European Jewish immigrants as a cheap but pleasing, bubbly refreshment. They called it "the workers' champagne." Some mistakenly believed that it had health-giving properties. Others simply regarded it as harmless, especially when compared with alcohol. Although there was no religious taboo on alcohol—only on drunkenness—Jews evolved a cultural aversion to its regular consumption, except for ritual purposes such as the one cup of wine that must be drunk on the Sabbath or the required four cups on Passover.[57] Many Lower East Siders also tenaciously held to the belief that kosher meat was better for the soul and perhaps the body as well. The high price of kosher meat in New York City drove immigrant women to organize a consumer boycott in 1902.[58]

So, too, must medical care be both available and affordable because Jews placed a high priority on consulting physicians. Eastern European Jews, as did southern Italians, continued traditional patterns after arrival, turning to folk healers and remedies before consulting a physician. However, for many this choice was dictated by poverty, not cultural preference as it was for many *contadini,* even after emigration.

Newcomers of all groups joined an urban work force that could afford to spend only a small fraction of its income for health care. A 1915 *Report on the Cost of Living for an Unskilled Laborer's Family in New York City* estimated that a family of five would need an annual income of at least $840.00 to live comfortably, of which $20.00 (2.3 percent) would cover routine health care. Experts, such as the Metropolitan Life Insurance Company statistician Louis Dublin, thought the figure much too low.[59] However, even $20.00 was more than most newcomers could afford. Income, more than ethnicity, limited Jewish immigrants' use of physicians.

Moreover, good physicians were at a premium because few had come to America from Europe; it would take a generation before the sons (and a few of the daughters) of the newcomers began to flood medical schools with their applications. At first, Jewish immigrants on the Lower East Side turned to neighboring Irish or German physicians, such as Dr. Harrie Abijah James, a popular obstetrician who became conversant in Yiddish to meet the needs of his trade. Later, German-Jewish physicians whose families had arrived mid-century became popular. Soon Russian and Polish Jews such as Alexander Aronson, Julius Robinson, and Paul Kaplan conducted busy practices.[60] Again, economic differences crisscrossed ethnic lines; eminent physicians such as Dr. Abraham Jacobi were hardly accessible to the impoverished.

By custom and tradition, eastern European Jewish women barred men—all men—from the birthing room. Only a midwife was allowed to assist in the birth. Recent feminist scholarship suggests a broad cultural role for the midwife that doctors could simply not fill. In addition to aiding in childbirth, immigrant midwives sometimes treated ailments or performed abortions.[61] Knowledge about pregnancy, child care, health, and sickness was passed among women from generation to generation, but the midwife did something more. She was a neighbor, a friend, and, as she went from house to house, she was a conduit for information and news, helping to bind the community of women together. No American doctor could do that. In 1891, there were 46,854 recorded births in New York City. Of these, 24,134 were attended by private physicians, while 22,720 were delivered by midwives. However, in 1905, an article in the *Charities and Commons* reported that a count of Manhattan birth certificates recorded the attendance of a midwife in 42 percent of the births. According to F. Elizabeth Croswell, an investigator for the Public Health Committee of the Association of Neighborhood Workers, who compiled the report, immigrant women preferred midwives because it was a time-honored preference, "in complete accord with the deepest, most sensitive prejudices."[62]

That would change for women, poor and rich, native and foreign-born, but not immediately. By 1920 half of all women in the United States would still be delivered by midwives, but like other women, the Jewish mothers of the Lower East Side were increasingly attracted to hospital rather than home births and physicians in place of midwives.[63] One bartering chip that physicians used to persuade women to relinquish midwifery was the promise of painless childbirth through the use of anesthetics or other pain-reducing drugs. Another was the value of forceps and other instruments in assisting difficult births, including gynecological surgery. And a third was the expectation that puerperal sepsis, or childbed fever, would be less likely to take their lives if they were in a hospital, or at least under a physician's care. Of course, not all physicians were equally skilled with the forceps or adequately appreciative of the value of antisepsis in saving mothers. Still, as one observer correctly predicted, "Just as the village barber no longer performs operations, the untrained midwife of the neighborhood will pass out of existence under the effective competition of . . . painless wards."[64] However, she did not go quickly or without her champions.

American physicians, increasingly committed to the medical model of childbirth that regarded parturition as a pathological process, battled against midwifery, a campaign criticized by social workers such as Lillian Wald, a social worker of German-Jewish background who founded the renowned Henry Street Settlement.[65] Always persuasive, Wald was appalled by the low professional standards for midwives in the United States as compared with western European nations. She convinced city officials to intervene in the war between doctors and midwives, and to strike a compromise between tradition and science. In 1906, the New York Bureau of Child Hygiene assigned its nurses to examine the bags of licensed midwives to check the quality of their instruments and to give them a silver nitrate solution for application to the eyes of newborns, thereby preventing *ophthalmia neonatorum,* gonorrhea that resulted in infant blindness.[66] Still, in the end, the physician prevailed. Rather than make midwifery more viable, physicians succeeded in supplanting tradition with science, and more women, including immigrant women, turned to urban dispensaries and physicians during childbirth, though many still preferred the services of a physician brought to their homes.

Physicians serving the Jewish immigrant community faced competition from others besides midwives and folk healers. Quacks became experts in preying upon newcomers. Immigrant patients often assessed doctors by the warmth of their manner and the bitterness of their medicine. "Doses before and after each meal, and a pill or two, preferably silver- or gold-coated, before retiring" greatly impressed Lower East Side patients.[67]

Lazy and rapacious physicians proved as damaging as quacks to the reputation of American physicians in immigrant communities. Immigrants were most critical of lodge doctors, physicians who ensured themselves a trade by contracting with one or more of the many mutual benefit societies serving immigrant communities. Though many of the physicians employed by eastern European Jewish lodges, or *landsmannschaftn,* were attentive and sympathetic to their patients, others grew so tired of climbing tenement stairs that they would call up to a sick patient, make the diagnosis, and suggest a prescription from the bottom of the stairs.[68] Their fees were tossed to them through an opened window. Eastern European Jewish immigrants, who considered a physician's bearing and comportment as critical to his credibility as the success of the therapies he prescribed, were frequently critical of "lodge doctors," even using the expression with a tone of derision. Yet, when sickness struck the home of Jewish immigrant workers, they often turned first to the lodge doctor, who spoke their language and understood their perspective on health and disease.[69]

Dr. Maurice Fishberg believed that eastern European Jews' willingness to obtain medical attention from physicians, coupled with their habits of nutrition and hygiene, accounted for low rates of Jewish mortality and resistance to particular diseases rather than innate racial characteristics, suggested by nativists such as E. A. Ross, Madison Grant, and Manly Simons. "'Race' alone may often be only a cloak for our ignorance, unless all the conditions of the *milieu* have been excluded."[70] Whatever the causes, statistics did indeed indicate that Jews manifested different patterns of longevity and morbidity than Gentiles in the United States.

Even as the emigration of Jews from Russia and other regions of eastern Europe was beginning, a study prepared as a *Bulletin* of the eleventh census in 1890 demonstrated that the health of native American Jews was superior to that of other Americans. The report was prepared by the U.S. Army surgeon in charge of vital statistics, Dr. John Shaw Billings.

In the introduction to his 1890 bulletin, *Vital Statistics of Jews in the United States,* Billings explained that the government wished to know the extent to which Jews were assimilable.[71] When it came to the Jews, was nature or nurture supreme in matters of health and vitality? Did environmental conditions level the playing field so that members of the "Jewish race," as Billings called them, would find themselves able to achieve biological parity alongside those with whom they must compete for life's resources? Billings's use of the term *race* suggests that he shared the assumption of many European and American scholars that the Jews were a separate race from other Caucasian Europeans. His rationale for conducting his study also suggests his belief that the innate biological differences among the races might well determine the ability of a particular race—in this case, the Jews—to cope with the physical tests posed by the American environment.

Billings's remarkable statistical snapshot of American Jewry was prepared with survey data acquired by Adolphus Simeon Solomons, a prominent Washington, D.C., Jewish spokesman who compiled the names of fifteen thousand Jewish families who had been in the United States for five years as of December 31, 1889. Their names obtained from rabbis and community leaders across the country, families were sent questionnaires.[72] American Jews had a mortality rate of 7.1 per 1,000, a rate slightly more than half of that for other Americans of the same socioeconomic status. Women had a higher death rate than men, 7.16 as compared with 6.47. Male Jews in their prime, aged fifteen to forty-five, whose mothers were born in the United States, showed a mortality rate per 1,000 of 1.98; those with German mothers, 2.86; Russia/Poland, 11.59; and England and Wales, 6.32. Jews were most likely to die from diarrheal diseases, diphtheria, and diseases of the nervous system, especially of the spinal chord, as well as diseases of the circulatory and urinary systems and ailments of the bones, joints, and skin. Jews were less likely to die from tubercular diseases such as phthisis, scrofula, tabes, and hydrocephalus.

At the conclusion of his report, Billings observed that it would be necessary to study the population of large cities where Jews had settled to evaluate birth and death rates. However, judging simply from the available data, Billings concluded that "the Jews in the United States preserve many of the peculiarities which have been noted among them in Europe," though he deferred further comment because "more extended and reliable data with regard to their birth and death rates in this country are highly desirable."[73] Billings left to others speculation as to whether the Jews' "peculiarities" that he identified would help or hinder their assimilation.[74]

Later studies using data from the peak period of immigration confirmed Billings's results. Metropolitan Life Insurance Company statistician Louis Dublin and Gladden Baker of the Yale Medical School published a study, "The Mortality of Race Stocks in Pennsylvania and New York," in 1920, using 1910 data.[75] The two investigators concluded that in New York, "the death rates for Russians of both sexes per 1,000 are appreciably lower than those for native stock at all ages" (see table 6.1).[76]

Contemporary investigators treated infant mortality as an important empirical measure of a group's health and vitality. The deprivation and environmental health risks of urban industrial life made death in the early years of life much more common in cities than in the countryside in the late nineteenth and early twentieth centuries. The U.S. Children's Bureau calculated a mor-

TABLE 6.1

Mortality Rates for Persons Born in Russia Compared with Native-Born of Native-Born Parents, All Ages in New York State, 1910[77]

	N.-B.	**Russian**
Male	15.9	7.5
Female	13.9	6.6

tality rate of 111 per 1,000 live births across eight cities.[78] However, for the children of foreign-born Jewish mothers, the rate was a low 54 as compared with all foreign-born whites at 127 and native-born whites at 94. For deaths during the first month of life, the neonatal mortality rate for infants of Jewish mothers was 28; for foreign-born whites, 46; and for native-born whites, 42.

Data from the 1910 census confirm both the lower mortality of Jewish immigrants generally and, in particular, the low mortality among children of Jewish immigrants in the United States.[79] Variables designed to test the influence of degree of exposure of the foreign-born to the native population such as length of residence, naturalization, and English language ability suggest that exposure alone did not cancel the Jewish advantage. Maurice Fishberg had believed that assimilation would lead to loss of the mortality advantage, predicting, "When the Jew assimilates with his non-Jewish neighbors, adopting their modes of life and habits, he gradually loses his immunity and his longevity, and in time does not differ as to health and sanitation, from the people amongst whom he happens to live."[80] However the 1910 findings appear to contradict Fishberg, at least for the first generation born in the United States. Why? Some contemporary critics argued that because Jews had been urbanized for many generations, they were better adapted to city life and could cope better than other groups with impoverished urban circumstances.[81]

More recent studies suggest that the extended period of breast feeding practiced by Jewish mothers may well be one of the most significant of such "adaptive" mechanisms responsible for low rates of infant mortality. Also, ritually required washing before eating and the laws of *Kashrut* have been cited by demographers as factors in keeping Jewish mortality figures low for the very young.[82]

A 1971 study by U. O. Schmelz, "Infant and Early Childhood Mortality Among Jews of the Diaspora," notes that although Jews had a slightly high mortality in the first month of life, they had a lower mortality after that, a conclusion that argues for nurture above nature in explaining the discrepancy between Jewish and non-Jewish populations. As did turn-of-the-century investigators, Schmelz emphasized the importance of child care practices and other environmental variables. Echoing the explanations of Fishberg, Schmelz suggested that Jewish mothers took greater care of their infants than did mothers of other equally poor urban-dwelling groups. Higher rates of breast-feeding, lower instances of factory work among Jewish mothers, and fewer cases of illegitimacy all appear as significant, though unproven, explanatory variables. As do others, Schmelz links culture, hygiene, morality, and longevity. Among the factors leading to longer lives for immigrant Jews, generally, he credits the religious injunction to wash hands before eating, the religious prohibition against drunkenness, and the acceptance of the physician's role in curing disease. Low rates of venereal disease, alcoholism, and even childbirth all suggest a portrait of a community that moved across national boundaries, carrying with them laws, customs, and practices grounded in religious beliefs and having the effect of making the population healthier and long-lived wherever it settled. More recent mortality analysis by Gretchen Condren and Ellen Kamarow using 1910 federal census data attributes low Jewish mortality to "a complicated interplay of cultural and structural phenomena" that historical demographers have yet to fully sort out.[83]

For members of the Jewish community in the United States, the debate over their health was more than academic. Eastern European Jews were frequently associated in the public mind with tuberculosis, or consumption, as the disease was often called. "It came to be regarded as 'a Jewish disease,' or 'the tailors' disease'," according to historian Irving Howe, because so many young Jewish consumptives earned their living with a needle.[84] Tuberculosis was neither peculiar to Jews nor to those who worked in the garment industry. And yet nativists, especially anti-Semitic nativists, stigmatized Jews as carriers of tuberculosis. In part, they equated the emaciated appearance of consumptives with the feeble appearance and behavior that many nativists attributed to Jewish arrivals. Physical descriptions of those suffering from consumption were often appropriated by E. A. Ross and others as convenient stereotypes of Jews' physical appearance.[85] Jews, according to Manly H. Simons, "are very tuberculous," a condition he attributed to an unhealthy life-style.[86]

For much of the nineteenth century, consumption was the foremost killer of humanity among infectious diseases. Once a diagnosis of consumption was confirmed, it was generally held to be incurable, striking especially those between the ages of fifteen and thirty-five, the most productive and otherwise healthy period of life. Tuberculosis struck the rich and/or famous such as Chopin, Napoleon, and Thoreau, the poorest, and those in between. If less spectacular in its effects on large concentrations of human population than epidemic diseases such as cholera and yellow fever, consumption was no less feared. Expressions of concern filled the popular press as well as professional journals. Much of this literature was aimed at identifying and isolating those individuals and groups who ran the most risk of developing consumption.[87]

By the end of the nineteenth century, physicians and public health professionals in this country had come to accept that consumption was caused by a rod-shaped bacillus capable of affecting any part of the body, and more than one part simultaneously, though the pulmonary form of the disease was the most familiar. It spread, they learned, primarily through the sputum of those infected, who would spray the air with bacilli when coughing or contaminate food shared by others. Before Louis Pasteur's insights gained wide acceptance, contaminated milk was an especially important source of infection affecting children. The word *tuberculosis,* meaning "full of tubercles," first appeared in the midnineteenth century but did not become the preferred name for this disease until the early twentieth century. Pulmonary tuberculosis was more commonly known as phthisis, from the Greek *phote,* a body shriveling up under intense heat, or consumption, from the Latin *consumere,* describing the wasting of flesh. Sometimes victims were described as in "decline." A later nickname was the White Plague, perhaps a play on the nickname for bubonic plague, Black Death, and a reference to the pallor common among TB's victims.[88]

Interestingly, even as public awareness was increasing, tuberculosis mortality rates were declining. Well before Robert Koch's 1882 discovery of the tubercle bacillus, mortality from tuberculosis began to diminish among all segments of the population; beginning in around 1840 in England and 1870 in the United States. Rates continued to decline until World War II, rose during the war, and then fell dramatically afterward, the result of antimicrobial treatment. The reasons are a matter of some historical controversy, but a persuasive case has been made that efforts to reduce the contact between those infected with pulmonary tuberculosis and those still uninfected dramatically shrank the opportunity for victims to spread their infection.[89]

At about the time that mortality from TB was beginning to decline, the TB mortality figures for Jews living in Europe were lower than for any other group. Tuberculosis was found less frequently among Jews residing in Russia, Austria-Hungary, and Romania than among Gentiles.[90] These patterns carried over to the United States, but American nativists predisposed to believe that Jews were tuberculous did not cloud their minds with contradictory data.[91]

Racists such as Manly Simons found Jews distasteful because they were generally diseased but equally hateful because they seemed unnaturally resistant to certain diseases including TB. According to Simons, "he [the Jew] has been so [tuberculous] for so long, that he exhibits great tolerance of the effects of the disease."[92] Scorned for living in squalor, the Jew ought to be equally reviled and resented, Simons seemed to suggest, for the natural immunities to tuberculosis his body manufactured as a result of his exposure to the disease. Even the advantages of natural selection were disadvantages for the Jew.

In 1902, Maurice Fishberg, while admitting that the evidence was hardly conclusive, believed that lower rates of mortality from TB among Jews had something to do with "the careful selection of carcasses in Jewish slaughterhouses."[93] Because tubercular animals would never have passed the rigors of inspection, certification as kosher would be out of the question. Dr. Kate Levy, who had taught clinical medicine at the Women's Medical College of Northwestern University in Chicago, similarly observed that although the ritual slaughterers, or *schochetim,* did not inspect carcasses "according to our modern knowledge of infectious diseases," the *schochetim* were able to discard the carcasses of animals whose lungs had been ravaged by TB.[94]

By 1911, Fishberg had changed his mind. Observation and data now persuaded Fishberg that consumption was not usually contracted via one's dinner but via one's neighbor, thus discounting the significance of *Kashrut* as a preventive of infection via ingesta. "In Eastern Europe, where the Jews follow the dietary laws, strictly adhering both to the letter and spirit of the sacred ordinance, there is more consumption among them than among their co-religionists in western countries who disregard the dietary laws in part or completely."[95] Nor did he accept the notion that a Jewish preference for cleaning house with a moist rag rather than a "dusting-brush" adequately explained the lower incidence of tuberculosis among Jews.[96] Fishberg concluded that the incidence of tuberculosis among Jews "depends more on their social and economic conditions [perhaps including living and working in clean, airy, uncrowded rooms] than on racial or ritual affinities."[97] In his linkage of crowded, unventilated living spaces to the spread of TB, Fishberg was actually anticipating the thought of recent scholars on the linkage between the isolation of TB patients and declining rates of TB.

Recognizing it as the basis of much anti-Semitism, Fishberg denied that Jews were a separate race, as they had been defined in Europe, deploring the view that the "'Semitic' blood which flows in their veins renders the Jews immune to the virus of infection and gives them an advantage in the struggle for existence when they meet the 'Arayan' in Europe."[98] Fishberg the anthropologist denounced such arguments as "fallacious" because "the Jewish 'race' is not as pure as is generally believed."[99] Why, then, the lower rates of TB? Fishberg and others did at times argue for biological adaptation. If the Jews possessed greater immunity to TB than did Gentiles, it was because centuries of living in the "adverse milieu" of overcrowded, dark, and poorly ventilated dwellings had resulted in a natural selection. "Those Jews who could not adapt themselves to a confined atmosphere succumbed to various diseases which thrive in such *milieu*, chief among which is tuberculosis, and they were eliminated from their midst, having had slight chances to perpetuate their kind."[100] Eighteen hundred years had been time enough for "weeding out a large proportion of those Jews who were excessively predisposed to tuberculosis infection." Thus Fishberg, as did Manly Simons, credited lower rates of TB among Jewish populations to natural selection.[101] However, Fishberg's belief in biological adaptation did not thoroughly exclude behavioral or cultural adaptation, which he treated as interrelated with biological adaptation. Even as he argued for natural selection as an explanation of the Jews' low mortality from TB, Fishberg continued to argue that "the rarity of alcoholism and syphilis among the Jews has a salutary effect in this respect." He considered it a matter of common knowledge that "chronic alcoholics are very prone to be attacked by tuberculosis; the alcohol lowers the vitality of the tissues and thus enables the bacilli to develop and grow more readily."[102]

Fishberg concluded that selectivity and adaptation even determined the course that TB would take when contracted by Jews. First, those who contracted it had a "more favorable prognosis" than other people. Second, galloping TB and acute miliary tuberculosis seemed to him rare among Jews. Few were "stricken with high fever and rapid extension of the disease with cavity formation within a few weeks or months." More frequent among Jews than among non-Jews of the "same social status" in Fishberg's experience were "cases of the extreme chronic type, running on for years, and still permitting the victim to make himself useful at some occupation," a pattern that Montefiore patient Bertha Pearl would have recognized all too well. The reason? Fishberg speculated: "It appears that thoroughly urbanized humanity does not offer a good soil for the growth and development of the tubercle bacilli, while the inhabitants of the plain, and less so the peasant or farmer in modern European and American villages, offer a virgin soil for these parasites. This is the only reasonable way we can explain the high rates of morbidity and mortality from tuberculosis of the rural dwellers who emigrate to cities. Being more predisposed to infection, they also more often supply cases of the acute fulminant or galloping type, as well as acute miliary tuberculosis. Among the Jews it is just the reverse."[103]

If Fishberg wavered in his choice explanation of lower rates of TB among Jews, it was because he could no more sort out the issues of nature and nurture than any other researcher of his era. However, it was also because he was taking special care to avoid backing into agreement with his nativist adversaries, who considered Jews a separate race lacking in health and vitality but naturally favored when it came to TB, as evidenced by markedly lower mortality rates.

Whatever the reason for these lower rates, even among Jews socioeconomic class was a factor. Data from Manhattan's fourth, sixth, eighth, and tenth assembly districts—mostly Jewish, immigrant poor—showed rates of TB 11.9, 12.0, 13.5, and 11.7 per 1,000 persons, respectively. The thirty-first assembly district in Manhattan, consisting of wealthier Jews, showed a rate of merely 3.0 per 1,000. Regardless of ethnic group, tuberculosis seemed to flourish most readily in the crowded, unhealthy conditions of the poorest sections of urban enclaves.[104] Whether rich or poor, victims of tuberculosis faced extended treatment, no chance of full recovery, and a problematical chance of being able to resume a normal life after treatment. If diagnosed at what physicians considered an early stage, the disease sometimes responded to a change in climate and diet, though advice about what climate and what diet differed from physician to physician. Doctors recommended different preparations to combat individual symptoms such as lung hemorrhage and cough, and opium was widely used to relieve the distress of the final stages.

The closing decades of the nineteenth century witnessed the rise of sanatorium care. A German innovation, the sanatorium was introduced into the United States by New York physician Edward L. Trudeau in the 1870s after he was diagnosed as tubercular.[105] Sanatoria applied earlier therapies of beneficial climate, fresh air, and supervised, nutritious diet to groups of patients in an institutional setting. The sanatorium could not offer a specific cure for tuberculosis, but it reflected a revision by the general public in its view of tuberculosis from being an incurable to a treatable disease. Physicians introduced various surgical techniques of collapsing seriously damaged lungs in the hopes of arresting the disease. It was not until after World War II, in 1947, that antibiotic therapy for TB began with the introduction of streptomycin and later, even more effective drugs such as isoniazid.

The peak immigration period coincided with the era of sanatorium care for tuberculosis. Many thousands of poor immigrant workers stricken with TB were in need of costly extended care. Some managed to enter publicly financed state institutions, although poor Jews deplored becoming wards of the government and feared that it would fuel the arguments of anti-Semitic nativists for immigration restriction. Many more found their way to private institutions, their expenses assumed by labor unions or mutual aid societies or industrial insurance policies. Jewish lodges, or *landsmannschaftn,* functioned as mutual benevolent societies, providing benefits to members with TB, but most poor shop workers lacked sufficient coverage to defray the costs of a disease that drained individuals and organizations of health care dollars even as it wasted the bodies of its victims. In response to the ubiquitous moan of Jewish consumptives, "Luft, gibt mir luft—Air, give me air," Jewish physicians prescribed sanatorium care. One Jewish physician, knowing that some patients could not afford even a short sanatorium stay, prescribed rest and some cough medicine, writing on the prescription slip, "Join the Cloakmakers Union."[106]

Some sanatoria were located in upstate New York, such as Trudeau's facility at Saranac Lake. Jewish philanthropists, especially renowned financier Jacob Schiff and Lyman G. Bloomingdale, whose family founded the department store and who had himself lost a daughter to TB, financed another, the Montefiore Sanatorium at Bedford Hills in Westchester County.

The Montefiore Home for Chronic Invalids was opened in 1884, even as the great wave of immigration to the United States was just beginning. Its supporters were philanthropists concerned by the plight of those for whom medicine offered little hope of a cure. The category included those with cancer, syphilis, the "opium habit," arthritis, chronic kidney disease, chronic melancholy or what modern physicians might diagnose as clinical depression, and tuberculosis. A clean, warm place to rest, regular nursing, healthful food, and some family assistance to relatives of the sick was all that could be done for incurables, but the expense was well beyond the means of poor Jewish immigrant laborers.[107]

Montefiore's first New York City facility consisted of twenty-six beds in a small frame house on Manhattan's Upper East Side. Later facilities were located in Harlem and, finally, in the northwest Bronx, then a green and healthful area on the fringe of the city. Doctors at Montefiore shared with others throughout the country the belief that fresh air, rest, good food, and regulated exercise was a successful regimen for treating patients in the early stages of TB. However, disagreement was rampant over the details. What was more important—cold air or fresh air? What was the best climate and topography—the high altitude of mountains or the low-lying desert?[108]

When the Montefiore trustees decided to build a separate tuberculosis sanatorium, they put their faith in altitude and opted for a site at Bedford Station in Westchester County, the "highest point in the county which was also entirely free from mosquitos and malaria," and only forty miles from New York City, a ninety-minute railroad ride. Thus, patients could be transferred easily and physicians from the home could come when needed "at slight cost to the institution."[109]

The Montefiore Sanatorium thrived as philanthropists such as the Lewisohn family contributed to the building of new pavilions, housing more beds. Such contributions also made possible purchase of a 136-acre farm. Farming and farm work was part of the therapy for some of the patients, and it served as a source of food for all Montefiore facilities. Trustees hoped that the experience of country living might persuade some of those released to settle away from the congested urban tenements of New York where some may have contracted TB.

Many of the patients were only too aware of the conditions that had contributed to their illness and said so in patient publications. There were two literary societies—one that conducted business in English and one in Yiddish, the language of recently arrived eastern European Jews disdainfully referred to as "jargon" by assimilated German-Jewish physicians. Patient Bertha Pearl, who wrote in English for the *Montefiore Echo* and recounted a familiar immigrant experience in her 1916 article was typical of many introspective Montefiore patients.[110]

At Denver's Jewish Consumptive Relief Society (JCRS), founded by and for "respiratory refugees" from eastern Europe via American factories and tenements, the pattern was similar.[111] The patients most often expressed themselves in Yiddish, their beloved *mamma loshen* (mother tongue). The *Sanatorium,* a JCRS publication that invited patient contributions, had a Yiddish section. The *Hatikvah* (The Hope) was a patient publication begun in 1923 at the JCRS that often translated into English pieces that had originally been written in Yiddish. And what did patients say?

Tuberculosis sucking the breath from a Jewish tailor's body. *Sanatorium,* 1907, Jewish Consumptive Relief Society. (Courtesy of the Rocky Mountain Jewish Historical Society)

Some were nostalgic for times past. Often they gave vent to maudlin feelings, those that came with the knowledge of how sick they were and the fear that they would never get well. Such expressions came easily because, in the world of Yiddishkeit, there is a strong element of lamentation over life's hardships. Typical is a poem, "The Invalid," by A. Druskin that appeared in *Hatikvah* and was translated from the Yiddish.[112]

Chained
Sickness holds me down
Tho free, I lie in bed a captive,
I cannot tear my eyes
From your bewitching mountains
That weave in majestic peace,
With blue transparent light
Their eternal mountain dreams.

I stretch out my arms to them
And it seems to me
I feel their cool breath
That brings relief
To my fever burning body—
But I drop my hands,
I cannot reach the far mountains.

And again I am an unawakened dreamer, tired
The sun before my eyes glares,
Stifled am I and crushed
Between my own four walls.

The very beauty of the environment to which Drushkin and so many others had come for therapy could be a cruel reminder of their imprisonment by TB.

Similarly maudlin are some of the Yiddish songs written about tuberculosis and its victims. One called *"Menschenfresser"* ("Man-eater") described how the lungs of victims were invaded by *mikroben* (microbes) that destroyed the lungs and brought death.[113] *"Das Schwinsüchtige Mädel"* ("The Consumptive Maiden") was written by Jacob Marinoff, a JCRS field secretary and lyricist, with music by L. Samoiloff, a former New York Metropolitan Opera composer. In this song, a young girl at the sanatorium plants flowers by one of the white tents, praying she will live long enough to see them blossom. But alas, the "seedlets heeded not, nor ear lent they to her request" and "to the poor afflicted one the flowers came too late."[114] Again, a tubercular patient reaches out to nature for comfort and receives none.

Not all patients shared this teary view; others defied the grim reaper's shadow with humor, an equally typically Yiddish response to dire situations:

> Just give an ear to me for a while, if you can,
> Till I tell you of some queer lingers in our Tubercular San.
> Some short and some long, most all of them tin,
> When they all begin coughing, O! L-rd
> What a din.
>
> The treatment here is superb, and four ply
> But if you're inclined to work you needn't apply.
> We rest six times a day, whether weary or not.
> Eat six times between rests and woe to him who says "rot."[115]

Hardly up to the standards of the great Yiddish poet Yehoash (born Sol Bloomgarden), who was also a patient at JCRS, but "Our TB Sanatorium" was intended to bring a smile to the faces of those who had so little to laugh about. Others published open letters. Many of the patients were young men and women from the sweatshops, and their letters reflect their interests. One letter written by "Nat" to a friend boasts of the pretty girls he is meeting in Denver and the fact that one of his fellow patients was part of Jewish prize-fighter Benny Leonard's entourage.[116]

Always at JCRS there were celebrations for the Jewish holidays, services, masked balls for Purim, and, of course, Seders for Passover. At such times, the line between patients and doctors lowered; Jewishness and a spirit of hope transcended lines of class and education. One such account at the JCRS, described the Seder as "conducted in good old orthodox Jewish fashion." The writer added how grateful the patients were to be "in an institution where they are treated as brothers and where they may maintain their self-respect and at the same time regain their health."[117]

As the years passed, so did the generation of eastern European immigrant Jewish consumptives who sought to preserve their past as they regained their health. Patient publications abandoned Yiddish for English exclusively by the 1930s, and those who did translations found themselves with many fewer to do.

While some immigrants suffered in body, others suffered the psychological stress of immigration. In his capacity as United Hebrew Charities medical examiner, Maurice Fishberg learned firsthand how mental strain bruised the body. In 1903, he described a typical case:

> A young man, eighteen years of age, born in Russia. Arrived at New York three months ago, intending to work hard and earn enough money to send steamship tickets to his parents and bring them to the United States. . . . His general appearance is worried, his look is rather anxious. He tells me that he suffers from "pain in the heart," insomnia, loss of memory, and inability to concentrate his mind on anything. He is, he claims, too weak to work. The cause of all this he knows. The climate has an injurious effect on his organism; the air is rather "too strong" for him. He knows he will die soon. All he asks is that the United Hebrew Charities pay his transportation to Russia where he may die near his parents. A physical examination reveals that, excepting minor bow-legs and slight anemia . . . he is a healthy man. His heart and lungs are normal.[118]

What happened to such young men? According to Fishberg, such an immigrant "sooner or later breaks down as a result of care, worry, and anxiety. His appetite vanishes and his digestive organs refuse to properly perform their function." Next various nervous disorders manifest themselves, the immigrant is "sleepless, suffers from headaches, pain along the spine and pains and aches all over." Soon, Fishberg observed, the recent arrival is "a total wreck physically and mentally."[119] In extreme cases, the condition advanced to "melancholia and hypochondriasis." He noted that those newcomers who had spent a longer time in the United States were more apt to suffer from neurasthenia and hysteria. Fishberg added that the symptoms he saw and the stories he heard were hardly confined to Jewish immigrants. In any case, the therapy differed little—minimal counseling followed by institutionalization if there was no improvement.[120]

Fishberg believed that even as immigration irreparably damaged some newcomers in mind and body, others and their descendants would be healed by coming to the United States. Fishberg concluded on an optimistic note: "The wretched physical appearance of some of the poorer people among us is the result of centuries of confinement in European ghettos. In the United States, where he enjoys freedom, the Jew recuperates physically to an extent unknown among other races. The physical and moral deterioration of the first generation of immigrants in a strange country is not to be observed among the Jewish children, even among those who are unfortunate and for that reason may be compelled to apply for assistance to a charity organization."[121]

Fishberg's insistence that the health of most Jewish immigrants improved once they were truly settled in America reflects his confidence in the culture and habits of body and mind that he believed travelers from eastern Europe packed with their Sabbath candlesticks and embroidered prayer shawls for the trip across the Atlantic. Whereas Italians had customs rooted in rural tradition, Jews had often already adapted to urban culture. In sickness and suffering, Jewish immigrants turned to their faith and folk remedies as did

Italians. However, even more aggressively than the Italians Jews also sought the help and advice of physicians, who were trusted and highly respected members of the community in the Old World and continued to be revered after emigration. Health care in eastern Europe began with the family. But in their homelands, Jews had also supported their own health care institutions, lest they become unmanageable burdens to their families or, worse yet, dependent upon the unlikely generosity of their Gentile neighbors or a hostile government, which might revile them for their misfortune. Institutions, outside the family were to be revered, then, not spurned. The tradition of the entire community, not just a family, caring for its own by creating health care institutions remained part of the Jewish response to disease in the United States. Also, because of the diaspora, Jews were long accustomed to being an ethnic minority. So along with their city ways, Jews arrived in the United States with long experience in making suspicious peoples and hostile institutions responsive to their needs.

7

"The Old Inquisition Had Its Rack and Thumbscrews": Immigrant Health and the American Workplace

"I knew that America's streets were not paved with gold, but little did I suspect that I would be asked to pave them." This joke shared in different languages and variations by immigrants of many groups and their children or grandchildren is a reminder that throughout American history newcomers were rarely so naive as to be lured to America by extravagant promises or expectations of great wealth easily acquired. What they expected was employment at higher wages than they had been receiving at home; what they got was a job based on the skills the market demanded and, often, a backbreaking one at that. All too frequently, though, the jobs that immigrants were recruited to do not only drained their energy but endangered their health. That was the case that Drs. Antonio Stella and Maurice Fishberg had made on behalf of southern Italians and eastern European Jews, respectively, in response to charges that immigrants were somehow inherently responsible for the diseases and disabilities that felled them in the United States. In the fast-paced industrial capitalism that characterized the American economy at the turn of the century, recruitment of workers by private industry was not accompanied by an assumption of responsibility for their health or the security of their families. But the severity of the problem and its redefinition as a public health concern eventually led to government investigation and involvement. In a congressional investigation of immigration and its consequences, problems of immigrant workers played a significant role.

A 1911 United States Immigration Commission report, directed by Sen. William Paul Dillingham, included a twenty-one–industry survey that found 57.9 percent of all employees were foreign-born.[1] That the United States had become dependent upon migration for its work force was never so obvious as when the economy stalled and foreign-born workers departed. It happened in the autumn of 1908. In a *New York Times* editorial titled "Disillusioned Immigrants," it was mentioned that from October 1, 1907, to October 1, 1908, "717,814 aliens returned, almost balancing the 724,112 who in this period landed on our shores." Although it had been what the *Times* characterized as a "'panic' year" in the American economy, the paper also blamed "false representations" disseminated by "private immigration agents" who advertised throughout Europe and were not scrupulous about accurately depicting conditions of work and life in the United States. The *Times* cited an example of a Bohemian glassmaker mentioned by Commerce Secretary Oscar Straus. Making only $0.75 per day in his home country, the craftsman was told that he could make $2.50 per day in the United States. When he arrived, according to Straus, he discovered that the cost of living in the United States was also higher—"no golden apple [was] ready to fall into his lap." Supporting Straus's desire for great candor, the paper firmly announced, "We need immigrants as much as ever, and this is still a land of wide opportunities."[2]

If the immigrant worker was so crucial to American economic prosperity, why was the workplace increasingly a public health nightmare haunted by the ghosts of immigrant men and women who had succumbed to disease or disability? How did the presence of millions of foreign-born workers affect health conditions on the job and the struggle to improve them?

The toll taken on the health of individual immigrant workers who arrived in the great wave of immigration between 1880 and the 1920s is part of a familiar visual record. You see it in the photographs of Jacob Riis, Lewis Hine, and other turn-of-the-century photographers. Here are hollow-cheeked Slavs staggering exhausted from mine shafts, perspiring Bohemians in steamy tenement apartments with both hands busily rolling cigars, stoop-shouldered

Russian Jews curving their spines over sewing machines in sweatshops, or emaciated southern Italians crouched low in open fields to pick crops. However, even these candid photographs cannot convey the agony of foreign-born match workers whose faces were grotesquely altered by "phossy jaw" or painters who vomited their lives away from lead poisoning, or sweat-soaked field workers in the Mississippi Delta, spending their last hours in the hallucinations of a malarial fever. Disease and disaster were the Scylla and Charybdis of all workers and their families, native and newcomer alike. However, the increasingly large foreign-born component of America's industrial work force faced some special health problems in the workplace.

Because newcomers from abroad frequently took hazardous, low-paying jobs that American workers spurned, injury and illness rates were higher among immigrant workers than among those born in the United States. The Pittsburgh Survey sponsored by the Russell Sage Foundation noted in its volume on work-accidents that between July 1, 1906, and June 30, 1907, 526 workers in Pittsburgh were killed performing work-related duties; 293 of them were foreign-born, almost 56 percent.[3]

However impressive the statistics they amassed, investigators could not help but be even more moved by the suffering they personally witnessed among immigrant workers. One industrial health reformer recalled visiting immigrant workers in the lead trades: "A Hungarian, thirty-six years old, worked for seven years grinding lead paint. During this time he had three attacks of colic, with vomiting and headache. I saw him in the hospital, a skeleton of a man, looking almost twice his age, his limbs soft and flabby, his muscles wasted. He was extremely emaciated, his color was a dirty grayish yellow, his eyes dull and expressionless. He lay in an apathetic condition, rousing when spoken to and answering rationally but slowly, with often an appreciable delay, then sinking back into apathy." Nor could this woman physician/reformer, herself the granddaughter of an immigrant from Northern Ireland, forget another of her more recently arrived interviewees: "A Polish laborer worked only three weeks in a very dusty white-lead plant at an unusually dusty emergency job, at the end of which he was sent to the hospital with severe lead colic and palsy of both wrists."[4]

Many immigrants were young and inexperienced when they arrived, with less than the equivalent of a grade-school education and little knowledge of English. All training was on-the-job, and those with poor language comprehension ran greater risks than the native-born of being the victims of the machines and tools with which they earned their living. Critics of immigrant labor often blamed work-related illness or injury on foreign workers' ignorance rather than on unfamiliarity with English and the ways of the American workplace. Unfamiliarity with language and customs also made it harder to find other jobs, so immigrant workers were less likely to quit unsafe jobs than native-born Americans. Industrial workers of the era were frequently exploited, even bullied by employers; immigrant workers were even more likely to be so victimized. Often, immigrant workers did not understand their rights in the workplace and therefore were less likely than other workers to get good care. In an era when the modern labor movement was in its infancy, workers had few rights worth knowing about. Those who advocated workers' rights deplored the lack of awareness of foreign-born workers, which critics feared would hurt all workers. As one writer lamented, "half a million of these people are entering the United States every year to be mistreated and exploited, to become helpless victims of industrial accident and industrial disease."[5] After the passage of extensive immigration restrictions in the 1920s, the growth in illegal immigration further complicated the picture, with ill or injured immigrant workers fearful that a doctor or a hospital might report them. Such reticence only furthered the financial success of quacks who preyed on newcomers.

Workers dying from their work—both native and foreign-born—was a problem not limited to the United States. Industrial countries elsewhere also experienced rising morbidity and mortality figures because of workplace health hazards. Still, the United States lagged far behind other industrial nations in defining the problem, investigating it, and involving the government in finding solutions. In her 1943 autobiography, industrial hygiene pioneer Dr. Alice Hamilton speculated on the reasons for America's tardiness: "When I talked to my medical friends about the strange silence on this subject [industrial illness] in American medical magazines and textbooks, I gained the impression that here was a subject tainted with Socialism or with feminine sentimentality for the poor."[6] Whether it was antiradicalism or sexism or a misguided hesitance to permit public interference in the affairs of private industry, the delay in imposing government industrial controls, even in data gathering in the United States, kept the extent and seriousness of industrial accidents and disease obscure well into the early decades of the twentieth century; industrial illness remained what one critic called a "matter of 'scientific conjecture.'"[7]

American businesspeople saw little reason to take responsibility for the health of employees. Why should they, when the sheer volume of newcomers granted employers in need of skilled or semiskilled labor such an edge? They could readily replace ill or injured workers with others—often even lower-paid foreign-born workers. Only as labor unions gained strength and demanded healthier work conditions, industry by industry, and the restrictive legislation of the 1920s curbed the annual pool of new workers desperately seeking work at any cost did employers find it advantageous to keep their work force healthy.

Some Americans were critics of the foreign labor being recruited. Oblivious to the differences in experience between English-speaking native-born workers and newcomers strange to the society, its language, and its customs, such critics blamed the victims of accidents and illness for what befell them. In a 1913 article, one writer said of Slavic workers, "This class of eastern European peasant lacks the intelligence and initiative either to avoid the ordinary dangers of rough labor or to keep in efficient health; and their employers have to pay the bills for teaching them." The same critic complained, "The Americans know how to take care of themselves. Most important of all, they wash their hands and faces when they stop work. The immigrants from Eastern Europe do not, unless someone stands over them and makes them do it."[8] Similarly, in 1922 immigration opponents Jeremiah W. Jenks, a professor of government and public administration at New York University, and W. Jett Lauck, a former academic, quoted an investigator of the New York State Bureau of Industries and Immigration to the effect that in fiscal year 1919–1920, there were 345,672 industrial accidents in the state, many of which were the result of the victims' shortcomings. The investigator, Jenks and Lauck gloated, believed that "a large proportion" of these accidents and "the consequent losses incident thereto were undoubtedly due to illiteracy." The investigator was certain of this, Jenks and Lauck wrote, because over 70 percent of those filing compensation claims needed the assistance of an interpreter.[9] From the perspective of these nativist authors, the need for an interpreter suggested an illiteracy in English that probably contributed, if it did not cause, the industrial accidents in which so many newcomers were involved.

Immigrants did have obstacles to overcome that many Americans callously dismissed as deficiencies in the quality of immigrant labor, but most of the dangers in the industrial workplace menaced native-born and immigrant workers alike. Although the federal government did not systematically or regularly collect data on industrial health and sanitation throughout the nation, the Public Health Service Act of 1912 formally broadened the responsibility of Public Health Service officers to investigate disease in both the laboratory and the field, a mission that had been implicit since Joseph Kinyoun first established his small laboratory.[10] Federal physicians joined the fray, justifying their participation by observing that new demands on worker productivity and new technologies to spur production combined to make health and hygiene in the workplace a public health concern. In the preface to a 1915 public health bulletin, "The Health of Garment Workers," U.S. Public Health Service Surgeon J. W. Schereschewsky made clear the necessity of more general concern with the health problems of the workplace, observing that "industrial conditions affecting the health of the worker can never be separated sharply from those of his total environment, nor is it possible to define where the field of industrial hygiene ceases and that of public health begins."[11]

Different trades affected different parts of a worker's body, but all took their toll on workers' lives by making them sick, by compromising their general condition and resistance to infection, or by leaving them with a disabling handicap. In the first category were diseases caused by exposure to industrial poisons, toxic substances that diminished the working capacity of the laborer. Such toxins could enter the body via the mouth and digestive system or through the nose and respiratory system, or by permeating the skin. Phosphorus poisoning and lead poisoning received much attention from American investigators in the early twentieth century.[12]

Phosphorus poisoning was the curse of American parlor match manufacturing, an essential industry in an era well before electric lighting and when gas or electric stoves became standard in American homes. Heads of parlor matches were made by dipping one end of a small wooden stick into a paste containing a mixture that included white or yellow phosphorus, a highly dangerous, toxic substance. Workers, whether they mixed the flammable paste, dipped the small wooden splints, or dried and packaged the matches, were all exposed to danger in one degree or another. Only excellent ventilation and adequate precautions to keep the workplace hygienic and workers' bodies scrubbed and protected could keep fumes or particles from invading match workers' bodies. Once poisoned, workers suffered "general anemia, lowered vitality, and predisposition to other diseases, and in some cases 'phossy jaw.'"[13]

Phosphorus necrosis, or "phossy jaw," was a disease caused by absorption of phosphorous through gums or teeth. Small particles entered the body through cavities in the teeth. Infection soon extended along the length of the jaw, loosening teeth and causing them to drop out, while the jaw decomposed and the disease spread throughout the worker's entire body. Early twentieth-century physicians treated the condition surgically, removing the jawbone to arrest the progress of decay. According to one 1911 industrial disease expert, "It is such a loathsome affliction that sixty-five per cent of all workers in American match factories are liable; and among women and children employed in such establishments the percentage is far higher and the pity of it is that such suffering should continue when it is preventable through the substitution of harmless phosphorus, and that the United States should continue to exact this toll when almost all civilized nations have already forbidden the use of white phosphorus!"[14]

While reformers demanded that Congress tax the importation of white phosphorus so heavily as to virtually exclude its export, match workers continued to suffer and die from their trade. When a nontoxic substitute for white phosphorous, sesquisulphide, was discovered by a French chemist, its American patent rights were purchased by the Diamond Match Company. In an extraordinary display of industrial generosity, Diamond waived its patent rights in 1909, allowing the entire industry to abandon white phosphorus. Congress then passed the Esch law, imposing a tax on white phosphorus matches high enough to cover the difference in cost between these matches and newer ones made with sesquisulphide. Finally the American match industry abandoned the killer that had illuminated the country's homes and helped cook its dinner.[15]

It was conversations with Jane Addams about "phossy jaw" at Chicago's Hull House and the European data collected by British physician Sir Thomas Oliver that inspired reformer Dr. Alice Hamilton to turn her attention to industrial disease. The same year that Diamond was leading the way to abandon white phosphorous, Hamilton was investigating another industrial killer, lead poisoning. Lead, a ubiquitous industrial poison at the turn of the century, enters the system slowly over time and is not easily or quickly eliminated. Though there are often no immediate symptoms, accumulation is toxic. In severe cases, the stricken report acute colic or encephalopathy, including convulsions and temporary blindness. The nervous system is also affected, as manifested by paralysis, especially wrist drop, and the mental debility characteristic of senility, quite striking when observed in younger victims. In early stages of the disease there are milder but still recognizable symptoms such as pale complexion, loss of weight, absence of appetite, stomach cramps, constipation, and goutlike discomfort.

In 1909, Illinois Gov. Charles S. Deneen appointed Hamilton director and chief medical examiner of a nine-member state commission to survey industrial disease in Illinois. The commission had only one year to plan and implement its investigation and to make recommendations to the state legislature. The imposed deadline caused Hamilton to narrow her focus to poisonous trades and occupations. She and a staff of young assistants devoted their attention to lead poisoning in the state's industries.[16] The Illinois Commission on Occupational Diseases discovered that twenty-eight industries in the state posed the risk of lead poisoning for workers. Of these, five accounted for the great majority of the state's lead poisoning cases: white lead manufacturing, lead smelting and refining, manufacture of storage batteries, manufacturing of dry colors and paints, and the painter's trade.

Although all workers were at risk in the lead industries unless they were "thoroughly acquainted with the proper means of guarding themselves against poisoning," Hamilton reported, it seemed to her that most at risk were "unskilled foreigners who enter upon the work utterly ignorant of its dangers or with only a vague, unintelligent dread."[17] In the report of the Illinois commission, Hamilton recounted the story of a young Bulgarian worker who found employment emptying pans of dry white lead in a white lead factory the first week he arrived in Chicago: "He was given no respirator and had no idea that he had a right to ask for one. Nobody told him the white dust on his hands and mustache was poisonous. He had only one suit of clothes and wore his working clothes home. He had a severe attack of lead poisoning at the end

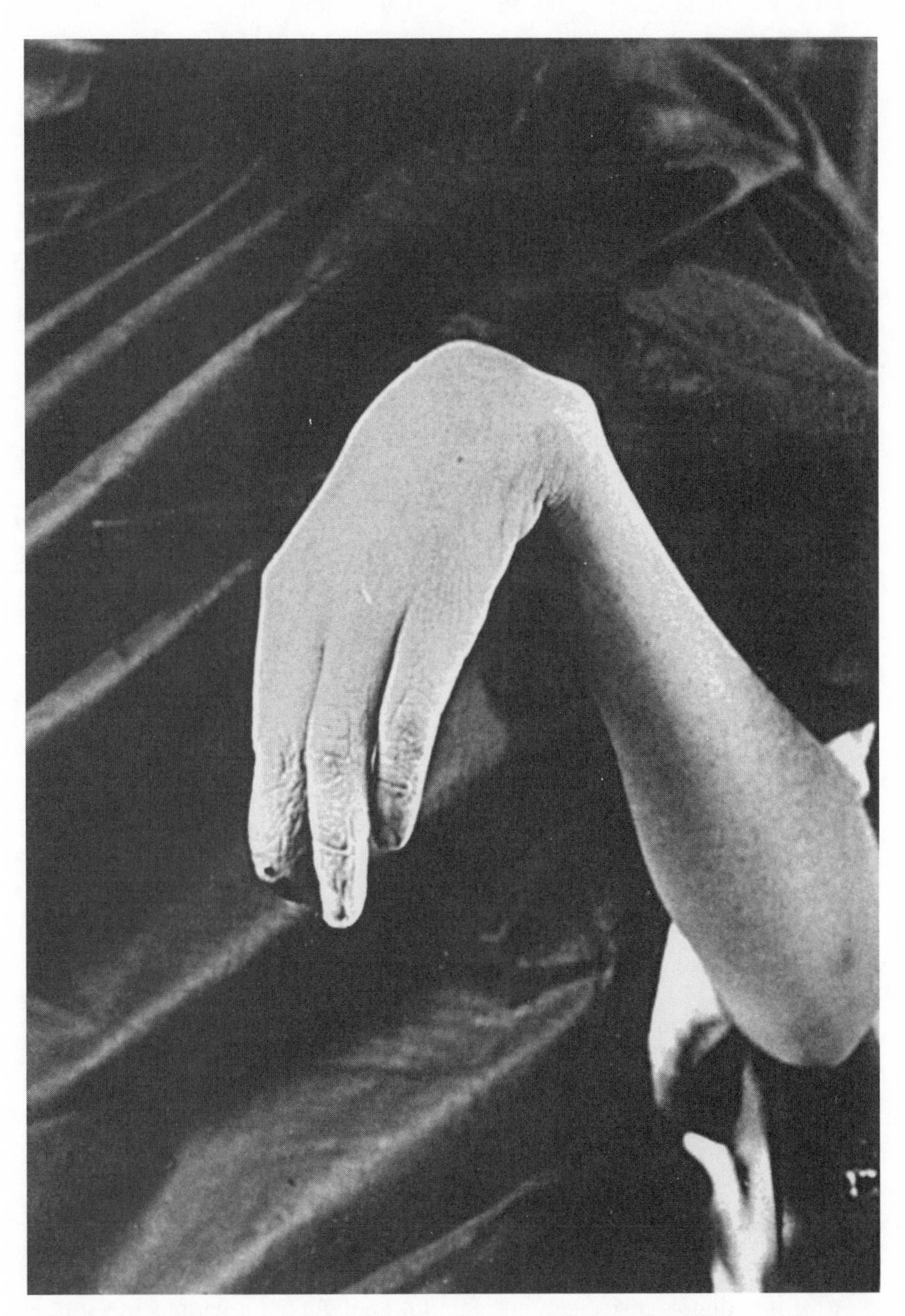

Wrist-drop, a paralysis caused by damage to the nervous system, symptomatic of the lead poisoning common among painters and other industrial workers at the turn of the century. (Courtesy of the Mutter Museum of the College of Physicians of Philadelphia)

of five weeks."[18] In a storage battery factory, the investigators found a Russian Jew who was told to make red lead paste. They found him "utterly ignorant of the substances he was handling and used to moisten his fingers in his mouth as he made the paste." After only ten days' work, he was severely poisoned. Hamilton was deeply disturbed by how employers neglected to tell foreign-born workers what they needed to know for their protection from toxic substances. She reported to Governor Deneen, "We have found almost no effort in the lead works to instruct the foreigners in the care of their persons and in the avoidance of danger."[19]

The paint industry was the most frequent origin of lead poisoning cases. Over thirty thousand men worked in the trade. Painters were exposed to the toxic substance when they chewed lead-smeared tobacco, ate lead-spotted food, breathed dry lead dust, mixed white or yellow lead with putty or paint, or sandpapered coats of lead paint after drying and inhaling the lead-laden dust swirling around them.[20] In Michael Gold's novel of immigrant working-class life, *Jews Without Money,* young Michael recalls a glimpse his father's encounter with "painter's disease":

> It was summer. My father worked on a scaffold in the sun. One day he grew dizzy with the painter's disease—lead poisoning.
>
> Paint is made with white lead. When a house painter mixes oil and turpentine with the dry pigment, its lead is released in fumes which the man must breathe. Or this free lead also enters through the skin. It eats up the painter's stomach and nerves, and poisons his bones.
>
> My father suffered from painter's nausea. One summer night he came home after his accustomed hour. His pale face under its tattoo of green and red paint was twisted grotesquely, like a Chinese dancer's mask. He stripped off his overalls in the kitchen and sank into a chair.
>
> "Quick! give me the bucket, Katie!" he groaned to my mother. She brought it and he vomited.[21]

Gold's fiction echoed the suffering of many thousands of workers and their families. Painter and union organizer Philip Zausner recalled in his autobiography, "The East Side immigrant painter . . . knew nothing about the poisonous effects of his trade. When as a result of his hurried lunching with lead-covered hands he conveyed the poison into his bowels and got a sudden attack of lead-colic, he didn't know why he was in pain. Maybe it was the extra large piece of watermelon which he ate with bread for supper last night—maybe that had caused his bellyache."[22]

After her Illinois investigation, Hamilton conducted a nationwide survey for the U.S. Department of Labor that exposed the national dimensions of the lead-poisoning problem.[23] The study included twenty-two of the twenty-five factories in the country manufacturing white lead. At one of the factories, a New Jersey plant that had denied Hamilton admission, she found a way to obtain information from workers. Throughout the country, most of the workers in the industry were unskilled and many were immigrants. "In St. Louis, East St. Louis, Cincinnati, and Philadelphia many Negroes are employed, and also many Slavs. In Omaha, Chicago, Pittsburgh, Perth Amboy, and Camden Slavs predominate. Italians are found in almost all white-lead factories, but not in large numbers except in West Pullman, Philadelphia, Brooklyn, Long Island City, and Staten Island. A few Greeks are found in St. Louis, West Pullman, Chicago, Omaha, and Pittsburgh. Very few American workmen, aside from Negroes, and not many immigrants from North of Europe are found in the factories of the Middle West." While Hamilton found more "American-born and English-speaking" workers on the Atlantic seaboard, "these men say they have taken up the work only as an emergency job during a temporary lack of other employment." She concluded, "It is not work that attracts experienced or intelligent men, except for a few skilled jobs in the mill."[24]

As for actual cases of lead poisoning, Hamilton discovered among the workers in white and red lead factories a total of 388 cases of lead poisoning between January 1, 1910, and May 1, 1911, a period of sixteen months. Precisely who got sick and the extent to which these data were accurate is difficult to determine, because "When one considers that there is no compulsory registration of cases of lead poisoning in the United States, that the victims of this disease are largely foreigners with no fixed abode, that there are no union sick benefit records to consult, and that the insurance companies can give no help, one can readily see that it is impossible to gain any accurate knowledge of the full extent of lead poisoning in this industry throughout the country."[25] Even physicians who saw patients in the better-run factories could not always be helpful because "There are always some suspicious foreigners or Negroes who have an unreasoning distrust of the company doctor."[26] It may well be that newcomers feared that if his diagnosis was negative, the company doctor would recommend dismissal. Whatever their reasons for being "suspicious," Italian, Polish, and Hungarian workers already stricken with lead poisoning found ways of eluding record keepers at hospitals and dispensaries as well. Hamilton offered an example of twenty-five cases of lead poisoning, "mostly Poles and Italians," that were collected from the hospital and dispensary records of a town in which there was one white and one red lead factory. Thirteen were traced to the factory; "the other 12 could not be found at all." The experienced physician speculated, "In all probability they, too, worked in this plant, but they are useless as far as our factory records go." Newcomers desperate to keep their jobs may have even received some assistance from factory doctors who empathized with their plight. Hamilton mentioned "a certain Polish doctor" who claimed to have seen from three to five cases of lead poisoning every month from among the Slavic workers in a particular white lead factory, but "he could not remember any names except of the last three," meaning that only three cases "instead of some fifty-odd, could be added to the records."[27] Clearly, Hamilton had good reason to suspect that her 388 cases of lead poisoning were far fewer than really existed.

Although Hamilton led the way in being skeptical of her own data, there could be little doubt about the suffering of individual victims. She admitted that the majority of interviews excerpted in her report were Irish or Scots or American because "they could give the fullest information [because they spoke English]," but southern and eastern European immigrants were not completely omitted. The first immigrant worker's interview in Hamilton's report was that of "V.O., Italian, speaking no English." Through an interpreter, Hamilton learned that V.O. had worked for thirty-six days in the stripping gang of a new factory, where "Nobody gave him any instructions nor explained to him the dangers of the work and the precautions he must take." He told the interviewer that he "never saw a respirator and did not know there were such things or that he had a right to ask for one." He got only cold water to wash with and no soap or towels, nor did he have a place to eat his lunch clear of the work area. Then one day, "an agonizing colic came upon him suddenly and he fell to the ground." He did not know that there was such a sickness as lead poisoning or that he had it until an Italian doctor told him about his condition and sent him to a public hospital, where he remained for two weeks. Subsequently, he returned home, had a relapse, and went back to the hospital. Three months later he was still not strong enough to work.[28]

Painters were as vulnerable as those who worked directly in lead factories. Other countries had somewhat diminished the risk to painters by providing "warm rooms in which they may wash their hands, change their clothes, and eat their lunches, as in Germany; by abolishing the use of white-lead paint as in France, or by using it only for exterior work, and by doing away with *dry* sand-papering of lead paint as in England."[29] The United States lagged behind these other industrial nations in imposing controls upon an unhealthy trade, although the increasing use of machinery for men and women did somewhat diminish the danger. Still, health education of painters and other workers in the lead trades and care for those in need of it continued to trail technology and productivity.

Dusts of various kinds—whether in the mines or the factories or stone quarries—contributed mightily to industrial mortality. The danger of the so-called "dusty trades" was not only that inhaled dust might contain poisonous particles but that even nonpoisonous dust could "impair the lungs and the delicate membranes of the air passages." Many mistakenly believed that the high death rate from tuberculosis could be directly traced to the textile dust workers inhaled at their trades for most of their lives.[30] Some also mistakenly thought that wool sorting, hide handling, and hair and brush manufacturing exposed workers in those industries to anthrax from sheep and cattle. Still, there can be little doubt that lungs impaired by dust particles compromised workers' health and made them more vulnerable to TB and other infectious diseases.

Dusts of all kinds were major health hazards for industrial workers. "Stone and marble[-]cutters, glass-cutters, diamond-cutters, potters, cement-workers, plasterers, molders, and others exposed to mineral dusts show abnormally high death rates, especially from respiratory diseases, that from consumption being about five times as high among stone and marble cutters as among farmers and farm laborers," according to one study.[31] In their study of silicosis, historians David Rosner and Gerald Markowitz observe that the germ theory of disease had so taken hold by the early twentieth century that all respiratory diseases were mistakenly diagnosed as tuberculosis. Symptoms—a hacking cough, difficulty breathing, and expectoration of blood—were blamed on bacteria. Only investigations among workers such as the stonecutters of Barre, Vermont, many of whom were northern Italian immigrants, revealed that the respiratory disease silicosis was a separate disease.[32]

Silicosis is a disease of the lung resulting from the inhalation of fine particles of silica dust. Silica is a mineral most frequently found in sand, quartz, and granite. Extended periods of inhalation destroy the lung's capacity to function, much as black lung does to miners. Silicosis was a widespread occupational disease in its own right, but it also compromised workers' health so that many in weakened physical condition also contracted tuberculosis. Diagnosis was difficult because some workers suffered from tuberculosis contracted from other workers, others of silicosis contracted from the pollution of their air by silica dust, and still others of silico-tuberculosis, or "stonecutter's shadow." Silico-tuberculosis was not a separate disease. Tuberculosis was contracted by workers during a period of deteriorated health caused by silicosis and accelerated in its devastating effect by silica fibrosis in the lung. Morbidity rates of silicosis and silico-tuberculosis skyrocketed when pneumatic tools that produced high levels of dust replaced older implements.[33] Some workers, such as the Italian quarry workers in Barre, Vermont, denied the degree to which they were sickened by their craft.[34] Others simply waited too long to seek medical advice, a characteristic especially common among immigrant workers who had difficulty explaining their condition. Historians Rosner and Markowitz poignantly begin their book with the story of Polish immigrant George Silinski, who "spoke very little English" and had given up farming to work in a Buffalo area foundry as a sandblaster between 1901 and 1918. By 1933 the forty-eight-year-old Silinski was dead of silicosis; the ash of his lung tissue taken at autopsy was 16.5 percent silica.[35] The Italian granite cutters of Barre, Vermont were often sunk to their graves by even greater weights of "deadly dust."

Also dangerous was the exposure to metallic dusts in factories. Metal grinders, polishers, tool and instrument makers, brass workers, jewelry workers, printers, and engravers also risked their health in an era long before adequate ventilation systems were required by state laws.[36] Those who bent their backs over grinding stones took harmful dust into their lungs with each breath, even as those did who worked in mines and quarries.

Disabling handicaps were frequently the result of extended exposure to dangerous conditions. Lifelong injuries were the physical reminders of what workers sacrificed for their livelihoods. Some were as obvious as a missing limb, some invisible, such as premature loss of hearing, but no less impairing. Deafness was quite common among engineers and fire fighters in train locomotives, boilermaking, bridgebuilding, shipbuilding, and other trades where hammers crashed against iron and steel. Young boys could become hearing-impaired for the rest of their lives while they were still apprentices to boilermakers and riveters.[37]

If factories, mines, and quarries could be hazardous to the health of workers, especially newly arrived immigrants unfamiliar with English and inexperienced in the ways of the industrial workplace, so could the home. Industrial homework was a handmaiden of the factory system. In the eighteenth century, farm families and artisans' households took in work. However, not until the middle of the following century did urban families begin to rely upon homework as their sole source of income. By the early twentieth century, immigrant men and women from southern and eastern Europe contributed their homework to a variety of labor-intensive industries. Several characteristics of homework explain the kinds of physical and emotional stress it placed upon workers. Homework was most common in industries where there could be clear divisions of labor by well-defined specialty, with each homeworker performing only a fraction of the total labor on each item. Such homework tasks were generally repetitious and paid by the piece at rates lower than those paid factory workers for the same task. Speed, then, was essential to keep one's wages livable. Especially useful for seasonal industries where flexibility in the work force enhanced profit, homework was typified by long hours of fast labor.

Industries requiring sewing parts of garments, making artificial flowers, sorting coffee beans, shelling nuts, making wigs of human hair, cigar rolling, lace work, and hatmaking were heavily peopled by immigrant homeworkers at the turn of the century. Men engaged in homework in the cigar trade and in the early days of the garment industry, before the advent of the system in which apartments and lofts were turned into small workshops—sweatshops—where several families might work together or strangers sat elbow to elbow, day in and day out. Most frequently, though, immigrant homeworkers were married women with children whose spouses' incomes were insufficient to

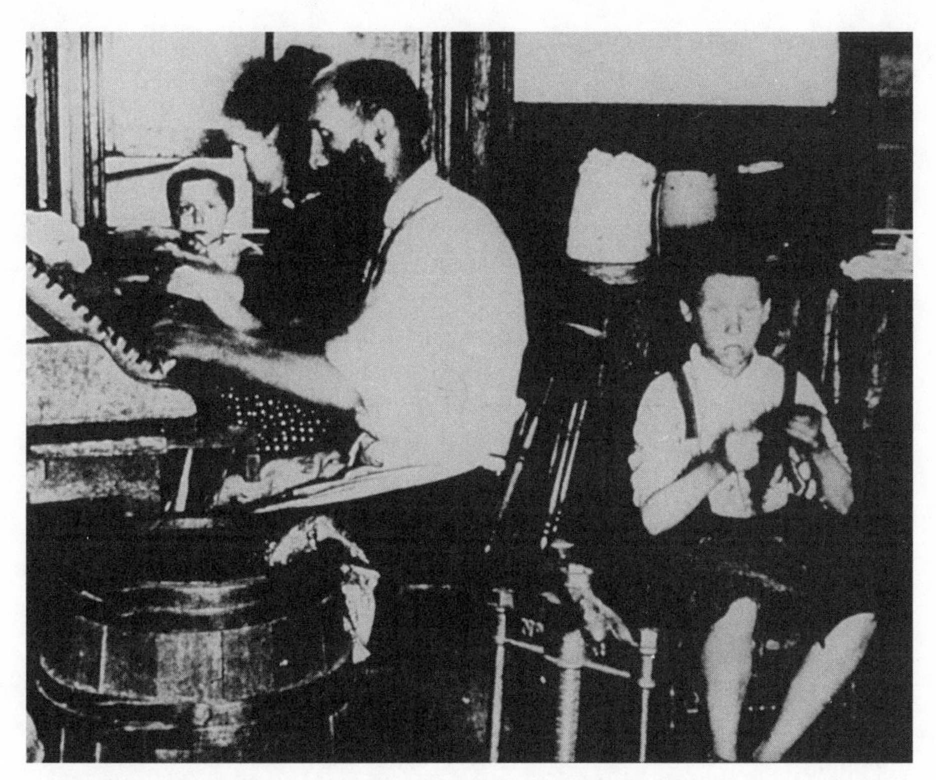

Jacob Riis's photograph of a cigarmaker working at home, c. 1900. The repetitive motion often resulted in cigarmakers' neurosis, characterized by pain and numbness in the shoulder, arm, and fingers. (Courtesy of the Museum of the City of New York)

meet the family bills. Homework was especially common among ethnic groups—such as southern Italians and eastern Europeans—with taboos on the employment of married women outside their own households.[38] As Jacob Riis and other reformers observed, tenement apartments were unhealthy places to live. Homework merely worsened conditions.

In his ramblings through New York City's immigrant neighborhoods, Riis observed that Bohemian immigrants manufactured cigars in tenement apartments on New York's Lower East Side, where "Men, women and children work together seven days in the week in these cheerless tenements to make a living for the family, from the break of day till far into the night." He denied that the tobacco workers were "less healthy than other in-door workers," or that they were particularly prone to tuberculosis, the scourge of the tenements.[39] Consumption was a leading cause of death among cigar workers, although quite likely the close working quarters were more responsible for the spread of the disease than any particular aspect of the actual cigar makers trade. Not so with "cigarmaker's neurosis." This neuropathy from repetitive motion slowed workers' ability to roll and often ended their ability to earn a living at their trade. Possibly caused by the repetitive motion of rolling cigars and the workers' cramped, sedentary position for many hours each day, cigar makers' neurosis was characterized by severe pain in the shoulder, arm, and head. In especially acute cases, cigar makers might lose muscle control in their hands.[40] The Cigar Makers International Union of America sought care for union members who needed it when the trade was moved from home to factory, but cigar makers who plied their trade at home had little resort except to private charities.

In their struggle to unionize the cigar industry, leaders of the Cigar Makers International Union argued that unsupervised homeworkers suffering from tuberculosis rolled a product that spread the disease to smokers. The union scored a brief success when the New York State legislature passed a law prohibiting homework in 1884, partly in response to the union's demand that the legislators, some of whom may have been cigar smokers themselves, act to protect the public. However, in less than a year, the New York Court of Appeals overturned the law on the grounds that without clear evidence that tenement dwellers' health was endangered by the presence in their midst of cigar workers, the law exceeded the police power of the state in its interference with lawful economic activity. The decision set a precedent for declaring homework illegal only when it could be shown to have detrimental effects on the public health.[41]

The Court of Appeals's message was not lost upon those with a special interest in women's health in the workplace. Women dominated homework. Although precise data are difficult to find, census data and homework licensing data suggest that "as many as 250,000 women may have been engaged in homework in New York City in 1910."[42] They were usually married women, twenty-five to forty-five years of age, with young children and husbands in their domiciles.[43]

The industry employing the most homeworkers was the garment industry, hiring over a quarter of all industrial workers in New York City in 1905. Garments were manufactured by a combination of factory production and homework. Women homeworkers were usually responsible for such labor-intensive work as hand finishing. Jacob Riis, an ardent opponent of homework as yet another condition that made tenement dwelling unhealthy, argued somewhat exaggeratedly that the sick women and children of the Jewish Lower East Side spread their disease through what they manufactured in their homes:

> The health officers are on constant and sharp lookout for hidden fever-nests. Considering that half of the ready-made clothes that are sold in the big stores, if not a good deal more than half, are made in these tenement rooms, this is not excessive caution. It has happened more than once that a child recovering from small-pox, and in the most contagious stage of the disease, has been found

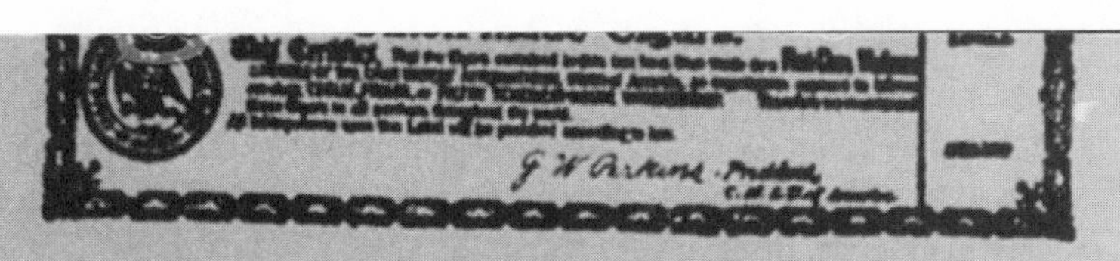

The picture on the other side
REPRESENTS A
TENEMENT HOUSE CIGAR
FACTORY.

BEWARE OF CIGARS MADE IN THOSE
FILTHY PLACES.
· THEY BREED DISEASE ·

The above Blue Label of the
C.M.I.U. OF A.
on a box containing cigars
is the only safe-guard
against
Tenement House Product.

Cigar Makers International Union of America (C.M.I.U. of A.) postcard urging smokers not to buy cigars made in tenement house sweatshops, c. 1900. (Smithsonian Institution)

> crawling among heaps of half-finished clothing that the next day would be offered for sale on the counter of a Broadway store; or that a typhus fever patient has been discovered in a room whence perhaps a hundred coats had been sent home that week, each one with the wearer's death warrant, unseen unsuspected, basted in the lining.[44]

Jewish and Italian women predominated among garment homeworkers prior to 1900. After the turn of the century, unmarried Jewish women worked in sweatshops and larger factories, but the wives of employed Jewish workers increasingly confined their homework to looking after boarders who rented space in immigrants' apartments.

Italian women and their daughters were now the workers who finished American-made garments—perhaps as much as 98 percent of them—in their homes. They continued to dominate the artificial flower trade as well. In 1913, Mary Van Kleeck, the secretary of the Committee on Women's Work of the Russell Sage Foundation, published a study of artificial flower makers.[45] Van Kleeck conducted her survey in lower Manhattan in a section of blocks

A consumptive mother doing piecework in tenement room, c. 1900. (National Library of Medicine)

bounded by Christopher Street running diagonally along the northwest, Canal Street on the south, the Hudson River on the west, and a broken line along Sixth Avenue, West Fourth Street, and West Broadway on the east, a district that adjoined one in which most of the flower factories were located. Location was important because as Van Kleeck observed, "It is quite natural that the workers should live near the shops, as they are obliged constantly to carry large boxes of finished flowers from their homes to the factories, and their earnings are too small to make it worth their while to lose time and spend carfare in a long journey."[46] She studied 110 families having 371 family members who worked on flowers. Almost all were Italian because of what the investigator described as "the attitude of Italians toward women and their prejudice against their employment outside the home," especially for married women.[47]

Van Kleeck regarded home manufacture as "a most threatening aspect of the sweating system" because "the labor of young children is utilized, and advantage is taken of the urgent need of their mothers to earn money without leaving their homes and children," a need that discouraged labor militancy, leaving the work force, "cheap and docile."[48] However, Van Kleeck emphasized, the harm was not to the workers alone, because "goods manufactured in these crowded tenement homes may carry disease not recognized as a result of the home-work system."[49] Van Kleeck saw flower makers already suffering from tuberculosis or "bad cases of skin disease," who handled the flowers "with no thought of the possibility of infection."[50]

Certainly the workday was grueling and the financial rewards meager. One Italian family of ten (parents and eight children) that Van Kleeck found living in Greenwich Village was typical of those supplementing a family income of $16.50 per week. The father earned $7.00 a week selling lunches in a saloon, while the eldest daughter made $6.50 in a box factory and the sixteen-year-old son contributed $3.00 earned as a wagon boy. Four children were in school and two babies at home. All but the babies helped make flowers at home. The mother worked irregularly during the day, joined by her children after school, and the older daughter and son in the evenings. The family made three-petaled violets at seven cents per gross, earning a weekly wage of $3.00. Their three-room apartment where they lived and worked cost $12.50 per month.[51] The congestion and long hours took their toll. Italian children had the highest mortality rates from scarlet fever and whooping cough in New York City and a mortality rate from measles that was five times that of all New Yorkers. Italian homeworkers themselves had a mortality rate from tuberculosis fourteen times that of some upper-income neighborhoods of the same city.[52] Forging epidemiological links between homework per se and disease is problematic because it is so difficult to isolate the work experience from the larger milieu of poverty in immigrant neighborhoods. However, in the minds of many contemporaries the connection was irrefutable.

And what did the government do about these flower makers and the sorry conditions of their lives and health? Not nearly enough, according to Mary Van Kleeck. New York State law required special licensing if a tenement was to be used for homework. In New York City, it was the duty of the Labor Department to check Board of Health records and Tenement House Department records to see that there were no outstanding orders against the property. If not, an investigator was sent, whose recommendation combined with the records search was sufficient for a license, provided that, as the law stated, "such building is free from infectious, contagious, or communicable disease, that there are no defects of plumbing that will permit the free entrance of sewer air, that such building is in a clean and proper sanitary condition, and that the articles specified in this section may be manufactured therein under clean and healthful conditions."[53] Van Kleeck was scandalized that "The number of persons per room, the habits of the family, their cleanliness, or the reverse are not important factors in securing a license if the building be moderately sanitary."[54] She reported that even the New York Commissioner of Labor admitted that the supervision of manufacturing in tenement houses

Italian artificial flower makers engaged in homework in New York tenement, 1908. Reformers blamed the congested conditions and long hours for high rates of morbidity from such contagious diseases such as measles, scarlet fever, and tuberculosis.
(U.S. National Archives)

was not "up to the standard contemplated in the statute," and that a study done for the federal government on the men's ready-made clothing trade had mentioned the inadequacy of New York licensing laws in health matters:

> It has been proved impossible, in spite of all existing laws merely regulating tenement-house manufacture . . . to guarantee to the consumer that clothing made or finished in homes is free from disease and vermin. . . . The New York state laws on this subject are looked upon as models for this class of legislation, and every effort is made for their enforcement, yet it has been found in this investigation that work was being done in homes in the city of New York that, while structurally sanitary, were insanitary from other standpoints owing to the presence of filth or vermin or diseased persons, or that they had become unsanitary because of the low standards of the dwellers in them.[55]

Mary Van Kleeck believed that homework ought to be prohibited, and she felt certain that she was not alone. She proclaimed, "Many are contending that as a health measure manufacture in New York City tenements should be absolutely prohibited by law."[56]

Perhaps Van Kleeck was correct. However, the condescending characterization of the health standards of homeworkers as "low" suggests that even well-intentioned investigators may have been more concerned with disease in the workplace because of its effect upon consumers than because of the suffering of homeworkers. Reformers were revolted by the sight of homeworkers who appeared to be suffering from contagious diseases sorting coffee beans, later to be used in fine restaurants. Even more shocking was the sight of other ill women workers sitting at their tables and cracking nuts with their teeth, shelled nuts that would later be sold to an unsuspecting public anxious to enjoy them.

As did the antislavery reformers of an earlier era, opponents of homework often attempted to play on the stomachs and souls of readers. One reformer could not resist writing about how one homeworker mother had come to the aid of her child who was suffering with whooping cough. To relieve a frightening choking spell, "the mother thrust her finger down his throat in an effort to relieve him," only then to wipe her mucous-covered fingers "on the pants on which she was at work."[57]

Expressions of concern over public health became standard fare for reformers and trade unionists. Because women usually purchased the family's food and clothing, appeals were directed at them. The *Women's Labor League Journal,* published in Chicago by the Woman's International Union Label League and Trades Union Auxiliary, criticized the unhealthy quality of bread baked in nonunion bakeries that employed many newcomers. "In Chicago a few years ago, an investigation headed by the commissioner of health disclosed the fact that in many bake shops there was little or no ventila-

A poster urging consumers to avoid disease by purchasing goods with a union label. (Smithsonian Institution)

tion and a number of bakers were employed in one shop which poisoned the air that was absorbed by the dough." Ladies were urged to look for the union label because they could be "assured that bread bearing the union label was made in a sanitary shop by clean healthy men."[58]

Women immigrants who ventured beyond the threshold of their apartments to help support their families were even more vulnerable to industrial diseases and accidents than homeworkers. Jacob Riis estimated that "at least one hundred and fifty thousand women and girls earn their own living in New York," but even he admitted "there is reason to believe that this estimate falls far short of the truth when sufficient account is taken of the large number who are not wholly dependent upon their own labor, while contributing by it to the family's earnings."[59] Many Italian and Jewish immigrant women worked in the shops of small contractors in the garment trades, called "sweatshops." Here they sat around tables in small groups, each repeatedly doing one of the tasks necessary to finish a coat, jacket, skirt, or pair of pants. In larger shops or factories, there were long rows of machines and no one woman made an entire garment. Each completed one part and passed it on to the next woman, while the boss or a foreman shouted at them to speed their pace. One worker recalled, "This sort of thing created a spirit of competition for self-preservation that ended only when the worker, too weak to compete longer with a stronger sister, broke down."[60]

Ernest Poole of New York's Charity Organization Society described sweatshop conditions in 1903: "Go tonight at nine or even ten o'clock down through the ghetto. You will find scores of small coat shops still lighted. These are nonunion shops, and a glimpse into one of them reveals the task system running at full speed. The room is low and crowded. The air is close, impure, and alive with the ceaseless whir of machines."[61] A young woman pieceworker in Chicago, working for Hart, Shaffner and Marx, recalled that the heat of the shop was exacerbated by the large, heavy winter coats that rested in workers' laps, making them sweat as they flicked at flies. The roar of the sewing machines was deafening, but a break in concentration could be costly: "You had to be careful not to stitch your fingers in."[62] Jacob Riis "guessed" that the number of such shops in New York was in the neighborhood of twelve thousand. One young woman quoted by Riis said that she had never made over six dollars per week. Another said that she worked from 4:00 A.M. to 11:00 P.M.[63]

Unfamiliarity with English and the ways of the workplace contributed to the exploitation. In her report on Italian women in industry for the Russell Sage Foundation, Louise Odencrantz mentioned a visit to a Cherry Street shop where she saw two Italian women standing and bending as they sorted through bales of dusty paper. The women worked longer hours than the legal limitations of female labor allowed, but "Neither could speak a word of English and neither knew that they were violating any law because they worked ten and a half hours every day from seven in the morning until six at night." The women knew only that "at the end of a week's work the Italian owner, who was a friend, handed them each a five dollar bill."[64] In an ostrich feather shop on Twelfth Street, Odencrantz found thirty Italian women, "some of whom could not speak English." They scraped, sewed, steamed, and curled feathers and plumes for forty-nine and a half hours per week for which each received $10 to $12, though they often supplemented their wages with "earnings from overtime or home work."[65] The abuse led to increasing enthusiasm for unionization in many trades.

Unionization in the garment trades was spurred by the danger to the health and safety of male and female workers. The United Brotherhood of Cloakmakers had been formed in 1900 but grew slowly. By 1906 it had still organized only 2,500 of the 42,500 cloakmakers. Ironically, it was following partially favorable settlements in the Shirtwaist-makers' strike of 1909 and the Cloakmakers' strike of 1910 with its famous Protocol of Peace crafted by Louis Brandeis that an industrial disaster reminded the entire world about the dangers of industrial life to the health and well-being of workers, especially the women and men who toiled in the garment shops.

The Triangle Shirtwaist Company was a nonunion shop, where the health and safety of workers was a low priority. At 4:35 P.M. on Saturday, March 25, 1911, a fire swept the upper floors of the ten-story building at the northwest corner of Washington Place and Greene Street, where the Triangle Shirtwaist Company was located. By two in the morning, fire chief Croker reported that 146 men and women of the 700 employees then at work had died in the inferno. More than a third of those who lost their lives died when they jumped from windows to escape the flames. The *New York World* screamed, "They jumped with their clothing ablaze. The hair of some of the girls streamed up aflame as they leaped. Thud after thud sounded on the pavements." Those who later walked through the charred ruins of the factory found sights equally ghastly. Bodies were burned to the skeleton and, "There were skeletons bending over sewing machines."[66]

Eyewitness accounts tugged at the heartstrings of readers. William G. Shepard, a United Press reporter who happened to be in nearby Washington Square on the afternoon of the fire, phoned in details even as the tragedy unfolded. He reported, "As I looked up I saw a love affair in the midst of all the horror." A young man was helping young women to jump from a window in the building rather than be burned alive. One of the women put her arms around him and kissed him before he helped her jump to her death. Then he jumped too, "His coat fluttered upward—the air filled his trouser legs. I could see he wore tan shoes and hose. His hat remained on his head. Thud—dead, thud—dead—together they went into eternity."[67] Another in the crowd recalled the death leaps as well. Louis Waldman, later a socialist labor attorney, saw "girl after girl appear at the reddened windows, pause for a terrified moment, and then leap to the pavement below, to land as a mangled bloody pulp." Waldman said that he and others saw the dead as "not so much the victims of a holocaust of flame as they were the victims of stupid greed and criminal exploitation."[68] Reports of the industrial tragedy spread, reaching the very villages of Russia from which some of the workers and their families had come. Parents who had daughters working in America and who had little notion of the country's vastness were wracked with worry that their child had been consumed in the flames.[69] For those in the streets below and those on the other side of the globe the tragedy was a lasting reminder of what the life of a garment worker was worth in America.

Why had so many escaped and why had so many died? The former was reported by the *New York World* as largely due to the bravery of workers, "elevator boys" who transported hundreds to safety until the flames severed the cables. An inadequate fire escape that consisted of a single ladder running down to a rear narrow court and "one narrow door giving access to the ladder" contributed to the high death toll, as did the insufficient amount of fire equipment on the scene and its inadequacy. Ladders reaching to only the seventh floor did little good to rescue workers who had fled to the roof. However, the main reason for the large number of deaths was the practice of

The friends and relatives of the Triangle Shirtwaist fire victims identifying bodies in coffins in a makeshift morgue at the Twenty-Sixth Street Pier. (Courtesy of UPI/Bettmann)

locking workrooms from the outside to keep the young machine operators from sneaking away from the job early.

At a memorial meeting for the Triangle workers held in the Metropolitan Opera House on April 2, 1911, the organizer for the International Ladies' Garment Workers' Union (ILGWU) and the Women's Trade Union League, Rose Schneiderman, told the audience, "The old Inquisition had its rack and its thumbscrews and its instruments of torture with iron teeth. We know what these things are today; the iron teeth are our necessities, the thumbscrews are the high-powered and swift machinery close to which we must work, and the rack is here in the firetrap structures that will destroy us the minute they catch on fire."[70] Schneiderman's passion on the opera house stage did not echo from the legal bench. While the owners of the factory, Max Blanck and Isaac Harris, collected almost $65,000 in insurance for the fire, none of it was shared with workers' families and they were acquitted by a jury when they were brought to trial on manslaughter charges in December 1911. Joseph J. Asch, who owned the building, proved that his structure had met fire regulations when it was opened in 1901 and that he had never been asked by city authorities to make modifications in the ensuing ten years. He settled twenty-three lawsuits. Settlements averaged $75.00 per life lost.[71]

While many families of the dead lost not only their loved ones but breadwinners, existing law offered little help or hope that such a tragedy would not happen again. However, public outrage over the fire and continued union protest resulted in the formation of the New York State Factory Investigating Commission, which included the American Federation of Labor head Samuel Gompers as a member as well as Francis Perkins, the well-known reformer, who would later serve as secretary of labor in Franklin Roosevelt's cabinet. The report of the commission and the tireless efforts of state legislators Alfred E. Smith and Robert F. Wagner, Sr., resulted in some of the most progressive legislation protecting the lives and health of workers in the nation, including a new workmen's compensation law. Meanwhile, the ILGWU and the Amalgamated Clothing Workers Union (ACWU) led the way in improving wages and conditions in garment factories. David Dubinsky, whose leadership of the ILGWU would make it one of the most powerful unions in the nation, recalled that he had joined the Socialist party months before the fire, but that he was "just a listener at its rallies until the horrible fire in the Triangle Shirtwaist Company."[72] The tragedy left its legacy for the health and well-being of workers.

Part of that legacy was an increased commitment to know the health problems of both male and female workers in the garment trades. The Joint Board of Sanitary Control of the Cloak, Suit, Skirt, and Dress and Waist Industries of New York (hereafter Joint Board) was formed as a result of the Protocol of Peace that ended the Cloakmaker's strike. The Joint Board requested that the

U.S. Public Health Service undertake an investigation of workers' health in the industries covered by the protocol as well as hygienic studies of shop lighting and air, the latter "as affected by the use of gas-heated pressing irons." Such industrial investigations, encouraged by the 1912 Public Health Service Act, were carried out under the supervision of Dr. Joseph W. Schereschewsky, who the PHS put in charge of all the service's industrial hygiene work.[73]

The 1915 final report had a long, detailed list of conclusions and recommendations. Schereschewsky noted, "Garment workers, as a class, exhibit a large number of defects and diseases, only about 2 percent being free from defects. That these defects or diseases are of such character as to interfere with individual efficiency is shown by the large number of workers (over two-thirds) who do not feel perfectly well."[74] The workers had many ailments and physical malformities, but "Tuberculosis is undoubtedly the most serious disease prevalent among garment workers, being over three times as prevalent among males as females." Even among women workers, the rates of TB infection were markedly out of proportion, as demonstrated by "the fact that among the females examined it is about three times and among males nearly ten times as common as among soldiers in the Federal Army."[75] The relationship between actual shop work and TB among the garment workers was unclear because the possibilities of infection were enhanced by so many aspects of the impoverished, crowded conditions in which most of the workers lived. Still, it seemed to contemporary investigators that "bad postural habits," "the presence of suspended matter (woolen 'fly') in the air of workshops," and "the low rate of metabolism induced by the inactivity of a sedentary occupation" only enhanced chances of contracting TB.

Recommendations called for a broad program of educating workers in habits of good hygiene such as the elimination of "promiscuous spitting," but also removing industrial health hazards such as "[woolen] 'fly' from the air of workshops." To lessen dust in the air, "Dry sweeping should be rigidly suppressed." The report exhorted factory owners to do their part by paying for and providing the means of keeping the workplace healthful. For example, the report stipulated, among many things, coating floors with dust-resisting oil, suction cleaning, reducing the vibration of machines by means of "blocks of rubber or similar dampening substances beneath machine supports," and attaching "nozzles or other dust-exhausting devices at the knives of cutting machines," and providing of adjustable seats with backs to improve posture.[76] If the owners were to improve the work environment, the unions must be prepared to pay for the care of workers, both preventive health maintenance and treatment of the ill. Sick benefit funds were recommended for those already stricken by TB and "periodical examinations" for the healthy so that "all workers who apply may receive an opinion as to their physical condition and proper advice be furnished them." The unions were urged to establish a "special dispensary for garment workers, convenient to the clothing district, where special attention will be paid to the correction of ocular defects, dental prophylaxis, and diseases of the respiratory and digestive tracts."[77]

The federal recommendations furnished to the Joint Board were not supported by federal enforcement power. Educating workers, negotiating contracts with employers that required healthier work conditions, and providing unionized workers with health care, such as that organized in 1911 by the International Ladies Garment Workers' Union Health Center, were tasks that would occupy unions and their leadership for decades to come. The Schereschewsky report represents an early, if very incomplete, federal involvement in the health of the industrial workplace where millions of immigrants were transforming the size and health problems of the national work force.

Among those who thought the pace of change glacial were the three female reformers who organized the Workers' Health Bureau. Grace Burnham, Harriet Silverman, and Charlotte Todes—three college-educated Jewish female reformers with a radical perspective—sought to persuade workers that occupational health and safety were class issues that demanded a higher place on the agenda than even most labor activists would admit. Therefore, the Workers' Health Bureau, which lasted from 1921 to 1928, sought workers' control of health and safety conditions in shops and factories and offered unions health and safety data derived from the research of engineering and technical experts. The organization was so successful in attracting attention and support within the labor movement that some union leaders in the American Federation of Labor feared it as a potentially competitive organization, accusing it of fostering "dualism." Others thought its research important but membership in the bureau an expensive luxury in an era when union membership was not robust. Still, during its short life the bureau demonstrated how medical expertise could be a weapon in the class struggle that some radical unionists perceived, and, at the very least, central to improving the health and well-being of all workers.[78]

Rural workers and their health problems were often obscured by the diffuse distribution and migratory patterns of the agricultural work force, especially immigrants. Farm workers were often invisible to urban reformers, and the absence of labor unions has left little possibility of reliable statistical surveys of immigrant workers' health in the fields of rural America. An exception occurred when violation of the laws against peonage awakened foreign governments as well as officials in Washington to the plight of foreign-born workers in the American South.

Vigorous recruitment of foreign-born labor into the South began during Reconstruction but mostly had lapsed and disappeared by the depression of 1873. However, by the end of that decade, the effort resumed, spurred by the growth of the Southern textile industry. When cotton resumed its place as the centerpiece of the Southern economy, the demand for a large, steady supply of cheap labor went beyond the capacity of the ex-slaves to fulfill. Immigrant labor, especially that of southern Italians, seemed a solution. As early as the 1880s, *padrones,* or labor brokers, were coaxing workers from America's congested northern cities to do construction work on Mississippi levees. By 1885 there was an Italian agricultural settlement at Friar's Point (Coahoma County).[79]

While southern Italians were used to working on land in the heat and humidity, they faced unfamiliar health hazards. If the *padrones* promised that few would contract tuberculosis in the sunny fields of the South, they undoubtedly said little about the risk of malaria. In the low-lying swampy fields of the Mississippi Delta and elsewhere where cotton production was high, so was the probability of being bitten by the female *Anopheles* mosquito and suffering paroxysms of chills, fever, sweating, anemia, and an enlarged spleen. Malaria was chronic and relapsing; without proper care, it could be fatal.

Malaria was mentioned in the forty-two volume 1911 report of the United States Immigration Commission. After studying Italians engaged in rural pursuits in thirteen states, some 4,142 families, the commission concluded that "the rural community has had a salutary effect on the Italians, especially those from the southern provinces of Italy. It has frequently taken an ignorant, abject, unskilled dependent, foreign laborer and made of him a shrewd, self-respecting independent farmer and citizen."[80] However, in Yazoo Mississippi Delta communities, such as those at Rosedale, Greenville, Shelby, and Friar Point, or Sunnyside, Arkansas, malaria was the price paid for economic opportunity.[81] The report observed, "Aside from the various local diseases, chief among them malaria, the Italians have suffered very little from sickness. At Sunnyside malarial fever proved so disastrous during the early days that two-thirds of the colony deserted to other regions."[82] However, the report also blamed the victims: "The Italians are very careless about their health; they will not boil their drinking water, nor take the necessary care of themselves during the damp weather. The women are so busy that they give the children very little attention, leaving them to their own resources."[83] What the report downplayed, quite likely for political reasons, was the extent to which Italian immigrants in the delta were the victims of an economic disease—peonage—before they were felled by malaria.[84]

Peonage, or debt servitude, developed in the post–Civil War South as a labor system substitute for the black slaves that had been freed by the Thirteenth Amendment to the Constitution. Until 1906, most peonage cases involved former slaves or their progeny. However, when labor agents began to recruit agrarian labor among immigrants, with promises of high salaries and working conditions healthier than those in urban factories, new cases arose.

The debt into which newcomers were plunged by the cost of transportation, food, and other necessities often trapped them.[85]

A federal investigation of peonage was launched during Theodore Roosevelt's administration, led by Mary Grace Quackenbos, the first woman to serve as a United States attorney, a special assistant to Henry Stimson, then the U.S. attorney for the Second Federal District of New York.[86] The observations of federal investigators, Quackenbos, and others shed some light on the health hazards faced by Italian peons.

Long hours of labor in the hot sun and the wet conditions of the Mississippi Delta took their toll. A March 1909 report compiled by federal agent Michele Berardinelli, filed after a trip through Mississippi, noted that at an early stage of Italian arrival, "these peasants, from a healthful section of Italy were compelled to meet the problem of adaptation, and before they got used to the climate a large number of them took sick with Malaria Fever and many died."[87] Similar descriptions of illness and overwork pervade reports, though no systematic records of deaths and injuries were kept.

Moreover, there is evidence that some plantation doctors were corrupt, giving their allegiance to profits and planters rather than patients. The Immigration Commission's report commented that at the Sunnyside, Arkansas, Plantation, "The resident physician . . . declares the present death rate to be less than 1 per cent among Italians, proving that the health of the community has greatly improved in the past ten years."[88] The data are suspect. In her September 28, 1907, report on the Sunnyside Plantation, Mary Quackenbos sketched an unflattering portrait of the plantation physician, Dr. H. H. Norton, who was himself owner of a large plantation near Sunnyside. Quackenbos described him as "a young Southerner with an assured income of from $5000 to $10,000 yearly, in the lucrative plantation business; he told me that illness of the Sunny Side Tenants alone yielded a return of from $4000 to $5000 yearly." Norton's charges ranged from $2.50 to $5.50 a visit according to distance, a fee structure he defended on the grounds that all plantation doctors from Memphis to New Orleans charged $1.00 per mile. Childbirth yielded Norton $25.00; illnesses $12.00 or $15.00—amounts that Quackenbos described as "sometimes equal to a whole month's food" for an immigrant renter.[89] Norton settled his bills with the plantation company each month, with the plantation retaining 20 percent of the fee and 10 percent interest on the full amount of the bill as profit. The company also charged immigrant workers $6.00 and 10 percent interest for a "cheap filter" to make the plantation's bad water drinkable.

Quackenbos was irritated that Norton could not speak Italian and was therefore "strange to these foreigners." Equally peculiar to her seemed the small bottles of "chill tonic" that Norton regularly sold to the Italians at $0.75 per bottle plus $1.25 for the office visit. Quinine pills were extra. She concluded that "sickness is expensive in the Delta lowlands."[90]

Recommending that an Italian physician be hired to speak with his patients in their native tongue, Quackenbos argued that office visits be limited to $1.00 per visit.[91] The Italian consul in New Orleans offered to provide "free of any cost an adequate supply of the best quality of genuine quinine to be distributed *free* to the tenants," with renewals as needed.[92] Quakenbos also thought that two Catholic nuns, Sisters of Mercy, ought to be allowed residence at the plantation "to care for the sick, to teach English" and "to listen [to] and adjust all complaints coming from tenants against the Company." Perhaps taking a barb at one of Sunnyside's owners, Leroy Percy, Quackenbos wrote to him that she thought the Sisters could be supported by the tenants "when the doctor's bills, interest and commissionary charges are reduced." It would be a good investment, she predicted. The result would be "health, happiness and unity which would promote intelligence, education, and citizenship."[93]

Anxious not to be indicted and convicted on federal peonage charges, directors of the O.B. Crittenden Company, owner of Sunnyside, promised to hire such a physician and to take no profit from his fees, to give him a house, and to contribute an additional $25.00 per month toward living expenses.[94] Quackenbos reported that she had contacted the Italian ambassador to help find such a physician and that the company agreed to pay this new physician when the tenant had no money and to charge the tenant 6 percent instead of 10 percent interest on the amount.[95] A month before Quackenbos issued her report, Percy expressed his desire to comply with her recommendations in return for the "friendly interest and assistance of the Italian consul," who Percy hoped would not discourage Italian workers from seeking employment at Sunnyside. Percy described the consul's offer of free quinine as "a liberal one" and agreed to "cooperate with him by distributing it free among the tenants."[96] Throughout their correspondence with Quackenbos neither Percy nor any other member of the company acknowledged, much less expressed, concern for the physical suffering of foreign-born workers at Sunnyside.

Some Italians who managed to flee peonage headed north, but many stayed. One of these was Joseph Rocconi, who worked on the Red Leak Plantation, just south of Sunnyside. Recording his emotions in a typescript autobiography, Rocconi described life on the plantation as "only rich of toil, illness and pain." He recalled the suffering from "marsh-fever," probably malaria, which he attributed to "woods and marsh lands, with mosquitos and flies." However, not until the boll weevil infestation of 1911 and the flood of 1912 did Rocconi stop trying to succeed as a cotton farmer in the delta and move to Alabama. When he and his neighbors left, "Many went to Chicago, several to Memphis, others even to South America."[97]

Some returned to Italy. Unhappy Italian officials looked on as their compatriots arrived home in far worse physical condition than they had departed.

Among those returning in 1904, 49 of 503 immigrants had malaria, a disease that was second only to tuberculosis among returnees.[98]

Neither the efforts of Mary Grace Quackenbos nor of any other federal attorney succeeded in gaining convictions of Leroy Percy or other plantation owners who kept Italian immigrants in peonage. Percy, especially, succeeded in quashing investigations. By 1911, he had also succeeded in becoming the United States senator from Mississippi and as a member of the United States Joint Immigration Commission signed the majority report, which undoubtedly pleased him because the forty-two volume report included only seven pages on peonage, the practice of which the commission could find only "sporadic instances," leading to the conclusion that there was "no general system of peonage anywhere."[99]

Whether they were recruited abroad or after arrival for American industry or agriculture, immigrant workers encountered common obstacles on the playing field of opportunity. Dreams of success in America foundered for reasons of ill health or injury. Factory and farm were kind to neither native-born worker nor immigrant in the late nineteenth and early twentieth centuries. However, the vulnerabilities of the foreign-born made their encounters with the American economy particularly hazardous to their health. In her autobiography, Dr. Alice Hamilton documented an all too familiar pattern of what happened to newcomers stricken by ill health in the workplace. Investigating factories where workers enameled bathtubs, she observed, "Because enameling tubs was notoriously hard, hot, and dangerous, most American men shunned it and I found in Pittsburgh and the surrounding towns, in Trenton [New Jersey] and in Chicago, foreign-born workmen, Russians, Bohemians, Slovaks, Croatians, Poles." When she inquired about the workers' health, a foreman told her, "They don't last long at it. Four years at the most, I should say, then they quit and go home to the Old Country." Hamilton asked, "To die?" In a matter-of-fact voice, the foreman replied, "Well, I suppose that is about the size of it."[100]

8

"There Could Also Be Magic in Barbarian Medicine": American Nurses, Physicians, and Quacks

Most workers and their families who arrived in the United States did not go home to meekly die. Since the arrival of the Irish and Germans in the 1840s and 1850s, critics of immigration had warned that immigrant workers felled by disease or disability might remain in America unwilling or unable to care for themselves. What if immigrants, whether incapacitated by working conditions in the United States or their own innate physical or emotional insufficiencies, required health care beyond their own capacity to provide? The critics need not have fretted. New arrivals engaged in the push and pull of acculturation, even those from wholly different medical traditions, found ways of caring for their own among strangers.

In 1900, the Chinese Benevolent Association, a powerful fraternal society based on the regional origins of its immigrant members, built their own medical facility, the Tung Wah Dispensary, in San Francisco's Chinatown. This facility, incorporated into the Chinese Hospital in 1925, reconciled patient preferences and assimilationist pressures by hiring both Western-trained physicians and Chinese herbalists. Hubert Dong, born in 1907, was treated at the dispensary for acute appendicitis (perhaps circa 1918). In his memoirs, Dong recalled:

> The first serious illness descended upon the Dong family and I was the victim. The magic Chinese herbs that had cured anything from colds to broken bones did not seem to work this time. I was suffering from an acute attack of appendicitis. Luckily our father was intellectually flexible and progressive. He had been the first Chinese father to call upon a Caucasian doctor to attend the birth of his children, instead of just having a mid-wife. He was also astute enough to recognize the limitations of Chinese herbal medicine, so he called Dr. Waters, an occidental surgeon to treat me. I was being carried down the stairway at #3 Chinatown Lane on my way to the hospital, all our relatives and friends gathered around the ambulance shaking their heads with their queues swaying back and forth—quite concerned and making dire predictions that I would never return home. Hospitals and operations meant sure death to the Chinese traditionalist. But I returned a few weeks later, looking better than ever to prove that there could also be magic in barbarian medicine.[1]

Years later, Hubert Dong synthesized his cultural experience with a modern medical education and became a physician at the instituition where he had once been a patient.

The horror that Hubert Dong's neighbors expressed as he was removed to the hospital speaks to the significant function that ethnic and religious hospitals played in providing desperately needed health care to newcomers, so vulnerable and defenseless in the face of physical illness in a strange land. The Chinese were hardly alone in their suspicion of America's hospitals, where they would be out of the sight and control of family and friends. Immigrants patronized health care institutions and health care professionals with whom they could feel comfortable, where treatment and care would be offered in a familiar language and with a compassion born of empathy. By contrast, progressive reformers, especially those interested in the broader issue of public health, treated health as an important round in the bout to acculturate immigrants on terms set by the native-born, not the newcomers.

The health care field became a cultural battle among old and new Americans and, at times, even among newcomers themselves. At stake for the foreign-born were their physical well-being and cultural identities. At stake for their American-born hosts was the public health, so crucial to the growth and prosperity of a newly industrial, burgeoning America. An ever-present subtext was pluralism's proper role in shaping the culture of American medicine. The battleground was the hospital, the home, the clinic, the corner drugstore. The disputants were, in addition to immigrants and their advocates, physicians, nurses, and pharmacists.

The hospital was an important arena in this cultural tug of war precisely because even in the United States it had a problematic relationship to healing the sick. At the end of the nineteenth century and in the first decades of the twentieth, hospitals in the United States were just beginning to shed their reputation of being cold, impersonal institutions where the poor were taken to die under the care of strangers.[2] Hospital social service departments were multiplying as a dedicated corps of social workers labored to bridge the abyss between their institutions and the surrounding communities. However, at the same time, hospital stays were lengthening to accommodate the more complex medical procedures available.

For the foreign-born, the hospital remained a frightening place. A Polish physician observed that "Poles are decidedly opposed to going to the hospital under any circumstances, or to allowing their children to be taken . . . They actually believe that patients die of neglect or are killed by 'black medicine.'"[3] A Russian physician explained that among the reasons that Russians shunned hospitals was that their inability to speak English "makes them feel strange and helpless." An Italian physician reported that many Italians thought "people are abused or neglected, or killed in hospitals." The physician was confident that the "younger generation" (perhaps he meant second generation in America) was abandoning such ideas, but not fast enough to suit him. He explained that he still had to deal with many cases at home "when I know recovery would be much more certain in a hospital."[4]

The foreign-born's distrust of hospitals was an obstacle both to their own well-being and to the broader public health. Reformer Michael M. Davis, Jr., who published a comprehensive study of immigrant health in 1921 supported by the Carnegie Foundation, understood both sides.

Born in 1879 into a highly assimilated New York City Jewish family of Sephardic origin, Davis received a bachelor's degree in chemistry from Columbia University in 1900 and a Ph.D. in political economy in 1906. As a young, idealistic social worker and educator of immigrants and the poor at the People's Institute of New York, he became enamored of Progressive reform. By 1912, Davis was director of the hundred-year-old Boston Dispensary, which he turned from a professional dinosaur into a state-of-the-art model of medical administration. His intention was to integrate the social needs of patients with their medical requirements and thus to provide a more holistic perspective on the patient, one he thought few physicians appreciated and some even hoped to scuttle.

Davis was a Progressive reformer suspicious of profit seekers in the medical profession, reared as an assimilated Jew, and married to the daughter of a Unitarian minister. He was sympathetic to the immigrant poor but not completely empathetic with their desire to hold on to the past. He was Americanization's advocate as long as it was undertaken with a modicum of sensitivity to the feelings of newcomers.[5]

According to Davis, immigrants who came from large cities with hospitals were less resistant to entering an American hospital than were those to whom the entire notion of institutional care was alien. However, he found most newcomers were somewhat hesitant about hospital care. Hospital stays were long in the early twentieth century, involving several weeks of convalescence after a surgical procedure or serious illness. Patients and their families often suffered anxiety during these separations. From interviews conducted by his staff, Davis created a series of statements that summarized how newcomers felt about hospitals, ordered according to frequency and importance, as best he could determine:

> "A strange place."
> "A place in which I cannot understand what people say, nor be understood."
> "A place where doctors practice on you, especially young doctors."
> "A place where people die."
> "A place where I cannot get the food I like or am used to."
> "A place where either I have to take charity or pay more than I can afford."[6]

Davis was moved by what he heard about the problems of immigrant hospital patients. He mused, "Imagine an Italian or Pole who lies ten days or three weeks in a ward amid strange surroundings, unable to speak English. He receives food which is unfamiliar and often distasteful, however well prepared from an American point of view. Perhaps he is without friends who can talk to him even at the necessarily infrequent periods when ward visitors are allowed."[7] Davis regarded the scenario as not only "pitiable" but perhaps even detrimental to recovery. A patient whose spirits had been so dampened might be discouraged about getting well and possibly not even understand the physician's directions about care after discharge and resumption of a normal lifestyle. Davis's solution was the establishment of a more active and interventionist hospital social service department. He believed that accommodating diet and language, for example, should pose no real difficulty for a hospital. Culturally sensitive dieticians could respect Jewish and Catholic dietary restrictions.[8] Interpreters would make certain that names, addresses, and medical histories were recorded accurately, without omission. Properly comprehending a patient's complaint of physical discomfort or pain would become routine.[9]

Well before Davis's 1921 study, immigrants and their descendants tackled the problems of language and diet and religious observance by creating hospitals to care for their own. Catholics, mostly Irish or German, had been building hospitals since the middle of the nineteenth century. However, newer Catholic immigrants such as the Italians or Poles were not always welcome or comfortable in institutions run by earlier arrivals. They wanted their own hospitals to provide sick brothers and sisters with hospital facilities, catering to their distinct cultural as well as medical requirements.[10]

Some immigrant hospitals began as small institutions founded by a single individual. In the Italian neighborhood of Philadelphia, Dr. Giuseppe Fabiani established such a hospital in 1904 to care for his compatriots who had emigrated to the United States. Born in San Pietro a Maida, in the province of Catanzaro, in 1864, Fabiani had trained in medicine and surgery at the University of Naples and received his M.D. at the Scuola Superiore of Florence. After military service and civilian practice in Naples, he emigrated in 1902. After a year of residency, his credentials were accepted for licensing in New York, New Jersey, and Pennsylvania. The Fabiani Italian Hospital opened its doors two years later.[11]

Other groups also opened hospitals, often with broad-based support and financing resulting from vigorous campaigning within the community. The Chicago Polish community opened its own hospital, St. Mary of Nazareth Hospital, in 1894.[12] The case for contributions to the project was made in an article soliciting funds in the Polish newspaper *Dziennik Chicagoski:*

> While I visited a sick friend in St. Elizabeth's Hospital on Sunday, I noticed a boy about six years old, to whom the attending Sister was saying something in English. One of the patients told me that the boy was Polish, so I walked over to his bed. The child, his head bound, his face pale and eyes sunken, was sitting up and asking for something. The nurse, leaning over him with all the tenderness of a guardian angel, endeavored to understand what the child wanted, but not knowing the language she was helpless. The child was impatient and began to cry; he was suffering, evidently. I asked the Sister in German what the trouble was. She told me that the child was suffering from some sort of insanity. Then I asked the boy in Polish: "What do you want, my child?" "Water," he answered, "water," and stretched out his arms to me. . . . The child was suffering from thirst for perhaps as much as half an hour, while the Sister, who would do anything to relieve him, could not understand what he wanted and thought that he was suffering from some kind of insanity. . . . How can one place a child in an institution where no one can understand what he needs and what he is suffering?[13]

Funds for the project came from wealthy philanthropists but also from modest contributions generated by benefit performances, such as the "beautiful drama, based on the life of the common people entitled, 'The Hut Beyond the Village' (Chata Za Wsia), . . . staged by prominent Polish artists from Chicago," and "the dance after the play."[14]

The Fabiani Italian Hospital in Philadelphia, c. 1904. (Courtesy of Balch Institute for Ethnic Studies)

When Hubert Dong had his appendectomy, his Chinese neighbors may have thought that the youngster was being taken to one of the small "hospitals" that the Chung Wah Kung Saw, or Six Companies, had built for members as early as the 1860s. Each consisted of little more than several bare rooms, furnished with straw mats, where the aged or ill could await death. Such hospitals were not places to get well. Moreover, these "Halls of Great Peace," as they were named by the Chinese, operated in direct violation of San Francisco health codes. Still, local officials, while appalled by sanitary conditions in the facilities, permitted them to remain open, quite likely because they relieved municipal officials from the social and financial burden of providing the Chinese with health facilities. It was simpler to merely dodge the Chinese need for institutional care while assuring the Caucasian community that it would be protected from Asia's scourges.

During the 1870s leprosy scare, San Francisco public health officials debarred Chinese lepers from public facilities, ruling that the Chinese Six Companies should be compelled to maintain the welfare of Chinese lepers and bear the expense of sending them back to China. From August 1876 to October 1878, Chinese lepers were relegated to the small Six Companies' hospitals until nervous public health officials decided that it was in the public health in-

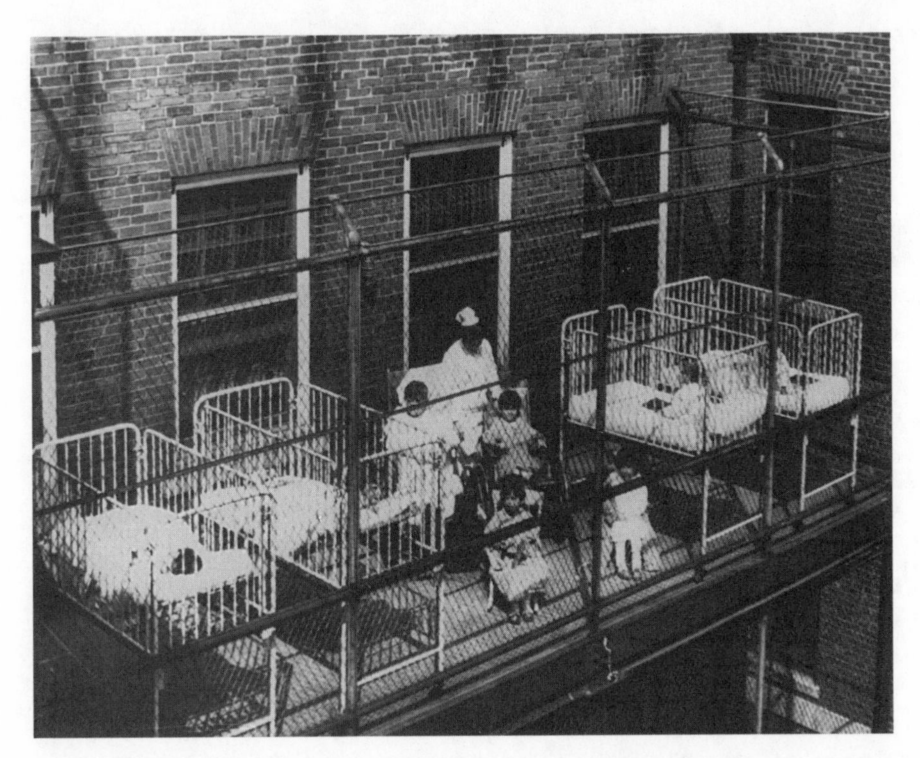

A sun porch for sick children at the Jewish Hospital of Philadelphia. (Courtesy of Balch Institute for Ethnic Studies)

terest of the larger community for the lepers to be isolated at the San Francisco City and County Hospital.[15]

By the 1890s, there had been several abortive plans for a Chinese health facility that would be more than a place to die. On one occasion, land was purchased on the southern outskirts of the city in the name of the Chinese consul general. Blueprints were drafted and fund-raising was proceeding with wealthy merchants on both sides of the Pacific. With construction imminent, municipal officials decided to forbid further work, objecting to plans to provide treatment drawing on Eastern as well as Western systems of medicine. However, it is likely that the hospital's location was as much of an obstacle as what might go on inside it. Some Caucasians regarded the facility's location on the perimeter of the city as threatening, preferring to keep the Chinese ghettoized and surrounded.[16]

Extended institutional health care, such as that prescribed for tuberculosis patients, tested both the financial resources of immigrant communities and their ability to provide an atmosphere in which therapy could be combined with cultural sensitivity. Consumptive immigrants, their lungs filled with tubercle bacilli and their pockets emptied of their earnings, often went west. By the 1890s Denver, Colorado, with its clean, cold, dry climate had already become a "Mecca for consumptives," many of them poor, Jewish, and from the cities of the East and Midwest.[17] Word of mouth and books such as *Rocky Mountain Health Resorts* (1880) argued that the city's high altitude made it a perfect place for cure-seeking consumptives.[18]

Under the leadership of Rabbi William Friedman, a committee of Denverites pledged $3,000 and incorporated the Jewish Hospital Association of Denver to build a hospital for the care of respiratory diseases that, though built by Jews, would be available to all.[19] On December 10, 1899, the hospital opened, and appropriately enough for an institution dedicated to serving all the suffering, the hospital's first patient was not even Jewish but a twenty-five-year-old Swede from Minnesota, Alberta Hansen.[20]

National Jewish was founded by assimilationist German Jews, many of whom practiced Reform Judaism. Not all those the institution hoped to serve were satisfied that the hospital was sufficiently sensitive to the spiritual needs of the growing number of eastern European Jewish TB victims. Their religious practices were orthodox and they needed care that combined therapy with opportunities for traditional religious observance. Unable to find a satisfactory synthesis of care and cultural climate at National Jewish, some of the newcomers turned to a new facility founded by others like themselves, the Jewish Consumptive Relief Society (JCRS).[21]

Twenty eastern European Jewish tradespeople—including a tinner, a furrier, a silk weaver, a tailor, a house painter, a cigar maker, an actor, and a photographer—all victims of tuberculosis, met in October 1903 to found an institution that would meet their needs. It was a humble beginning. Among themselves, the twenty could raise only $1.10 for their Denver Charity for Consumptives.[22] However, a change of name to the Denver Appeal Society for Consumptives to eliminate the word *charity* because of its condescending connotations, coupled with an advertising campaign launched in Yiddish and English, gave the idea momentum.[23] Most contributions came from workers and their labor unions. A letter of February 28, 1908, thanked the Workingmen's Circle of New York for their contribution of $3.00.[24] A similar letter written several months later expressed appreciation to the president of the Rochester, New York, Workman's Circle for a personal donation of $1.35.[25] Agents of the JCRS combed Jewish communities around the country for donations, entreating philanthropists and placing *pushkes* (collection boxes) in stores, homes, and meeting halls for whatever coins a community could spare.[26]

The guiding light of JCRS was its secretary, Dr. Charles Spivak. Spivak was born Chaim Dovid Spivakofski in 1861 in Krementshug, Russia. He was a revolutionary socialist in his youth and was forced to flee Russia in 1882 to escape being apprehended by the Russian secret police for his political activities. Arriving in New York City at the age of twenty-one, Spivak held various jobs from road paver on Fifth Avenue to mill worker in Maine to farm laborer in New Jersey, but eventually he sought an education and a medical career. In 1890, he graduated with honors from Jefferson Medical College in Philadelphia, and by 1895 he was chief of staff of gastrointestinal diseases at the Philadelphia Polyclinic. In 1896, his wife's poor health, very likely tuberculosis, caused the family to move to Denver, where Spivak continued to teach and practice at the University of Denver and Gross College of Medicine.[27]

Spivak's own immigrant background attracted him to the idea of a sanatorium that would offer care to those with traditional cultural and religious needs. The JCRS was formally dedicated on September 4, 1904; several days later, six men and one woman became the society's first patients. They were housed in wooden tent cottages. Within a year, the patient population expanded to twenty and the tents increased to fifteen. By 1917, the patient population was 146: 115 men and 31 women. Since 1904, 2,974 patients had "chased the cure" at JCRS.[28] Open to consumptives regardless of race or religion, as was National Jewish, the JCRS accepted those at all stages of the disease at no charge. JCRS catered to the special religious needs of those who practiced Orthodox Judaism, including careful observance of dietary restrictions.

The cultural tensions that existed in New York City and elsewhere between assimilated German Jews and newer eastern European arrivals were reflected in the relationship between National Jewish and the JCRS.[29] Both sides gave as good as they got. Shortly after the JCRS was founded, it was attacked by National Jewish's Rabbi Friedman for extending "an open invitation to the Jewish communities of the United States to send their penniless consumptives to Colorado" and by bringing "incurable consumptives," posing a "dangerous menace to Colorado."[30] The JCRS defended its decision to accept cases of tuberculosis in advanced stages less on medical grounds than on humanitarian grounds, suggesting that the Jewish spirit of *rachmones*, or mercy for the poor and suffering, was a higher standard than a physician's prognosis derived from passionless medical science.[31] Rabbi Friedman countered that it was more humane not to raise false hopes in those with advanced cases by bringing them to Denver. Instead, he argued, "Far better for the advanced consumptive to bear the ills he has, surrounded by family and friends, than to fly to greater suffering."[32]

An equally thorny issue was the debate over the observance of the Jewish dietary laws of *Kashrut*. This dispute cut to the heart of the differences between the highly assimilated German Jews who supported and administered National Jewish and the eastern Europeans at JCRS. National Jewish did not observe the dietary laws because it was believed that tubercular patients needed a combination of milk, meat, and butter at all meals to restore their health. Dr. Moses Collins, an administrator at National Jewish, had maintained that following the Jewish dietary laws was "inadvisable for medical reasons."[33] For many orthodox Jews, abandoning religious law was a step toward a fate worse than dying of consumption. The JCRS contended that patients need not abandon belief and tradition to recover their health. In an article entitled "Kosher Meat in Jewish Hospitals and Sanatoria," JCRS's Dr. Adolph Zederbaum argued that *Kashrut* and better health were consistent: "It is mere nonsense to claim that a sick person cannot thrive on Jewish meat and delicacies . . . at the Sanatorium of the JCRS dairy articles never appear together with meat." Moreover, he observed that by forcing patients to forsake dietary laws, more damage might be done to their health and general welfare than any positive nutritional value that might result by violating the laws.[34]

The conflict between competing tuberculosis sanatoria in Denver, each representing a different religious and cultural perspective within the Jewish community, suggests that illness not only precipitates struggles between natives and newcomers; it irritates existing tensions within ethnic groups over assimilation. Beyond that, it often widens the abyss between institutions and the individuals those institutions purport to serve.

American Catholics, too, realized that the hospitals they had built earlier in the nineteenth century to care for Irish and German patients were inadequate to meet the cultural needs of newly arriving immigrants, especially the millions from southern Italy.[35] Approximately 640 Catholic hospitals were established during the period from 1870 to 1910. Immigrants entering the United States through Ellis Island and settling in New York City could be treated at fourteen Catholic general hospitals.[36] As had been true earlier in the nineteenth century, Catholic hospitals were intended to defend against Protestant evangelization of the flock by serving as a vehicle for sustaining and communicating Catholic culture. However, given the variation of Catholic culture from one ethnic group to another, hospitals in cities such as New York and Chicago became not merely Catholic institutions but ethnic institutions as well.

Gone were the days when most Catholic hospitals were intended to be simply charity or chronic care institutions, substitutes for Protestant church–subsidized or state-run almshouses.[37] As were hospitals, generally, Catholic hospitals at the turn of the century were intended to be acute care institutions. Often, they were founded and run by orders of religious women, who supported their institutions through private and public contributions. When the flow of needy immigrants ebbed and the costs of institutional and technological modernization mounted by the late 1920s, Catholic sisterhoods began to relinquish their control over hospitals to lay and diocesan control.

Funding by a particular immigrant group gave each Catholic hospital its unique identity. Columbus Hospital in New York City offers an example of such an institution. It was founded in 1894 by the Missionary Sisters of the Sacred Heart and their mother superior general, Frances Xavier Cabrini, or Mother Cabrini as she came to be known. Cabrini and her order took over the Cristoforo Colombo Italian Hospital on 109th Street, which had fallen into debt. In September 1892 she moved her sisters and ten patients to a rented facility on the Lower East Side of Manhattan at East 12th Street.

Maria Francesca Cabrini was born in July 1850 at Sant'Angelo Lodigiano, a small town in Lombardy, twenty miles south of Milan. By 1889 Cabrini had taken vows and was headed to the United States with some of her sisters to work with the missionary institute of priests founded by Giovanni Battista Scalabrini to serve Italian immigrants. Originally intending to send his priests to Brazil, Scalabrini was asked by Archbishop Michael Corrigan of New York to provide "*good* Italian priests" for New York City's exploding Italian population.[38] The sisters started the Missionary Sisters of the Sacred Heart as a new community designed to share in the work with New York's Italian immigrant community. Providing health care in a Catholic environment was well within the broader mandate.

An Italian presence on the hospital's medical board suggests which Catholic community Columbus Hospital was intended to serve. So do the hospital's annual reports, especially data on patients and funding. In 1896 the hospital, now in a new facility on East 20th Street, had 615 patients. Of these, 434, or over 70 percent, were Italian. Financially, Columbus was very much a charitable institution, with 530 free patients and another 50 who could pay only part of their health care tab. Only 35 patients, or 5.7 percent, paid "full board" that year.[39]

Cabrini was not reticent about asking the Church and the Italian government to contribute to her mission. Although her requests were often turned down, she persisted. One of her favorite arguments was that unlike other facilities, hers was "exclusively Italian." In 1907, she requested subsidies for her immigration work from the Italian commissioner general of emigration, Leone Reynaudi. She noted that in 1905, 90 percent of Columbus Hospital's patients were Italian, explaining somewhat apologetically that non-Italians were treated only reluctantly. "The few non-Italians were emergency cases which we could not refuse, or paying patients who were admitted to help defray expenses for the free patients." Anxious to persuade Reynaudi of Columbus Hospital's attention to the Italianness of their Italian Catholic patients, Cabrini stretched the value of cultural sensitivity to mortality data well beyond what the medical board might have approved: "How efficacious was the care administered to our compatriots by our sisters, who know their eating habits and serve food according to their taste, is shown by the low mortality rate, which in the past was a little over 2 percent."[40]

Although she may have made extravagant claims for the food at Columbus

Hospital, Mother Cabrini understood some of the same truths about therapeutics that Dr. Charles Spivak knew about at the JCRS. Adequate health care of the foreign-born must include sustenance for the spirit as well as cures for the body.

The rapidly increasing eastern European Jewish population needed general hospitals as well as TB sanatoriums. Including both, over sixty Jewish hospitals opened after the first one in Cincinnati in 1850. Over half were established between 1880 and 1930.[41] But efforts to build an adequate Jewish hospital did not progress quickly or smoothly in all cities. Still, by the 1890s Boston was the only one of the fifteen largest cities in the United States that did not have a Jewish hospital. Indeed, as late as 1898, "only about 36 of Boston's 1,550 doctors were Jewish," and many were quite young. Although eastern European immigrants swelled Boston's Jewish population to almost thirty thousand by 1902, it paled beside New York's size and diversity.

Jewish immigrants in Boston had the same needs and suffered the same indignities as those in larger cities. As one recent volume by a physician observes, otherwise decent doctors who were neither conversant in Yiddish nor even especially informed as to the customs and problems of Jewish immigrants sometimes even diagnosed neurasthenic symptoms as a uniquely Jewish ailment, dubbed "Hebraic Debility."[42] Even Richard C. Cabot, a distinguished physician and friend of Michael Davis, admitted to having treated Jewish patients as invisible men and women, much as Caucasians do African-Americans in Ralph Ellison's *Invisible Man.*

> As I sit in my chair behind the desk, Abraham Cohen, of Salem Street, approaches, and sits down to tell me the tale of his sufferings; the chances are ten to one that I shall look out of my eyes and see, *not* Abraham Cohen, but a *Jew;* not the sharp clear outlines of this unique sufferer, but the vague, misty composite photograph of all the hundreds of Jews who in the past ten years have shuffled up to me with bent back and deprecating eyes, and taken their seats upon this stool to tell their story. I see a Jew,—a nervous, complaining, whimpering Jew,—with his beard upon his chest and the inevitable dirty black frock-coat flapping about his knees. I do not see *this* man at all. I merge him in the hazy background of the average Jew.[43]

Boston needed a Jewish facility. The aloofness of the non-Jewish population and later the Jewish community's own internal bickering were barriers, but eventually, even Boston built a lasting Jewish hospital—Beth Israel—which opened to its first patient on February 4, 1917.

Like Columbus and other Catholic hospitals, Jewish hospitals had a significant number of dispensary patients. Throughout the nineteenth and into the first decades of the twentieth century, the dispensary was on the frontline of medical care. However, immigrants were turning to dispensaries that were really outpatient facilities of urban hospitals, unlike the earlier freestanding institutions of the nineteenth century.[44] In their early years, dispensaries were similarly organized. There was one full-time employee, usually an apothecary or house physician who ran the facility and might perform surgery and even prescribe drugs for some patients. By 1850 there was both a resident physician and a druggist-apothecary. Some also had a physician junior to the resident who made house calls on the very ill.[45] At the turn of the century and especially in the 1920s, urban immigrant communities found freestanding dispensaries useful. However, hospital clinics had the additional advantages of being both well equipped and able to move the patient to ward care quickly if that became necessary.

Social reformer Michael Davis was a staunch advocate of clinics and an expert in their management, having been the driving force behind the modernization of the Boston Dispensary.[46] Davis and physician Andrew Warner, superintendent of Cleveland's Lakeside Hospital, coauthored a book for health care professionals. The authors made their case for the dispensary's role among the foreign-born as well as among native-born Americans on the grounds that the family physician system was breaking down under the strain of medical specialization and "the greater heterogeneity and mobility of our population."[47]

> The old neighborhood relationships have largely broken up, and this affects the close touch of a physician with a family. Furthermore, the influx of numbers of immigrants has made our population heterogeneous. The new peoples have brought few physicians of their own races with them. The young men of their race enter only slowly into the medical profession, and the immigrants have had to depend largely upon native physicians already in practice. The rise of a family physician system among such immigrant groups was hardly to be expected and has not occurred, even when the financial condition of the immigrants was such as to enable them to employ private doctors.[48]

Three years later, in his volume on immigrant health, Davis observed that while newcomers might complain about American clinics, they used them.[49]

Immigrant complaints against dispensaries were similar to their objections to hospitals. Those unaffiliated with their own ethnic group seemed distant and strange. Some feared that the doctors "practice on you."[50] A Polish woman complaining of "blurring eyes and splitting headaches" told one of Davis's associates that the dispensary physician had been helpful, but the facility was far from her home and the carfare was expensive. Because neighbors refused to come with her after the first visit, she got lost. "Then, of course, she cannot speak English and does not like to explain to everyone why she cannot pay. So she had rather stay at home and suffer."[51]

Not all immigrants shared the Polish woman's frustration and resignation.

A "Jewish doctor" said of dispensaries, "the people flock to them." Part of the reason may well have been the Jewish admiration of physicians: "They know they will receive the care of a specialist, a professor, and they are only too glad to avail themselves of the opportunity. There the mother can go into the clinic with the child and is permitted to talk and explain herself freely."[52]

Data from the Rush Medical College Dispensary in Chicago confirmed its popularity in Davis's view. (See table 8.1.) Out of 2,535 cases in a sample, 1,055 were persons from one of six immigrant groups:

Believing that immigrant feelings of alienation could be overcome with sensitivity and adjustment, Michael Davis urged dispensary administrators to act to humanize dispensaries by limiting waiting times and providing more privacy. He recommended use of signs and pamphlets in different languages, "foreign-speaking doctors," and interpreters. He urged placing skilled personnel at admission desks to make newcomers feel secure and welcome. Particularly, he believed that dispensaries should be located throughout the neighborhoods, with hospital outpatient facilities near transportation lines at central points in the city.

As did many other Progressive reformers, Davis ultimately hoped that "local dispensaries or health centers should be interrelated as to staff and administrative methods."[54] The dream of well-equipped centers that brought health care to the neighborhood was realized only briefly in cities such as Milwaukee, Cincinnati, and New York. The dissolving of distinct immigrant neighborhoods as newcomers climbed the ladder of socioeconomic mobility diminished the likelihood that health centers could remain close to those they were designed to serve.[55]

TABLE 8.1
Rush Medical College Central Free Dispensary, 1,055 Cases by Nationality[53]

Nationality	Number	Percent
Polish	343	32.5
Jewish (defined by Davis as a nationality)	333	31.6
Italian	182	17.3
Bohemian	115	10.9
Lithuanian	57	5.4
Greek	25	2.4
TOTAL	1,055	100.00

In June 1911, a sick Marya Tallarico sought medical attention at New York's Mt. Sinai Hospital. She found a place in the waiting room where she sat until she fainted. She claimed that when a physician finally attended her, he began by scolding her for marrying a non-Jew after she told him her last name and explained that her husband was Italian. In contrast, Mrs. Tallarico wrote, "The grandest and greatest thing Metropolitan [Life Insurance Company] did was send me a nurse who gave me courage and strength. For her kindness and fresh smile, I am alive and thankful."[56] Marya Tallarico was expressing her gratitude to her visiting nurse, preferring the care of this stranger, whom she came to trust and allow into her home, to the impersonal bustle of a hospital, even one run by her own kind.

The home as well as the hospital, then, became an arena where the immigrant's health care was mixed with a heavy dose of acculturation. Of all the images that have come to characterize the immigrant's milieu, few are as familiar as the famous photograph of a visiting nurse avoiding the many flights

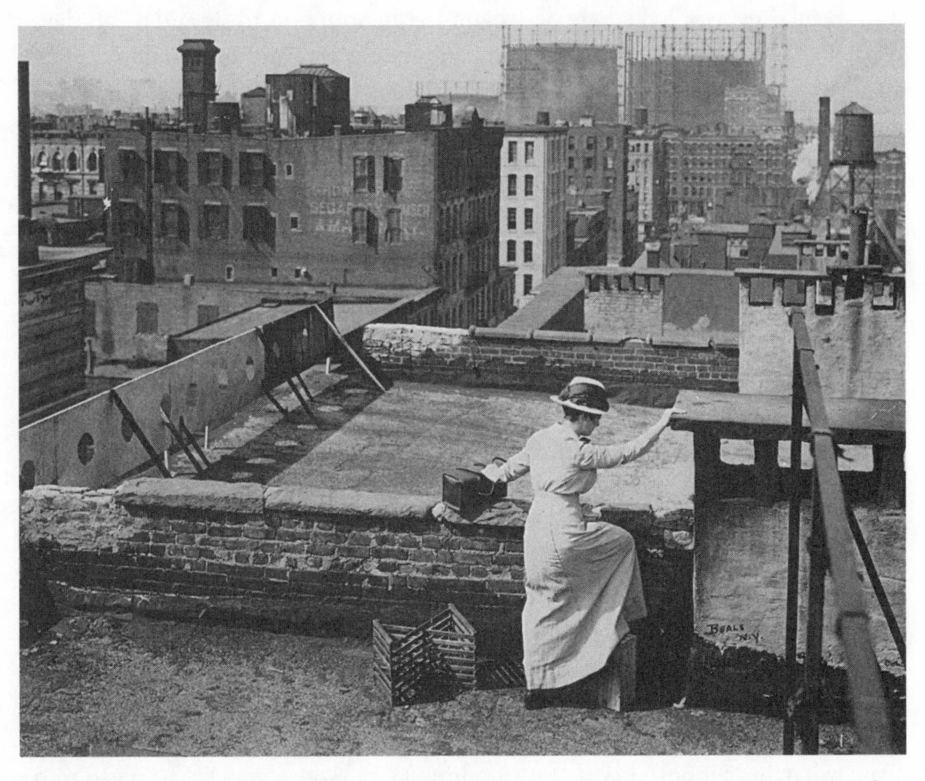

A New York visiting nurse, c. 1900, avoiding going up and down each tenement's stairs by stepping onto a milk carton, lifting her skirt, and taking a shortcut over the rooftop.(Courtesy of Visiting Nurse Service of New York)

of tenement stairs by crossing from one rooftop to another, carefully lifting her long skirts to facilitate the maneuver. Agencies providing nursing care for the poor began to spring up in various parts of the United States, but especially in northeastern cities in the late nineteenth century. There, where the high concentration of immigrants, poverty, disease, and filth was the greatest, wealthy and public-spirited women often perceived it a duty incumbent upon them to sponsor one or more nurses to care for the impoverished sick at home. These nurses were supposed to minister to the ill, but beyond that they were expected to educate the poor and foreign-born in good health habits and proper hygiene, all the while serving as a model of moral rectitude.[57]

On a more practical level, visiting nurses were cheaper than hospital care. In an era when many hospitals were charitable institutions supported by the rich, to fulfill their duties toward the poor, when hospital costs were rising, advocating visiting nurse services was a good deed at bargain prices. In 1909 New York City had fifty-eight different organizations sending 372 visiting nurses into homes across the city.[58]

The nurses, many of whom were motivated by strong feelings of altruism and social duty, worked a long day: eight to ten hours, six to six and a half days a week. They visited eight to twelve patients per day and confronted a wide variety of ailments, often infectious and highly contagious diseases, almost always complicated by poverty, and among immigrants, lack of familiarity with the language and culture of their new country.[59]

At times nursing services were fortunate enough to recruit foreign-born nurses or nurses of foreign-born parents, who spoke the language of the neighborhood to which they had been assigned. However, there was a small pool of such nurses in every city. In 1921, as the children of southern and eastern European immigrants were passing through school and pursuing careers, Michael Davis observed, "Comparatively few foreign-born girls take nurse's training, either because other vocations take less preparation and yield a quicker return, or because national customs hold personal service in lower regard than clerical work or teaching. Among nationalities where marriage is considered the only respectable profession for women, any kind of paid work is regarded as demeaning."[60] Davis also mused about whether such nurses were even an advantage. He felt that they sometimes were lacking in "conviction in their message and confidence in their pupils' ability to change." Even when such nurses really were persuaded by the health lessons they taught to newcomers, Davis fretted that their conviction was often "accompanied by contempt of, or impatience with, their compatriots." Moreover, looking at it from what he supposed was the immigrant's perspective, Davis considered that perhaps a family inclined to adopt American ways "would prefer to learn directly from a native [-born] American."[61]

For nurses who had little or no link to immigrants in their districts, language was a significant barrier. As interpreters, nurses frequently pressed into service a neighbor or child in the household being visited. Rarely did nurses have the advantage of an experienced interpreter, although they were prepared to pay for one if necessary. Always, nurses reported they were concerned that the ideas and perceptions of the interpreter would become intertwined with the patients' own description of their feelings or symptoms. Therefore, some nurses told Michael Davis that they favored a school-aged child of about fourteen because of "her schooling in this country and because she will give a more accurate interpretation than an adult who has learned the language in a general way and who is also apt to add his own interpretation and suggestions to what is said."[62] The sensitive Davis cautioned that even these ideal interpreters presented problems, if not for the nurse, for the family. Such an arrangement might well hasten the stress fractures in immigrant family life by causing parents and children to reverse roles whenever the visiting nurse crossed the threshold, an occurrence that became ever more common as an increasing number of organizations sought to do their philanthropic duty to the poor and the foreign-born by sponsoring such visits.[63]

Visiting nurses were surprisingly popular among immigrant patients. Newcomers who needed nurses managed to express their sentiments in a manner the nurses understood: "In many cases the appreciation of the patients is made evident in various little ways. A call[,] a plant or a length of knitted lace brought to the house as a gift, visits on which the new baby is brought in to show progress in growth and the small baby accomplishments so numerously evident to a mother's eye, an invitation to a christening party or a circumcision celebration."[64]

Philadelphia nurses, with support from the Bureau of Municipal Research, helped in the annual summer struggle against disease in the most congested neighborhoods of the city, the immigrant wards. Nursing Superintendent Margaret Lehmann explained with pride that her nurses helped reduce the summer infant death rate by 40 percent. It was never an easy task to "teach an ignorant mother how to care for her baby . . . especially when she has had several others, and seems to think she knows all there is to know, concerning each small thing, but the nurses never forgot that patience was the keynote to it all." Nurse Lehmann was especially proud because her nurses' summer work had "dealt largely with the foreign element, who could speak almost no English, and [among whom] ignorance, poverty and superstition reigned supreme."[65]

Nurse Lehmann's condescending language reflects the double-edged quality of visiting nursing among the immigrants. Nurses and their supervisors worked long days to save lives and battle disease among immigrants, while often holding in contempt the very individuals they labored so hard to make healthy. At times, the methods used to coax compliance from Italian mothers

A visiting nurse comforts immigrant mother as police officer aids her in removing sick baby to the hospital, 1916. (Courtesy of Center for the Study of the History of Nursing, University of Pennsylvania School of Nursing)

were of questionable taste. In 1914, the nurses cooperated with the Child Federation in its Baby Improvement Contest, being held "in the Italian quarter of the city." For nine weeks four of the nurses visited the home of each baby entered in the contest. Each would record on a specially prepared chart "the care of the infant, the housing conditions, the intelligence and cooperation of the mother." Not surprisingly in this impoverished neighborhood, mothers responded eagerly to the prospect of cash prizes, leading nurse Lehmann to comment awkwardly and in a self-serving manner, "The eagerness with which many of these Italian mothers responded to the instructions was of particular interest, and was attributed not entirely to the money prizes offered by the Child Federation, but also to the fact that the Visiting Nurses were so well known in this section that confidence in them was already established." She proudly boasted of the 2,730 visits that her nurses made in connection with the contest.[66]

The preoccupation with the immigrant's supposed ignorance of American ways and need for health education was a constant in Visiting Nurse Society reports. In 1916, a student nurse from the Philadelphia General Hospital echoed the theme in a description of her day: "Before going out I make my program for the morning. My first visit will be to Bessie Rucci, a young Italian mother whose baby boy was born yesterday. Her sister tries to care for her, but owing to her ignorance of American customs, she needs some one to teach and help her. And so the Visiting Nurse plays a great part in making the mother and baby well and strong by teaching them day by day the value of cleanliness about themselves and their homes, hygiene, the proper feeding, etc."[67]

Surely it must have occurred to nurses that many of the women they sought to "teach" had more firsthand experience with childbirth than most of the nurses, having already brought healthy children into the world. That was likely quite beside the point. Visiting nurses undoubtedly did believe that immigrants' standards of health and hygiene were well below those that they had been taught in American nursing schools. Moreover, to the extent that they could persuade others of the dire condition of those they served, they justified their existence and the philanthropy that supported their efforts. Perhaps the young Philadelphia nursing student spoke for others when she summed up her sincere, if patronizing, perspective on her routine among Italian immigrants: "So from day to day we find new symptoms, new types of humanity, we see things we did not know could exist, but I firmly believe that in time this ignorance will be overcome because of our constant, urgent teaching."[68]

Others managed to emphasize the contribution of visiting nurses made to the public health without demeaning the patients. One such episode that found its way into nursing reports was during an outbreak of typhoid.

> Tony was a little six year old Italian boy, very sick with typhoid fever. He had seven brothers and sisters, ranging in age from two to fourteen. When the nurse first visited them there were but three beds and Tony, the sick child, had two brothers sleeping with him. The nurse at once explained to the mother how dangerous this was to the other children, and helped her to clear out the first floor front room and improvise a bed with chairs and pillows until she could get a single bed, which she did within a day or two. It was remarkable how this simple Italian peasant woman could follow implicitly the nurse's directions, taking all typhoid precautions, and even was persuaded to allow the other children to have the antityphoid serum. As this happened in the summer the help of all the family was engaged to minimize the danger from flies. A screen was placed over the bed of the sick child, the windows were screened and the other children were taught the joys of "swatting the fly." The result was that our little patient was the sole victim in the large family, and he is now flourishing.[69]

It was precisely this spirit of assistance with respect and cultural sensitivity that inspired Lillian Wald to begin her nursing service in New York.

Lillian Wald, although the founder of the "nurses" settlement, later the

Henry Street Settlement, on New York's Lower East Side, did not introduce the concept of the visiting nurse in New York. But her creation of a visiting nurse service in the midst of one of the poorest immigrant neighborhoods in America, and the acclaim that her efforts to aid the poor received, brought high visibility and considerable philanthropic support to Wald and others who advocated bringing modern health care into immigrant homes. Unlike other home nursing services available to poor New Yorkers, Wald's had no institutional affiliation until 1909, when the Metropolitan Life Insurance Company subsidized some of Wald's nurses to attend to company policyholders. As Wald wrote, "We perceived that it was undesirable to condition the nurse's service upon the actual or potential connection of the patient with a religious institution or free dispensary, or to have the nurse assigned to the exclusive use of one physician, and we planned to create a service on terms most considerate of the dignity and independence of the patients."[70]

Wald and colleague Mary Brewster lived and worked as visiting nurses on the Lower East Side of New York soon after receiving their nursing degrees. Because so many of Wald's patients were newly arrived eastern European Jews, millionaire philanthropist Jacob Schiff and his mother-in-law, Betty Loeb, underwrote Wald's initial nursing efforts. Wald and Brewster jointly received $120 per month for living expenses and nursing supplies, with additional funds provided for particular projects as needed. Later, names such as Warburg and Lewisohn appear on Wald's contributor lists. Known for their generosity and their intense embarrassment at what they regarded as the primitive life-styles of Russian Jews, such German-Jewish philanthropists were only too happy to spend their money where it would save Jewish lives and hasten assimilation.[71]

Soon Wald and Brewster were joined by a cluster of dedicated nurses such as Lavinia Dock, a graduate of the Bellevue Hospital School of Nursing who had aided the injured during the Johnstown flood, served as assistant head of the Johns Hopkins School of Nursing and authored one of the first nursing textbooks, *Materia Medica*.

Aware that nurses, too, had class prejudices and cultural biases to overcome, Wald hoped they would try hard to be sensitive to those they served, reversing patterns of authority. In her autobiography, Wald explained, "The new basis of the visiting-nurse service which we thus inaugurated reacted almost immediately upon the relationship of the nurse to the patient, reversing the position the nurse had formerly held. Chagrin at having the neighbors see in her an agent whose presence proclaimed the family's poverty or its failure to give adequate care to its sick member was changed to the gratifying consciousness that her presence, in conjunction with that of the doctor, 'private' or 'Lodge,' proclaimed the family's liberality and anxiety to do everything possible for the sufferer."[72] As Wald understood intuitively, public acknowledgment of a family's poverty was "a great humiliation to people who are trying to maintain a foothold in society for themselves and their families."[73]

During their heyday, the Metropolitan Visiting Nurse Service and other visiting nurse services reached the immigrant in settlement houses, in public school clinics, and most of all in their homes. In the early 1920s, Michael Davis praised private nursing services for their effectiveness in dealing with immigrants and observed, "The most important single step made by health departments toward effective methods of work with the immigrant has been the employment of visiting nurses to do infant-welfare or tuberculosis work, school nursing, prenatal work, or general public-health nursing."[74]

Certainly, many of the immigrants who owed their health to the ladies who climbed the stairs and crossed the rooftops were aware of their debt. Although their persons might be the targets of tenement rats and dogs in the street, nurses were not menaced by those whom they served. To those on the stoops and staircases of tenements in summer's boiling heat, the broad-brimmed hat, the blue cotton dress, freshly washed and starched, and the black bag, an innovation of the Henry Street Settlement's Mary Brown, was a familiar and welcome sight. Few of the visiting nurses wore uniforms. Although some supervisors thought that uniforms would protect these unescorted women who went into the most impoverished neighborhoods, nurses argued that the appearance of a woman in uniform entering an apartment only served to stigmatize the occupants as charity cases in the eyes of their neighbors and discouraged those who needed help from summoning assistance.[75]

It was perhaps only logical that immigrants would prefer the visiting nurses to going to neighborhood dispensaries or hospitals, where they would have to wait in cramped, overcrowded waiting rooms and where the status of some as nonpaying patients put them at what they perceived to be a disadvantage in getting assistance. These newcomers may have intuitively understood that medical care was a cultural negotiation; debating matters of health and hygiene in the privacy of their own home might be the only advantage they had. One can only speculate on the role that gender played in the negotiation. Was it additionally empowering for immigrant women patients and their daughters to see the negotiation carried on at the other end by a woman? Perhaps the relationship between women immigrant patients and visiting nurses revolved around the issue of modesty, with immigrant women appreciating that the hands probing their body would not be those of a strange male physician.

Physicians, too, often made house calls. However, their ever-increasing prestige in early twentieth century American society had to be checked at the threshold of many immigrant households. There they often suffered diminished standing as compared to folk healers with whom there might be ties traditional and personal stretching across oceans and continents.[76] Moreover, while folk healers rarely demanded immediate payment in cash, often waiting for compensation until the patient was healed, physicians had a fee for services policy that seemed cold and impersonal to newcomers.

Physicians, these strangers in their midst, who arrived at an immigrant's doorstep only for acute crises, were at times treated no better than tradespeople. And as in the case of any service vendor, physicians could be fired as easily as they had been hired. If the immigrant was not satisfied with a physician's service, he or she felt little compunction about banishing one and summoning another. A Hungarian physician from Connecticut told one of Davis's field workers:

> There is no such thing as a family doctor among the Hungarians. In a case of pneumonia they might have any number of doctors even at the time of the crisis. A man will say to me, "We do not want you any more." "Why not? What is the trouble?" I would get the answer, "Oh, my wife she get worse. You no good. We get another doctor." At first this discouraged me, but I have gotten used to it now. I have seen six or seven doctors meet at a bed unexpectedly.[77]

A Polish physician from Chicago reported that in his ethnic community, calling in three or four physicians on the same case was quite common. The first doctor to be called made a diagnosis and left some medicine. However, the patient expecting the same startling results after several pills or spoonfuls that one might get from a charm or magic potion quickly lost confidence and called a second physician without notifying the first. Nor would patients pay for a house call to check up on a sick patient. "A doctor who wishes to know how a patient is getting along and makes a call without being sent for must do this at his own expense."[78] Even eastern European Jews, who honored and respected physicians, also believed them to be excessively rapacious. Lodge doctors, especially, seemed to some of their patients more interested in fees than cures.[79]

Still, whatever ambivalence they may have felt toward physicians, young Jewish immigrants and the children of the foreign-born poured into medical schools. According to one study using data from 1933–34, a decade after the end of the peak period of immigration, of 22,000 medical students enrolled at seventy-six "approved" medical schools, 4,000, or 18 percent, were Jewish.[80] Such data sent chills up the spine of assimilationist Jewish leaders such as Rabbi Morris Lazaron, who so feared an anti-Semitic backlash within the medical community that he conducted a survey to see how medical school deans felt about the influx of young Jewish students. He sent questionnaires to the deans of sixty-five medical schools. From the forty-four responses he received, he drew the somewhat problematic conclusion that "while there is limitation of Jewish enrollment [usually through quotas] there is no anti-Semitism involved in the admission of Jewish students to medical school."[81] However, the deans' letters suggest that at least some had their doubts. Dean Burton Meyers of the Indiana University School of Medicine deplored the "anti-Jewish reaction in Germany" but thought it might be echoed in the United States if "forty percent of the Doctors of Medicine were Jews."[82] Dr. C. C. Bass, dean at Tulane, said their teachers thought Jewish medical students had a "disagreeable manner and attitude in their classes," while being "less acceptable to patients with whom they come in contact in their studies," though he conceded, "Some of the most distinguished members of our Faculty are Jews. Some of our best students are Jews. Some of our most distinguished alumni are Jews."[83] Dr. W. McKim Marriott of Washington University in St. Louis saw this issue more as a problem of the marketplace and the tendency of patients to prefer the care of their own:

> The relationship between patient and physician is of such a degree of intimacy that the average person chooses as his physician a practitioner of his own race. With the Jewish physician limiting his activities almost entirely to the larger centers of population and to a great extent to Jews, the training of a large proportion of Jewish physicians fails to benefit the medical situation of the country as a whole.[84]

For still others, the "problem" was not the Jewishness of Jewish students but the manner and style of those who were themselves immigrants, the son of a "recently immigrated" Russian Jewish family who is "intensively aggressive and presents other personality difficulties," or students "who were born in Europe, or whose parents have recently migrated from Europe" and who "are apt to be an entirely different type, sometimes radical, sometimes asocial, often unstable."[85]

Nor were Jews the only immigrant group to suffer discrimination in medical school admissions. According to a 1935 report by the secretary of the Association of American Medical Colleges, "Italians are flocking into medicine at an alarmingly rapid rate." Whether they fell to quotas designed to limit the number of Catholics or to quotas specifically aimed at Italians, the children of the Mezzogiorno often found it as difficult as the children of Israel to get into Yale's School of Medicine and many other prestigious schools as well.[86]

If young immigrants suffering the slings and arrows of discrimination were having their doubts about the medical profession in America, so were those who were victimized by medical quacks. Quackery among physicians seemed to especially irritate health reformer Michael Davis, perhaps because it discouraged the trust in physicians and other aspects of American medicine that Davis was seeking to encourage among immigrants. Also, because so often the exploiters were from the same group as they were bilking, quackery struck at newcomers' most vulnerable point, their unfamiliarity with American language and custom. Davis explained, "This would-be friend [the medical quack] reaches the immigrant in his home through the foreign-language paper and like most things in this new land, is taken on faith."[87]

A 1920 Cleveland Health Survey reported that the same quack, writing ads in several papers, called to "my Rumanian brothers" or to "my sick Lithuanian brothers." In another ad he urged, "Sick Italians, don't be discouraged. Thousands of your countrymen have found health and happiness by going to see Dr. Landis." The nationality altered with the language of the paper. Sometimes the appeal stressed that the physician could address his patient in a familiar language: "You can hold a conversation with me in your own tongue . . . Here we speak Hungarian." Those charlatans in pursuit of the immigrant trade adjusted the tone of the ad to suit the temperament of the group being approached. The report noted that "the type of appeal is more vivid and dramatic in Italian, Hungarian and Polish papers, while in Swedish, Lithuanian or German, more matter of fact."[88] Publishers were criticized for accepting ads from known quacks, but, as the Cleveland study reported, newspapers depended upon quack ads for as much as 36 percent of their advertising.[89]

Regarding medical fraud as part of human nature, with fault enough for both perpetrator and victim, Davis admitted, "To be sure, 'quackery and the love of being quacked are in human nature as weeds are in our own fields,' and quacks conduct a certain amount of business among our native-born and English speaking people." When newspaper exposés curtailed the market among the native-born, medical quacks "turned their attention to the fertile field of our immigrant population."[90] Marveling that as many as 1,233 foreign-language newspapers were being published in the United States by 1918, Davis and his researchers contacted approximately a hundred newspapers for their study. They found over seven hundred medical advertisements printed in eighteen languages.

Davis bristled at the clever wording designed to avoid prosecution and the wide array of illnesses that ads promised to banish. A list of curable ills promised by quacks included:

> asthma or anaemia, backache or bad luck; colds, cancer, constipation, or catarrh; dyspepsia or drunkenness; eczema or ear trouble; fistula or fatigue; gout or general weakness; headaches or hernia; indigestion, infectious diseases, or impurity of the blood; liver complaint, lung trouble, or "any long-lasting chronic disease"; nervousness or night emissions; overwork; "special acute troubles," "special chronic troubles"; just "troubles," "too much food," teething, trachoma, tape worm; vertigo or yellow fever.[91]

There was also an abundance of obliquely worded ads promising cures to venereal diseases and others aimed at specifically female disorders, which included "dysmenorrhoea, infertility, or too frequent fertility."[92]

The largest number of quack ads, 147, were in Italian and included "all sorts, brief and lengthy, mild and lurid." But other groups, such as Greek, Hungarian, Polish, Russian, and Yiddish language newspapers, were well represented. The Davis team wrote letters to fifty-one different advertisers in the language of the advertisement to elicit further information. Thirty-four replied, usually with a typewritten form letter, sometimes with a questionnaire, often with testimonials, always with a preaddressed return envelope. The latter was for the money to pay for the remedy that would then be forwarded.

The investigators invented the Italian pseudonym Domeneco Pace and wrote to a Dr. Gibbon in San Francisco, describing no symptoms. The fictitious Signor Pace characterized himself as having been sick for a long while, having already spent much in physicians' costs and being anxious to know what it would cost for Gibbon to cure him. Gibbon simply responded, "Send me $10 (ten dollars) now for your medicine for your sickness and I will send it to you." Although Gibbons was careful to convey the cost in digits and prose, he never bothered to inquire about the illness his potent cure would drive from Pace's body.[93]

Eight respondents answered letters with booklets and pamphlets, such as material on Dr. Williams's Sanatorium showing fifty-six illustrations of individuals suffering from different kinds of cancer. There were also before and after pictures, demonstrating the effectiveness of miraculous therapies.[94] How could the foreign-born be expected to accept American medicine, consult physicians, and visit dispensaries and hospitals while they were the unwitting targets of these medical malefactors?

The Pure Food and Drug Act of 1906 had not curbed medical quackery. The federal government could only prosecute quacks engaged in interstate commerce, while most state governments had neither sufficient funding nor adequate staff to vigorously pursue prosecutions. However, the most significant obstacle to effective government intervention was the requirement that prosecutors prove defendants' intent to defraud, thorny legal underbrush not cleared until passage of the Food, Drug, and Cosmetic Act of 1938. Until then, though, investigations by reformers such as Michael Davis were essential in catching and convicting medical quacks.[95] Michael Davis and his associates were hardly alone among reformers in their concern with quackery aimed at immigrants. The Immigrants' Protective League (IPL) in Chicago, formed in 1908 by reformers connected to Hull House, also monitored medical fraud and made regular reports to the American Medical Society on North Dearborn Street. According to the IPL's files, Rev. Ivan Hundiak, editor of the Ukrainian paper *Dnipro,* regularly reported suspect physicians in his community. One such was John Smuk, whom Hundiak described in the IPL file as "not a competent doctor," with "nine degrees from different rather obscure places, but no M.D." According to the IPL investigator who spoke with Hundiak, self-proclaimed physician Smuk was "very ignorant in medicine and does more harm to the people who do not know him than good. He is only a chiropractor and a bad one at that, but practices general medicine and surgery." Smuk, himself a Ukrainian, had an office in a fancy downtown building. He did indeed engage in unlicensed surgery, operating "two or three times on some Ukrainian man [later identified as J. Stefaniuk] for tonsils and the man is not well yet." On each visit, Smuk cut away at the tonsils, bleeding his patient both literally and figuratively. When Rev. Hundiak reported Smuk, the charlatan learned who had done it and confronted his accuser, telling him to mind his own business and "That he [Hundiak] should not take his bread from him." Confident of his own popularity in the Ukrainian community and perhaps understanding how hesitant immigrants often felt about appealing to authorities in their new land, Smuk "talked to other Ukrainians about this to stir them up against Rev. Hundiak as if he [Hundiak] would be an informer." The report concluded with the pessimistic observation, "There are many people who have confidence in Smuk."[96]

Immigrants who could not afford a physician's house call or were unaware that the services of a visiting nurse might be available could yet turn to a local pharmacist. On their block, or around the corner, was an individual to whom newcomers could turn and explain their ills in a familiar language, where cherished customs and traditions were respected and observed.[97]

Local pharmacists or their employees most often spoke the language of the neighborhood, be it Chinese, Italian, Russian, Polish, or Yiddish. Moreover, pharmacists were respected as learned individuals but, unlike the physician, not so learned or removed from the community and its concerns as to be unapproachable. Although not an unqualified advocate of their influence, Michael Davis understood that these pharmacists offered a sympathetic ear, often one of a fellow immigrant. A young pharmacist born abroad but trained in America shared his perspective with one of Davis's associates:

> I can speak Russian, Polish, Yiddish and German. I know that neighborhood [in which the drugstore was to open] and the people there. I just bought out the old man who has been running that store. He has not made good, but I am sure I can. He only speaks English. He has had a Yiddish-speaking clerk, but the clerk was more interested in himself than in the business.[98]

When the field worker asked the young man whether he knew neighborhood physicians who might help him by sending patients there with prescriptions, he confidently answered that, as a pharmacist, he expected to surpass physicians in his local popularity: "I know one doctor, but the people will know me better than they know the doctors before long."[99]

Wherever immigrants lived, pharmacies to serve them soon appeared. Davis's team noted "an Italian store in Providence, a Greek in Lowell, a Finnish in Maynard, a Hungarian in Bridgeport." In smaller towns and cities, an ethnic druggist bore even more of the burden of medical care than in

A Chinese pharmacy in San Franciso's Chinatown, c. 1908. (Courtesy of Wells Fargo Archive, San Francisco)

larger cities, where there were physicians and often a keen competition among pharmacies for the local immigrant trade.

The testimony of a Hungarian pharmacist indicates that, like Chinese herbal pharmacists, he treated patients with drugs drawn from another, more familiar medical tradition. "Hungarians do not use as many patent medicines as Americans. They make more of their own brews from herbs in the fields of Hungary. Hence I carry a large stock of these."[100] However, unlike Chinese herbalists, the Hungarian seemed to be concocting remedies that drew on modern pharmaceutical understanding as well. The cultural orientation of the patient and the ailment determined the bottle on the shelf for which he might reach.

The Hungarian pharmacist complained that in the United States the "grocery and dry-goods stores carry drugs," a practice that he thought irresponsible and one which did not compare favorably with the more specialized role of the pharmacist in Hungary. He took great pride in his role as a trained medical professional, deploring the broader role that the neighborhood drugstore served in America, especially in ethnic enclaves. "Here in this country a druggist does everything: telephones, soda fountain, information bureau, doctor. In Hungary he is a skilled pharmacist." On the other hand, even this particular pharmacist was not above prescribing for minor ailments, because

"People do not like to bother with a doctor. He is a nuisance. He says not to eat this nor give the baby beans, or something else. Some medicine which the druggist says will cure you surely is so much simpler."[101]

Unfortunately, there were quacks among pharmacists, too. In the files of the Immigrants' Protective League is a translation of an article that appeared in a Bohemian newspaper about a Lithuanian, K. S. Ramashauskas, who was described as "a fake doctor and druggist," owning "a drug store or dispensary" on Chicago's South Halsted Street. In fact, Ramashauskas was neither a doctor nor a pharmacist, according to the article, but a complete charlatan who "prescribed in his dispensary carbolic acid for treatment of a shot wound which was incurred by accident by a boy of twelve years old, named Bruno Anderson. . . . As a consequence this boy died." Though Ramashauskas admitted at the inquest that he had been prosecuted forty-five times for violations of laws regulating pharmacies and dispensaries, he claimed that the boy's father had come to him and requested the carbolic acid; he denied prescribing it.[102]

Moral reformers took a special interest in venereal disease cures peddled by some drugstore quacks. The American Social Hygiene (later Health) Association (ASHA), established in 1913, led the way in investigating such practices, especially as they concerned the foreign-born. A report filed as late as 1931 by ASHA in response to the concerns of social hygiene advocates in Chicago mentioned the immigrants' vulnerability. "Frequently, persons infected with syphilis or gonorrhea, not realizing the seriousness of these diseases, apply to pharmacists for advice or a remedy to cure their condition. This is especially true among certain foreign-language and Negro groups. They take it for granted that the pharmacists can give them something that will cure all their ailments."[103] In their own limited survey, the ASHA staff found that among 270 young men "selected at random on the streets, in pool rooms, and other public places," 27 percent thought the drugstore the best place to go for venereal disease treatment, while 23 percent of those interviewed in two clinics for the treatment of syphilis and gonorrhea said they had been to a drugstore for treatment before coming to the clinic. Although the ASHA report included no data on the prevalence of venereal disease in any particular immigrant group, it called for "suitable pamphlets in English and appropriate foreign languages," concluding with a plea for education, again specifically mentioning "foreign-language and Negro groups."[104]

Issues of medical care, morality, and ethnicity were intertwined in the minds of those concerned with quackery in pharmacies. On the one hand, immigrants seemed to reformers to contribute to the problem through their naïveté. On the other, it was frequently either the foreign-born or American-born of the same group who led in the exploitation of immigrants. If not their own, whom should they trust? Native-born reformers thought the best solution for the newcomers and the public health, generally, was rapid Americanization.

And Americanize they did. In the neighborhoods where they settled, on the blocks where they lived, in the hospitals and dispensaries and pharmacies of the city, immigrants made some compromises and not others, cautiously negotiating their break from Old World definitions of health and from traditional therapeutic practices. Physicians—responsible and quack—pharmacists, hospital administrators, and visiting nurses all took part in the cultural haggling. Avaricious quacks did it for profit. For the sake of the public health as much as for the foreign-born patients before them, health care providers had no choice but to become actors in the immigrant's Americanization drama, constructing a paradigm or blueprint of health care to which the foreign-born were expected to conform but often did not—right away.

In the process of cultural disputation physicians, pharmacists, and nurses learned even as they healed. Even as they cured their patients' bodies of illness, they could not neglect the cultural attitudes held by the community to which a foreign-born patient belonged. The balance between assimilation and cultural integrity was at stake, as Hubert Dong so clearly understood when he joined the staff of the Chinese Hospital to make certain that newcomers such as his boyhood neighbors could find therapy for their bodies and respect for their beliefs all under the same roof.

9

"East Side Parents Storm the Schools": Public Schools and Public Health

In homes and hospitals and dispensaries and pharmacies, natives and newcomers jousted over the past, debating whether Old World understandings of sickness and therapy would help or harm public health in the new. In the schools, though, they fought for the future. There could be little doubt that young immigrants and the children of the newcomers were part of the educational system's captive audience. According to the 1911 United States Immigration Commission's report on thirty-seven cities across the country, almost 58 percent of all public school pupils in a sample of 1,048,490 had foreign-born fathers.[1] The proportion was even greater in cities such as New York (71.5), Boston (63.5), Chicago (67.3), and Cleveland (59.6).[2] These were the young who would either become tomorrow's robust Americans or part of an enfeebled and eventually disaffected mass. These children might, by their

Title taken from a headline in an article in the *New York Times,* June 28, 1906, 4, p. 3.

very presence, undermine the country that opened its doors to them and their parents.

Reformers, educators, physicians, and nurses, especially the Progressives, wanted immigrant youth to be cared for and educated in health and hygiene every bit as much as they wanted the new arrivals to master reading, writing, and arithmetic. They had twin motives: altruism and assimilation. Progressive reformers and their allies were often genuinely concerned with immigrant well-being and were anxious to provide opportunities to the young through education. However, such reformers were often equally persuaded that opportunities could be extended only to those newcomers willing to abandon old ways for new. Whether or not the emphasis on conformity to the white Protestant model of success made public schools as much a self-conscious grab for cultural hegemony and social control as some have suggested will undoubtedly remain a subject of caustic scholarly debate.[3] However, there can be little doubt that reformers and educators committed to the same Progressive values saw schools as engines driving assimilation forward at a rapid rate. As Prof. Irving Fisher of Yale had observed in his famous 1909 "Report on National Vitality, Its Wastes and Conservation," the schools were essential to "conserving national efficiency," which "increases economic productivity."[4]

Health and hygiene were seen as critical to the Americanization of the foreign-born, particularly those sitting in classrooms. In school, young newcomers might be pried free of Old World myths and nostrums. Exposed to the gospel of health and hygiene offered in the classroom, young converts would prosper and pose no threats to the health and vitality of the country that welcomed them, albeit cautiously. Lillian Wald, the founder of the Henry Street Settlement, believed that "Education should aim to develop good minds, good habits, good characters and good citizens, and this cannot be accomplished unless the mental training and the methods of education are directly influenced by the physical life and constitution of the individual child."[5]

William H. Maxwell, who became the first superintendent of the New York City public schools in 1898 when Greater New York was created, was especially articulate on the subject. Maxwell, after all, would know; he was an immigrant himself. A Scotch-Irish Protestant, Maxwell emigrated in 1874 at the age of twenty-two.[6] When Greater New York hired him, it placed in his hands the largest incubator of new Americans in the country. In 1898, the system included 400,000 pupils, but by 1911, it had expanded to approximately 808,000 students, over 70 percent of whom were the children of foreign-born fathers. Nor were the teachers working in the system that Maxwell supervised a sharp contrast to their boss and their students. Of 14,904 Greater New York public school teachers, 7,029, or 47.2 percent, had a foreign-born father and 1,180, 7.9 percent, were immigrants themselves.[7]

Maxwell praised the public schools' unique ability during a rising tide of immigration to be an equalizer, drawing "all social classes together in a common effort for improvement." Even more important, though, the public school created a level playing field for immigrant children and parents. Maxwell exalted the public school because "It accustoms people of different creeds and different national traditions to live together on terms of peace and mutual good will. It is the melting pot which converts the children of immigrants of all races and languages into sturdy, independent American citizens."[8]

Many immigrants were as enthusiastic about American schools as Maxwell, regarding the free American school with such gratitude and reverence that they were only too pleased to turn over their children. Not uncommon was the reaction of Russian-Jewish immigrant Mary Antin's father, who "brought his children to school as if it were an act of consecration, who regarded the teacher of the primer class with reverence."[9] However, the adulation was hardly unanimous among newcomers. Immigrants' different cultural values, predeparture experiences, and class position in the United States and their former homelands, as well as aspirations and expectations of return to the old country, all shaped their perspective on American schools and education.[10] Suspicion was equally common, sometimes more common than reverence.

Not infrequently, embattled urban school administrators, teachers, and school health care personnel felt more reviled than revered by immigrant parents, wondering whether their good intentions of helping the newcomers and protecting the United States could survive the minefield of misunderstanding that separated them from the immigrant families whose children were the prize in the tug-of-war between old world and new.

In unbearably crowded, unhealthy, and impoverished urban neighborhoods distrust often flared most quickly in warm weather, when residents met on the streets or gossiped through open windows, hoping for an entertaining story and a breath of fresh air. On the morning of June 28, 1906, there were riots on the steamy streets of the Jewish Lower East Side. Parents, most of them eastern European Jews, "Tearing their hair and talking wildly in Yiddish," according to the *New York Times,* were storming neighborhood public schools protesting the "murder" of their children by physicians.[11] A dozen public schools were besieged, with mothers in the majority, but often accompanied by "men with long beards," all "emit[ing] wails of anxiety and distress."[12]

The catalyst in the civil disturbances was swollen adenoids. A small number of Lower East Side children had developed swelling of this lymphoid tissue in the nose above the throat. The blockage often caused youngsters to breathe through their mouths, a practice that physicians of the day believed could lead to insufficient oxygen reaching the brain. Allowed to continue, mouth breathing might yield a feebleminded person, and because there was an opinion in some quarters that arriving immigrants already included a disproportionate number of feebleminded, it is not surprising that school officials sought medical help for the youngsters. Though educators often claimed to see marked intellectual progress in pupils who had adenoidectomies, an undocumented advantage to be sure, the operation had the disadvantage of brief but heavy bleeding. Rumormongers spread word of children bleeding heavily, their throats slashed by knife-wielding physicians. From there the story grew.

The riots did not occur immediately but a week after a "corps of physicians and nurses" had visited P.S. 110 at Broome and Cannon Streets, in the heart of the Lower East Side, where there were "three or four classes made up of children not physically sound." To "save the parents the trouble and expense of carrying the children up to Mount Sinai Hospital," Miss A. E. Simpson, the principal, had arranged to turn one of her rooms into a "temporary hospital." However, as the principal explained to a reporter, "No child was touched . . . without the consent of its parents." As for the children, the reporter learned that "they became distinguished in the eyes of their fellows, as Tom Sawyer was when he charged a fee for a view of his sore toe. They boasted of their ordeal."[13]

Now, a week later, the young heroes were being characterized by friends and neighbors as victims and the benefactors as accomplices in atrocities. During the disturbances, one man quite literally pulled out his hair in anguish, and embarrassed children struggled against their parents' efforts to drag them out of class. At one school, where the newspaper reporter covering the story for the *Times* felt obliged to note there was "only one Christian in the entire school, and the parents of 98 percent of the pupils speak only Yiddish," mothers and fathers gathered at the doors an hour after school opened, calling their children's names. "Puzzled [and perhaps amused] that their fathers and mothers screamed outside," the children "sat calmly at their desks and twiddled their pencils." When the principal finally admitted the parents to see that their children were safe, she described how "the mob rushed in and began to climb the stairs shrieking." Miss Rector, the principal, surmised that the parents "expected to see their sons and daughters dead and covered with blood. If a woman didn't happen to find her child at once, she set up a wail as if she had stumbled across its body." Rector wisely dismissed classes until the excitement died down, reporting that her classes had "marched out in good order."[14]

As might be imagined, P.S. 110, where the surgery had occurred, was at the center of the storm. However, Principal Simpson, whose well-intentioned efforts on behalf of her pupils' health had unraveled, was not about to let the incident pass without identifying the real culprits to the press. She recalled that four years earlier there had been a similar riot at Gouverneur Hospital in New York when parents stormed it after rumors spread that their children's eyes were being removed. In fact, the New York City Board of Health had sent school children to the hospital to have their eyes examined. Then, according to Simpson, the origins of the unrest had been "low-class doctors" who "openly complained to me that their practice was being interfered with by the city authorities." Now, she was certain, greedy, jealous physicians feeling deprived of their fees were at it again.

Simpson reported that she had been visited by at least twenty-five parents, all claiming that a strange man had come to their apartments and warned them of danger to their children. Such door-to-door agitation suggested to the *Times* reporter "a deliberate and coordinated attempt" to instigate disturbance by community physicians, who perceived school officials as endangering their professional interests by going beyond health education into health care, just as they criticized urban dispensaries for occasionally treating patients who might have been able to pay for private practitioners' services. The timing of the sieges, all beginning between 10:00 and 11:00 A.M., seemed equally "remarkable" and additionally hinted at some degree of collusion and coordination.[15] Not wishing a blanket indictment of all physicians who served this impoverished neighborhood, Simpson made a point of emphasizing, "none of the better class of physicians objects to the children's receiving hospital treatment."[16]

As the reporter wisely observed, recent reports of Russian pogroms had undoubtedly fueled fears that Jews in America were now to be subjected to terrors they thought they had left behind. He might have mentioned, too, that the episode was a peculiar reversal of the blood libels traditionally leveled against Jews in Christian countries to justify anti-Semitic atrocities.

To quell the disturbances, police reserves from eleven precincts, seventy-five men in all, were required. Although the unarmed parents were no match for the police, the anxiety level was high, as suggested by the testimony of Police Captain Byrnes of the Delancey Street Station, who "worked so hard that he had to stay in bed for an hour after it was all over."[17] Byrnes's men used slats from nearby construction sites to paddle the bottoms of rioters. Other militant protests by Jewish immigrants had left policemen with little patience or sympathy for those who were making the warm June day even hotter, such as the member of New York's finest who told a *New York Tribune* reporter, "There's goin' to be no spoilin' of these children. . . . We're not sparin' the rod any. They want riot over here all the time—bread riots, meat riots, coal riots—and now the wimmin and childer is havin' one for themselves. . . . They'll git it." And with a swing of his stick, the officer did indeed whack an "obstreperous" teenager who had gotten in range.[18]

Thanks to the cool-headed Miss Simpson and her colleagues at other Lower East Side schools, as well as the police, the episode concluded almost as suddenly as it had begun. When the children reached the streets, their parents required them to open their mouths "to see if anything usually visible was missing." Thus reassured, the parents retreated and the children returned to their desks in the afternoon. Only three arrests were made, the most serious being Max Scott's for having threatened to kill a physician leaving one of the schools. A policeman found Scott "pointing a revolver at the physician's head" while an enthusiastic crowd unsuccessfully encouraged the seventeen-year-old youngster to shoot.[19]

The arrests aside, the episode only slightly disrupted school schedules. The morning's instruction was lost and commencement exercises scheduled for the afternoon were postponed until evening, when students received their diplomas under the watchful eyes of police patrolmen. At one such school, P.S. 62 at Essex and Division Streets, parents paid little attention to the police. They were engrossed in a school production of the court scene from Shakespeare's *Merchant of Venice.* The *Times* reported that although many had to be "turned away at the door" because of limited seating at the commencement, "there was no sign of panic."[20]

Unfortunately for school officials and police, what upset Yiddish-speaking parents did not play well in Italian either. The next morning as the Jewish Lower East Side went about its business peacefully, three public schools "were forced by a howling surging mob of women to let their children out . . . long before the regular hour for recess."[21]

The origins of the rumors that "maddened the mothers who come from Sicily and Calabria" appeared to have been the previous day's newspaper reports. However, as the mothers gathered at the schools, "One woman confided her suspicions to another, suspicions grew into fears, and fears into agonized certainty." Commencement ceremonies were disrupted and a few children had even come armed. They carried small bags of salt in their pockets, and "any doctor who approached them was to have salt thrown in his eyes and then be stabbed with knives which they [the children] had brought hidden in their blouses." One principal, Miss Underhill, reported that while there was chaos in the streets below, "the children upstairs were saying their pieces and singing their songs, utterly unmindful of their agonized parents in the streets." However, by eleven o'clock even the patient Miss Underhill "decided she had had enough" and released the children on an extended recess." When the youngsters marched out, they were "kissed and hugged or slapped, according to parental temperament," reported the *Times.*[22]

In the Italian neighborhood, too, there was evidence of outside agitation. Near Public School 106, a man was seen haranguing Italian women on a street corner, telling them that while such operations generally were not harmful, six children that he knew of had died from such surgery. Later, an ambulance passed on its way to treat a case of heat prostration, but the women suspected that the bodies of children were now being removed "by the cartload" and they stormed the school. The school janitor, looking on, claimed to have been menaced by a man with a stiletto, but "Luckily one of his friends had come up and threatened the Italian with an iron bar," allowing the janitor to escape. Meanwhile, the teachers inside the school "piled up against the door and with a long, steady heave managed to slam it to." Firemen and mounted policemen arrived, but nothing calmed the mob until someone noticed that their children were peacefully filing out through a side door.

The scenes at other area schools were similar—women storming schools brandishing knives and stones. Frank Mulvanna, a school janitor at another of the affected schools, was jabbed in the body by a stiletto-wielding mother demanding her child; he managed to disarm her and force her out the door. Eventually police reserves restored order.

In the Jewish quarter of the Lower East Side, it was as if nothing had ever happened. Children played in the broiling streets as their perspiration-soaked parents bustled about trying to eke out a living; pushcart peddlers and housewives bargained and exchanged gossip in Yiddish, Russian, Polish, German, and English. Still teachers at the various schools expressed relief that the summer holiday had come so the "madness" could be forgotten.

As the Lower East Side riots suggest, the school doctor's or nurse's office was oftentimes a cultural battlefield where newcomers, school officials, social reformers, and health care workers wrestled with each other over what was best for immigrant children, best for the public health, best for the United States. Public school officials and their allies among Progressive reformers, including many physicians, saw the relationship between healthy bodies and sound minds as best pursued in the public schools. They intervened, if at times somewhat self-righteously, as the child's advocate but also assuming the mantle of the public's best defense against dangerous germs bred among the foreign-born. To immigrants, the provision of medical care to public school students—their children—often seemed an unwelcome intrusion rather than an altruistic act that could make children healthier and increase opportunities for success. Many could not comprehend how their children's health or personal hygiene was the business of American school teachers. Local physicians, meanwhile, were neither all simply heroes nor villains. Some aligned themselves with reformers, but others resented the medical care being offered to the school children, condemning it as a "policy of Socialism" that undermined their authority and income.[23]

By the time that ferries began shuttling back and forth to Ellis Island in 1892, the concept of compulsory education for all children at the public's expense was still only slightly more than a half-century old, but already public schools were the focus of debate over whether or not health care as well as health education was a legitimate aspect of education's mission. Almost from the very beginning public schools had served a public health function. Concentrations of students in a disciplined environment made schools useful in vaccination programs. In 1850 Massachusetts began the practice, using schools as the venue for mass vaccination against smallpox. However, compulsory smallpox vaccination programs for public schools were controversial. Some immigrants objected to being jabbed for reasons they neither understood nor accepted. More significant, though, was opposition from some school administrators, who approved of vaccination but objected to programs imposed by public health officials that might undermine the confidence of suspicious immigrant parents, curbing attendance, or, at the very least, disrupting school routine with nonacademic interventions. In New York City, Board of Health officials were able to vaccinate 17,505 children in the schools during the 1875 school year, but only after pitch battles with school principals. In other cities and states, court decisions were required to uphold Board of Health decisions to exclude the unvaccinated from public school classrooms. By the late 1890s, most urban health departments were either overseeing vaccination in schools or insisting that school boards include it among their duties.[24]

Vaccination at school was a precursor to school medical inspections. At first, such inspections were designed to make schools safe environments for youngsters, free of danger and disease. However, by the turn of the century, in part because of the arrival of millions of immigrant children from a diversity of cultural backgrounds and health environments, sentiment was growing among educators and reformers to expand from the inspection of school buildings to the medical inspection of students and increasing health education. Soon after, though, the debate moved to medical care. Was identifying a sick or disabled child the end of public responsibility, or was therapy at public expense a logical extension of medical inspection? Should routine medical examinations at school be undertaken to promote preventive care and establish standards of wellness for the schoolchild? And if students in need of care could not get it at school, what other agencies in the community might provide it?

Buildings, not bodies, were the objects of the first school inspectors. Too often schools were sufficiently unsafe and unsanitary that students were being asked to improve their minds at the risk of their health. Because spending for education was usually a local responsibility in the nineteenth century, one that drew minimal funding, school buildings were dirty and dilapidated, with too few toilets serving too many students. Vermin-infested barns, stores, and abandoned family homes were divided into classrooms, an economical way to avoid building structures especially designed for educating the young.[25] According to public health historian John Duffy, as late as the 1860s, in almost 50 percent of the nations public schools, boys urinated and defecated in the open air on one side of the building and girls on the other. Few doubted that such conditions contributed to making school environments incubators of disease.

Change came first to cities in the Northeast. By 1888 the Massachusetts legislature had passed a far-reaching sanitary code for school buildings that quickly became the model for other states. A year earlier New York required separate privies for male and female students.[26] There was a growing awareness that the schoolhouse must not remain off the board of health's map if the cause of public health was to be responsibly served.[27] The school environment was within the domain of public responsibility, but the health of the individual school child? That was more problematical for Americans suspicious of state intervention.

Europeans preceded their American cousins in the matter of school medical inspections of students. In France, an 1833 law charged city school committees with keeping schools clean, and an 1837 royal ordinance made it the special duty of maternal school (kindergarten) supervisors to concern themselves with child health. By the early 1840s Paris ordered that every school in the city be visited by a physician to inspect the school environment and the general health of the school children. Although the physicians were unpaid volunteers, the precedent was set that the health of the school child was a community responsibility, rather than just a parental prerogative.

In the United States, medical inspection to ascertain the health of the individual schoolchild began in 1894 in Boston, although two years earlier Dr. Cyrus Edson, New York City's sanitary superintendent, had appointed Dr. Moreau Moore medical inspector of schools.[28] In Boston, the Board of Health selected 50 physicians, divided the city into 50 school districts, and began inspecting. By 1897 New York's program was also operating; the New York Board of Health appointed 134 medical inspectors to the schools, headed by a Dr. A. Blauvelt, formerly assistant chief of the Bureau of Contagious Diseases, who assumed the post of chief inspector at an annual salary of $2,500. In the meantime, in 1895 Chicago was divided into nine districts with one medical inspector assigned to each district, giving each inspector responsibility for more than twenty square miles. Clearly, cities were progressing and watching one another carefully for hints at how best to meet what they now perceived as a public responsibility.

The inspection in various cities began as a hunt for victims of contagious diseases among school children. However, there was a gradual movement toward more thorough physical examinations revealing a broader range of illnesses and disabilities. It was not long before state legislatures, aware of the trend at the municipal level, began to enact legislation providing for this broader approach to health inspection. By 1899, the Connecticut legislature passed legislation mandating eyesight tests in all the state's public schools. The State Board of Education was now required to develop test cards and forms and instructions for their proper use to school authorities who would supply them to schools and teachers. In every school students were required to read the eye chart during the fall semester. Those who strained and squinted received a note to their parent or guardian briefly stating the defect. Arranging for further examinations or acquiring eyeglasses was left to the guardian. Clearly, legislatures understood that they were treading sensitive turf and remained wary of the conflict that overzealousness could generate among parents and physicians or local medical societies, who did not appreciate the competition with their services. Limiting school responsibility to identifying noncontagious ailments and waging treatment created trade for private practitioners, curbing their opposition and school inspection.

By 1910, then, the United States was no longer behind the rest of the world in providing health inspection for school children. However, the increasing population of many schools, especially in immigrant enclaves of large cities, left medical inspectors needing to be both expeditious and thorough, a dilemma not unlike that facing Public Health Service physicians on Ellis Island. In cities around the country, school children now had an annual examination similar to the one they or their parents stood upon entering the country. Forms issued by the New York City Health Department described the "routine inspection":

> At the beginning of each term the Medical Inspector makes a routine examination of each child in the schools in his charge. The eyelids, throat, skin, and hair of each pupil are examined. The inspector is not allowed to touch the child, but the latter is required to pull down its own eyelids, open its mouth, show its hands, and, in the case of girls, lift up her back hair. Individual tongue depressors are furnished by the Department.[29]

New, unprecedented responsibilities for the schools and the Progressive preference for rationality and efficiency gave birth to forms and procedures. As were the federal physicians on Ellis Island, school medical inspectors were part doctor, part bureaucrat. All cases of disease were recorded on preprinted index cards "with the proper data in appropriate columns," while code numbers were always used to designate the disease. Children requiring further examination were seen separately by the inspector at his office, much as newly arrived immigrants had been detained and examined in special rooms adjoining Ellis Island's great hall or in the island's hospital. Moreover, those found to have contagious diseases were removed from their classes and sent home with an "exclusion notice to the parents," while several other forms were scribbled—a record for school authorities, a record for the board of health, and a record for the school inspector. Documentation from the board of health or, at the very least, a physician had to be presented prior to a child's readmission.[30]

Children sometimes recalled being poked or prodded by physicians but little more, unless they were excluded from classes because of a communicable disease. However, school medical inspectors' memories of the encounters are rich in insight, especially the recollections of S. Josephine Baker. Her rough-and-tumble encounter with Typhoid Mary as a young public health physician in New York City did not dissuade Baker from a career in public health. That career in public medicine included a stint as a school inspector from 1907 to 1908. She was then asked to organize and administer the Bureau of Child Hygiene, which she did from 1908 to 1923, serving also as a consultant to the federal government's Children's Bureau, launched in 1912.

Baker recalled in her autobiography, "When I entered the room the children all arranged themselves in a line and passed before me in solemn procession, each child stopping for a moment, opening his mouth hideously wide and pulling down his lower eyelids with his fingers." This line inspection gave Baker and her colleagues "an opportunity of looking at the child's hands (skin diseases), his teeth, throat and eyes all at the same time." The astute young physician was not unaware, though, that "For the children's purposes it was a beautiful opportunity for making a face at teacher unscathed and they made the most of it."[31] In words very similar to those of Ellis Island physicians who believed themselves masters of the "snap-shot" diagnosis, Baker boasted, "After a year or two at this sort of inspection, I reached the place where I could pick out the undernourished children and those with other obvious physical defects almost as soon as the door was opened and before the children passed before me."[32]

Baker was appalled at the magnitude of the health problems that school inspection detected. There seemed to be so many cases of communicable disease that she feared depopulating the schools if all children with such ailments were excluded from classes. "Four out of five of these children, about 80 per cent, had pediculosis—the polite medical term for head-lice. One out of five, some 20 per cent, had trachoma, that highly infectious eye-disease which constitutes so serious a risk of blindness that the immigration authorities no longer allow any case of it to enter the country. . . . The infectious skin diseases—scabies, ringworm and impetigo—were almost as frequent. Those were not schoolrooms we inspected; they were contagious disease wards with all the different diseases so mingled it was a wonder that each child did not have them all. Many of them did: lice, trachoma, scabies, ringworm, all at once."[33] From the opening of school on September 15, 1902, to April 1, 1903, five and a half months, 5,381,616 examinations were completed, resulting in 57,986 exclusions. Exclusions for the quarter ending December 31, 1902, included measles, 18; diphtheria, 140; scarlet fever, 13; whooping cough, 61; mumps, 9; trachoma, 12,647; pediculosis, 8,994; chicken pox, 172; skin diseases, 662; miscellaneous, 1,823; total, 24,538.[34]

It appeared to Dr. Baker that the high percentage of immigrants and children of immigrants in New York's schools made an already serious public health problem more acute. She tried to make the newcomers comprehend why their children were being asked to leave the company of their schoolmates. "The children's parents were painfully astonished at finding that Giuseppe and Isador and Sonya were being kept out of school because of inconspicuous troubles to which no one had ever paid any attention in the old country." Frustrations with the dimensions of the problem made even the sympathetic Baker sound like a nativist:

> Our present rigid restriction of immigration had not yet begun, physical examination of immigrants was hardly yet existent, and the famous melting-pot of Manhattan Island had long since become a huge germ culture. The Mediterranean and Balkan slums and southern Russia were emptying their underprivileged into Ellis Island and, on reaching their new country, these underprivileged were likely to be a definite medical and public-health problem. No one had told them what to do about these seemingly minor ailments and they had never known themselves that anything could or should be done.[35]

However frustrated Baker remained with the problems she had encountered years earlier, she was still capable of recalling the almost farcical bureaucratic bungling occasioned by the absence of coordination between school truant officers and public health officials who had ordered the exclusion of children with communicable diseases. "It was a thoroughly insane situation. Not the least ridiculous detail of it appeared when the truant-officers, finding the schools emptied of pupils, began going around and ordering these children back into school. Here was one city department prohibiting the children from attending school and another city department commanding the parents to send them to school."[36] Baker speculated that by this time most immigrant parents must have been convinced that "all government officials in this new country were crazy." Clearly Baker thought them close to correct.

Thinking back upon those years of coping with sickness among the immigrant children of New York's poorer neighborhoods, Dr. Baker paid tribute to two other women known for their caring and expertise in matters of immigrant child health, Lillian Wald and Lina Rogers. Wald, the much beloved and widely known reformer, offered a solution to the problems revealed by school health inspectors such as Baker—the services of nurses. Wald lent the New York Health Department one of her most qualified nurses, Lina Rogers, who was appointed the first school nurse on November 6, 1902, and issued a gold-toned badge of office.[37]

Nurse Rogers, whom Baker recalled as "a dignified, attractive person who exuded capability and adaptability and all of the other required qualities," began her duties in a Lower East Side school filled with child health problems to see whether or not the presence of a nurse on site might make a difference. It did almost immediately, and a month later a corps of eleven more school nurses was appointed at salaries of seventy-five dollars per month.[38]

Rogers began with an assault on minor infections such as pediculosis, which required identifying the children with head lice and then educating whole families in the standard treatment, which involved a shampoo, a thorough combing with a fine-tooth comb, and then soaking the hair in kerosene to kill the nits. Baker recalled Rogers lining up the children for inspection: "I can still see the lines of little girls with their pigtails pulled forward over their eyes so that the nurse could look through their hair." And then a confession only amusing in retrospect: "All nurses periodically acquired lice themselves. Even I, who seldom had to do that work of inspection, have not avoided that infection on several occasions."[39]

Rogers and others experimented with different techniques of keeping skin infections and other such ailments from spreading throughout a school. She established a school clinic, supplied with simple remedies, and had infected children report each day for treatment. Sometimes there had to be special classes of children who shared the same infection, as in the case of trachoma. Children with trachoma sat in classes with other victims of the disease. They did their school work but "were not allowed any contact with the other children in the schools."[40] In serious cases, the children were sent to clinics for treatment by physicians and where surgery could be performed, if necessary. Lina Rogers's efforts left little doubt of the school nurses's public health value.

The public school teacher still bore the main burden of vigilance between annual inspections, alerting medical inspectors to suspicious signs of ill health in the student population. However, increasingly the school nurse played a crucial role not only in disease detection but also in health education, including the parents in the child's health care. One Philadelphia physician characterized the school health nurses of his city as "invaluable," especially in the "foreign, poverty stricken sections."[41] School nurses treated minor ailments in school, thus preventing loss of instructional time. Even more important, the nurse became the conduit between school and parents. Overworked parents, especially immigrant parents with scant knowledge of English, often were unaware of the seriousness of a child's illness or that a child had been excluded from school for health reasons because the parents could not read the exclusion notice sent home with the child. The school nurse visited the home and conferred with the parents, often securing treatment for the child by a family physician or a physician employed by a charitable institution, rather than permitting the child to continue unable to regain health and rejoin classmates. Because of the school nurses, New York school administrators reduced the number of excludable diseases from ten to five, requiring only those with measles, diphtheria, scarlet fever, mumps, and chicken pox to remain at home. For the school quarter ending in December 1902, 24,538 children were excluded. One estimate suggested that had the nurse service been in operation, only 460 children would have had to be excluded for health reasons. With the care of nurses, 24,138 students would have been allowed to continue classes.[42] Those with diseases such as mild cases of trachoma, which nurses were not allowed to treat on their own, were permitted to remain in school and report to the nurse each week to present a physician's certificate, properly dated, as evidence that the child was under continuous treatment.

Other cities were quick to follow New York's example in hiring school nurses, who were increasingly regarded as the "most efficient possible link between the school and the home," a provider of health care but also "the teacher of the parents, the pupils, the teachers and the family in applied practical hygiene." Those apprehensive about the health menace of immigrants celebrated the school nurse nevertheless for their own reasons: "Among foreign populations she [the school nurse] is a very potent force for Americanization."[43]

School physicians and nurses, many of them by background and experience sympathetic to immigrants, often faced obstacles as intransigent as disease among newcomers—culture. Dr. Jacob Sobel, borough chief of the New

A school nurse visiting a home to instruct student and parents in American ways of health and hygiene, c. 1900. (National Library of Medicine)

York Health Department's Bureau of Child Hygiene, reported to the Fourth International Congress on School Hygiene in 1913 on the difficulties of traversing the cultural divide. The culture and class of immigrant parents caused them to be uncooperative with health care professionals who shared their concern for their schoolchild's welfare but not their world views. Too frequently, the result was what seemed to Sobel "a fusillade of prejudice, tradition, and superstition, ignorance, distrust, apprehension, indifference, irresponsibility, poverty and antagonism."[44]

Sobel entertainingly invited his listeners "into the heart of the tenement district" to "listen to the answers which are given to the inspectors and nurses in response to their plea for early and proper treatment of the physical defects found in the school children and to their advice on prevention of contagion, on child diet, child care, hygiene and sanitation."[45] According to Sobel, physicians and nurses who determined that a child needed glasses often heard from the parents that eyeglasses "are a luxury, that they are worn for style, that they make the child look old" or "that the child will get used to them" and then will never be able to get along without them, thereby making the eyes weaker still. Jewish mothers, the doctor charged, were fearful that eyeglasses would interfere with a daughter's matrimonial prospects, "that she won't marry well," or "as one mother told me [Sobel] of her ten year old, 'if she needs glasses let her husband buy them.'" Sobel was mildly amused by some of what he heard, but much less so when in the same Lower East Side neighborhood he saw mothers attempting to cure "mucopurulent conjunctivitis" by "literally squirting breast milk into the eyes." Exclaimed Sobel, "What a woeful waste of nutriment! What an ideal culture medium for bacteria!" Equally appalling to him were the Bohemian mothers who treated ulcer of the eyeball by "dusting sugar into the child's eyes."[46] As for the ears, Sobel believed that many children had lost their hearing because their parents mistakenly thought that "running ears" like "running sores" allowed the poison to escape.

Tonsils, like the adenoids that had sparked the Lower East Side seven years earlier, also inspired misconceptions and apprehensions among immigrant parents. Said Sobel, "One of Ireland's buxom daughters, when told of the enlargement of her child's tonsils stated defiantly, 'Is that so? Sure God put them there and there they'll stay.' And in many cases they do—while you make a hasty exit."[47] Meanwhile, "The mother of the Jewish ghetto" offers as her excuse for opposing a needed tonsillectomy: "If the tonsils are taken out, the throat will be too wide and air will rush into the lungs too quickly and produce inflammation of the chest."[48]

In spite of the frustrations, Dr. Sobel concluded his presentation on an optimistic note, calling for more and better education and observing that an increasing number of immigrant parents were coming to understand not only the importance of medically treating disease and defects but doing it even before the child entered school, when possible. No longer did the word *operation* "carry with it the terror of years ago" for these parents of young Americans. Even care of the teeth had improved, although the hygiene was still imperfect: "Now we see the tooth brush and powder in many homes, albeit that at times one brush is called upon to do service for the entire family."[49] Sobel also placed his confidence in the children learning health and hygiene in school who would come home and be their parents' teachers. Demonstrating the kind of economic and cultural sensitivity all too rare, Sobel suggested that school health care workers not only emphasize to parents how proper correction of defects could later increase the child's "future wage-earning capacity," but engage in "a plain heart to heart talk in their native tongue."[50]

Medical examinations by physicians and the watchfulness of teachers and school nurses brought results. Contagious diseases were detected among the young that posed a danger to them, their classmates, and the larger public. Students in New York, Philadelphia, Detroit, Newark, and other cities were being excluded from classes because they were found to have such diseases as scarlet fever, diphtheria, measles, mumps, chicken pox, whooping cough, pediculosis, smallpox, ringworm, scabies, tuberculosis, trachoma, favus, acute conjunctivitis, and croup. Later, in the 1920s, public health professionals in New York and other cities made public schools battlegrounds for an all-out assault on diphtheria. In March, 1920, New York public health bacteriologist Dr. William H. Park approved giving the test developed by Dr. Bela Schick to children in 100 schools. A comprehensive program of tests and vaccinations worked and the number of cases declined. After a spike in diphtheria cases across New York City in 1927–28, a newly appointed Diphtheria Prevention Commission again turned to the schools. By the end of 1929, 292,000 children had been immunized. The death rate from diphtheria plummeted from 9.54 per 100,000 the previous year to 6.75.[51] Moreover, examinations by competent physicians revealed a large number of children in need of care for noncommunicable physical defects. Some, perhaps even most, had been produced by unhealthy home lives, especially among the poor, while others were either contracted at school or exacerbated by school conditions. Remaining were the problems of caring for those in need and preventive health care, preserving the well-being of those who were already healthy.

Aside from particular ailments or disabilities, poor diet, improper personal hygiene, and insufficient healthy exercise were all among the most common causes of disease and disability in immigrant school children. Many turn-of-the-century reformers mentioned the problem of hunger as it affected children, none more movingly than John Spargo in his 1906 volume, *The Bitter Cry of the Children,* in which he treated "the tragedy and folly of attempting to educate the hungry, ill-fed school child."[52] Praising the pedagogical advances in public schools, Spargo lamented the "nervous, irritable, half-ill children to be found in such large numbers in our public schools," as "poor material" for the fine public education that an enlightened society was making increasingly available. Spargo deplored the "physical and mental weakness and inefficiency" that was born of children going breakfastless to school. Differences over how many thousands of children went to school undernourished led Spargo to conduct his own study. He deplored the careless methods of board of education officials, whom he believed were trying to downplay the problem.

Embarrassed by their poverty, many children lied about their access to breakfast. A frustrated Spargo wrote, "The principal of a large school on the West Side reported that 'after careful inquiries' he had found only one little girl who came to school without breakfast, and she did so from choice, saying, 'Because I never used to have any breakfast in Germany, sir, and didn't want any.'"[53] Upon further investigation, Spargo discovered that the child's father was a dock worker and her mother kitchen help in a "cheap restaurant." Both parents left for work each morning before the child arose, leaving her only "some cold coffee and bread." The family owned no clock, so the child often arose and left for school without eating even that, fearing being late. Spargo recounted, "On the day of my interview with her she had spent her five cents for a cup of coffee with nothing at all to eat, as she had done for two or three successive days." Asked why she had purchased coffee, the child replied, "Because coffee is hot, sir, and I was so cold." When the father returned from work, he sent the child to a delicatessen for bologna sausage for their evening meal. The mother ate at the restaurant and returned home later. Essentially the child was fasting from the evening of one day to the evening of the next. To Spargo's compassionate eyes, what the children did was far more important than what they said. "There were also two boys, Syrians, who said that they had three meals each day but could never get enough to eat."[54] His own investigation and that of others in New York, Philadelphia, Buffalo, and Chicago revealed that of a total of 40,746 children examined, 14,121, or 34.65 percent, either went breakfastless or came to their studies with only tea or coffee in their stomachs.[55] John Spargo, Robert Hunter, Lillian Wald, and others concerned with the welfare of children wanted change. The public school's role gradually expanded to participate in this broader approach to the education of children.[56]

In response to the issue of hunger, some reformers advocated a program of low cost or even free school lunches for poor children. With many mothers out of the house, away at work, there was no one to ensure that children ate wholesome meals. Some were given a few coins and told to buy a midday meal. Proponents of school lunches maintained that too frequently the children spent their money on ice cream, although many immigrant children preferred pickles. Aside from the limited nutritional value of pickles, there were those among school officials who believed that craving for pickles suggested an addictive response to stimulants and might be indicative of incipient alcoholism.[57] In addition to such foolishness based upon no medical evidence, moralists among the reformers worried that children left to make their own decisions about lunch might gamble their money away in games of chance, while still others argued that keeping youngsters in school at lunchtime was an opportunity to teach cleanliness and manners.[58]

Provision of school lunches at cost had begun in English settlement houses. Philadelphia and New York had such programs by 1908. In New York the lunch program was spearheaded by reformer Mabel Hyde Kitteridge. Moved by her experiences volunteering at a church on the Italian East Side, Kitteridge had organized the Practical Housekeeping Center in 1901 on Henry Street, working closely with Lillian Wald, to teach the poor but especially immigrant women American techniques of housekeeping, including health and nutrition.

Beginning with lunches for students at vacation schools at P.S. 51, Kitteridge was soon asked to expand her experiment, which she did, creating the School Lunch Committee. Most of the Committee's work was confined to the immigrant enclaves of lower Manhattan, but by 1914, it was responsible for lunches at seventeen schools, eleven of them in immigrant neighborhoods, having a student enrollment of 24,087. Not all the youngsters ate lunch at school in spite of the low prices. At first, there were three lunch prices, at five cents, three cents, and one cent, respectively, but jealousies among the children over who could afford which meal caused the committee to eliminate all but the three-cent lunch. Ultimately, the committee went to a one-cent-an-item scheme but required children buying lunch to spend their first penny on soup and two slices of bread. Those fortunate enough to have more money in their pockets could spend it on sweets with some nutritional value, such as baked apples or custard. Cooking was done at a central location and the food was transported to schools, where some of the older children served it in return for free lunches, although the committee stationed paid supervisors at each school. School lunchrooms, where they existed, were neither equipped to serve meals nor spacious enough to handle the increased number of students eating in the building. Therefore, portable counters were assembled each morning and the youngsters ate standing up. The food suggested that Kitteridge and others were not insensitive to the cultural diversity of the students. At one school with a large Jewish population, meals were kosher; at another where the Italians predominated, meatless days were observed out of respect for Catholic belief.[59]

The program was popular, especially among school principals who testified that the well-fed children were better able to concentrate and that complaints of stomachaches had decreased, while student comportment at mealtimes was increasingly civilized. Still, municipal government refused to support the School Lunch Committee or to take over the program. Reformers and educators rallied to the cause. New York School Superintendent Maxwell went on record in his annual reports, calling for lunches to be publicly supported. He also solicited funds from philanthropists such as Andrew Carnegie and Solomon Guggenheim to help the School Lunch Committee. In 1913, the Board of Estimate appropriated a small amount to assist, but a withering opposition was mounted both within and outside the board by those who believed that feeding the poor shattered their moral character and served only to sustain their despicable condition. Milder opposition came from those who saw school lunches as a nuisance, a disruption that would encourage rambunctious students to engage in food fights and other naughtiness.[60]

Of course, the members of the Board of Estimate did not have to see the faces of hungry children every day. Those who did improvised, such as the teacher who administered her own milk fund, collecting a penny per day from each child for a warm cup of milk, served in the middle of the morning. Children who could not contribute still received milk, the teacher making up the difference from her modest wages. Other teachers saw that hungry children received the edible products of cooking classes three days per week. These efforts and others were praised by John Spargo.[61] These ad hoc solutions and Mabel Kitteridge's school lunch program continued from 1909 to 1919, when the New York City Board of Education finally received funding to assume the responsibility of serving school lunches.

Food for children came to be seen as fulfillment of a public responsibility, rather than as charity, removing the stigma for many immigrants. Still, cultural differences continued to dog well-intentioned but often insensitive efforts to feed students. Italians cherished the noontime *colazione* (meal). In Italy, families shared the meal together in the fields or at their tables. Now in America their children were being encouraged to eat away from family, at school. Children fortunate enough to have parents at home when lunchtime came often preferred the familiar meals of the *paise* (old country). Scholar Richard Gambino recalls that as a child in the 1930s, he preferred the fried eggs and potatoes on Italian bread he sometimes got at home to the "well-balanced meals" that professional school nutritionists had arranged.[62] In the previous generation, opposition to such intrusions upon family life were even more vehement.

Occasional immigrant opposition to school lunches was part of a more general opposition to the schools' health role. Foreign-born parents did not think it was the function of the school to embarrass them or their children by calling attention to discrepancies in standards of health or hygiene between newcomers and natives. Though the matter was serious, at times the exchanges across the cultural divide could be amusing. Even as she struggled to improve the health of millions of New York City schoolchildren, many of them foreign-born, Dr. S. Josephine Baker could not resist saving a letter she received from a mother whose child was sent home because he needed a bath. "Dear Teacher, Ikey ain't no rose. Don't smell him—learn him." And an angry response to a school inspection that perhaps had discussed the need for some minor surgery: "Dear Nurse: As for his nose, it don't need it. As for his tonsils, he was born with them. As for his teeth, he'll get new ones. Please mind your own business."[63] And as always, there were the misunderstandings occasioned by language differences. Baker recalled the indignant mother who created a storm protesting an unfortunate misunderstanding of an abbreviation. Waving a school nurse's form in Baker's face, the mother screamed, "My boy is as bright as any." Glancing at the form, Baker immediately saw the reason for the tirade; in place of "poor nutrition," the nurse had hastily scribbled, "poor nut."[64]

Most of the time, though, Baker and the many other health care professionals in the schools saw little that amused them. Lunch programs might tackle nutritional deficiencies, and the presence of nurses helped detect and control the spread of infectious diseases, but Baker and others still considered school health programs inadequate because so little was done about the problem of physical defects common in childhood. She deplored that New York City officials refused to appropriate the money for annual physical examinations for all children, although most children could count on an examination when they entered school, another about the fourth grade, and a final one prior to graduation. There was general public interest in curbing rates of infant mortality. Philanthropists such as Nathan Straus were willing to contribute significantly to establishing milk stations in the heart of immigrant neighborhoods so all babies could be guaranteed clean, wholesome milk, but as Baker knew all too well, "there was no comparable way to dramatize the older children's failing vision, decayed teeth and diseased lungs."[65] The health department experimented with forms that a family physician could fill out, hoping to get some of the parents involved in taking their children to the doctor, and, of course, bearing the expense. However, when parents realized that if they did not cooperate, the city would assume both cost and responsibility, the plan failed.

With so many examinations to administer, Baker feared that "our examinations were very superficial and of little account." Usually school physicians did not undress the children and "only too often there were innumerable layers of clothing to hide their bodies and make listening difficult." On one occasion, New York health officials were chastised by the Association for Improving the Condition of the Poor (AICP) for not undressing the children during exams.[66] Baker responded by instructing several female members of the staff to undress girls undergoing examination, but only if a nurse or the child's mother was present. Instead of quieting critics, Baker loosed an even greater wave of criticism upon herself and her colleagues when several newspapers ran headlines about schoolgirls being stripped by health department physicians with photos of the "insulted girls and much journalistic indignation." Undressing the children was abandoned, but Baker decided to check her staff's results by forming committees of specialists to examine the same children. Much to Baker's delight, the special investigators confirmed the results of her staff, leading her to conclude that however superficial the examinations might appear to some, sick children and those with disabilities were being diagnosed as best the state of medical knowledge and technology would allow.

Physical defects that Baker and others involved with school medical inspection especially hoped to identify and treat involved eyes and teeth. As early as 1905, Lillian Wald called attention to children's eyesight. She cited a limited study of 981 children, 30 percent of whom showed refractive error in one or both eyes. In June 1902, a special study was also done to detect the extent of trachoma in the New York schools. Sixteen trained oculists examined 55,470 children in thirty-five schools. They diagnosed 6,770 cases of contagious ophthalmia, or 12 percent, distributed as follows: "severe trachoma, 2,328, or 4.2 per cent; mild trachoma, 3,243, or 5.82 per cent; acute conjunctivitis, 1,099, or 1.98 per cent."[67]

Wald was well aware that in many communities eye exams were rather unskillfully performed by teachers. However, she took heart from a resolution passed at a meeting of the American Medical Association that called for "measures [to] be taken by Boards of Health, Boards of Education, and school authorities, and when possible legislation [to] be secured looking to the examination of the eyes of the children, that disease in its incipiency may be discovered and corrected."[68]

As Wald knew quite well, testing eyesight was one thing, furnishing glasses quite another. Many physicians opposed government providing glasses for the same reasons they opposed adenoidectomies performed in school clinics. Still, Wald and others tried. Baker recalled, "We examined the children's eyes as nearly annually as we could. We fought for duller paper in text-books [to cut the glare], better spacing of the print and better classroom lighting to avoid eye-strain. But the incidence of defective eyesight is just as high as it ever was."[69] Children who needed glasses and could not afford them were referred to one of the many charitable societies that hired physicians to conduct further tests and provide glasses. Such arrangements neither physicians nor their associations opposed. Only occasionally did parents oppose treatment.[70]

By the early twentieth century it was well established that mouth disease was a prime example of bacteriological pathology, but still few paid attention to the need for oral hygiene or focal infection until American physicians were persuaded by Dr. Frank Billings that infection beginning in the teeth or mouth could ultimately spread throughout the body.[71] Organized dentistry's aggressive approach transformed oral hygiene from a prudent precaution advised by dentists to their middle-class patients to part of the medical gospel being preached to the nation's schoolchildren. School physicians found dental caries and other oral diseases, often in advanced stages, to be among the most common health problems they encountered. However, many school health personnel were not trained to detect oral problems unless they were severe, and few understood dental hygiene. After a methodologically primitive but persuasive experiment by W. B. Ebersole in Cleveland's Marion School, one with a large component of poor and immigrant students, many public health advocates were convinced that there was a relationship between oral health and school performance. Bridgeport, Connecticut's Dr. Alfred Fones opened the first school to train dental hygienists for service in 1913. His graduates were prepared to teach children how best to keep their teeth clean through brushing and to provide the children with prophylactic treatments, largely tooth polishing and scaling. Fones fended off criticism that dental hygienists, most of whom were women, were minimally trained subspecialists who might compete with dentists for fees or join preexisting dental practices and undercut the existing fee structure. Instead he emphasized the use of hygienists in schools because only there could they reach "all the children of all the people."

Fones used graphic imagery to make his point: "The most conspicuous defect of the child is the unsanitary condition of its mouth. Like a pigpen or garbage drain slowly seeping its poison into the brook, which flowing into the reservoir, contaminates the water supply of a city, so do the products of abscesses and decayed teeth with decomposing food slowly but surely poison the human system." Moreover, Fones explained why proper oral hygiene for schoolchildren ought to be a more general public health concern: "Such mouths breed disease. Such children laugh and sneeze millions of germs made virulent and active in an ideal feeding ground . . . around and between the teeth."[72]

A member of the Bridgeport School Board, Fones used his persuasive powers to get ten of his first graduates hired by 1914. They inspected teeth daily, treated them prophylactically, and taught first and second graders about brushing. No other dental work was done on student teeth. His intention was to demonstrate the value of education and preventive care of teeth in improving children's health and school performance. Fifth graders were the control group in the five-year study. The results were dramatic. When the experimental group entered fifth grade, they had only one-third as many cavities in their permanent teeth as those in the control group. School retardation, the percentage of children more than two years behind their age cohort in grade, was a matter of great concern to all school health professionals. It declined by 50 percent in the experimental group. Seeming confirmation of Fones's belief that the mouth was the entrance for infections of all kinds, the incidence of diphtheria, measles, and scarlet fever also declined. A careful Fones did not overstate the significance of his results, but he did believe that his research had confirmed that the mouth was a significant "ingress" for communicable diseases and that "a clean mouth with sound teeth is the one most important factor for prevention" and worthy of a place in the program of hygiene enforced in the public school system.[73]

The concern with oral hygiene, including daily teeth cleaning, was also part of the larger mission of cleanliness that educators and health professionals saw as vital for children from immigrant neighborhoods, where cleanliness at times seemed neither a cultural imperative nor convenient and affordable. School teachers and nurses emphasized the importance of coming to school clean. A New York City school syllabus included a "Course of Study in Physical Training and Hygiene." Teachers were required to "lead pupils to cultivate habits of cleanliness," but the lessons would be no more than empty academic exercises if the children left the building with no place to go to take a bath.[74]

Reformers who had explored urban immigrant neighborhoods frequently recorded among their impressions the filthy appearance and fetid odor of both places and people. By the turn of the century, most urban officials and social commentators understood that if they were to improve the health and well-being of their poorest citizens, public baths were essential.[75]

How could the young be encouraged to avail themselves of public baths, indeed just soap and water? On New York's Lower East Side, reformer Jacob Riis noticed that the process of educating immigrant children in the virtues of cleanliness was well underway in the public schools, although in some cases teachers had to take soap and water to the children themselves.[76] Clearly there was much work ahead.

Defending his people against charges of uncleanliness as he did against charges that they were disease breeders, Dr. Maurice Fishberg vehemently denied that Jews were bodily unclean by choice, claiming instead, that youngsters learned cleanliness in religious ritual before they were ever exposed to the public school. "Before each meal an observant Jew must carefully wash his hands, and repeat this operation immediately after meals, and must then also rinse his mouth; and he must not walk four steps from his bed in the morning without careful ablution of his face and hands." Among Russian Jews whose piety had lapsed, the steam bath was a popular Russian institution that Jews were quick to seek out in American cities. Fishberg observed that on days when the Russian baths admitted only women, "they are crowded with women and children."[77]

Fishberg's admiration for religious ritual's inspiration of hygiene among eastern European Jews was not shared by Jacob Riis, who regarded the religious roots of hygiene as insufficient to ground cleanliness into the next generation: "Only the demand of religious custom has power to make their [Jewish children's] parents clean up at stated intervals, and the young naturally are no better."[78] Riis's contempt for Jews' personal hygiene may well have been merely another dimension of his stereotypical perspective on them. Again, commenting on child development among the schoolchildren he observed, "As scholars, the children of the most ignorant Polish Jew keep fairly abreast of their most favored playmates, until it comes to mental arithmetic, when they leave them behind with a bound. It is surprising to see how strong the instinct of dollars and cents is in them. They can count, and correctly, almost before they can talk."[79]

The Italian immigrants were also sensitive to nativist criticism of their personal hygiene, which in its most benign form was couched as well-meant advice to newcomers. In the guidebook for Italian immigrants published under the auspices of the Daughters of the American Revolution, they described the school health inspection with a warning, "The sickly child is always behind in his studies. Only well children make progress." Crucial to that wellness was cleanliness. Italian immigrants were instructed to "Bathe the whole body once every day."[80] Dr. Antonio Stella, a vigorous advocate of education for Italian immigrant children, charged that the uncleanliness that contributed to breeding disease among the young was largely the result of urban congestion bred of poverty, not an inherent opposition to healthy habits. Soap and water were important but little match for the bacteria bred in the filth of crowded tenements.[81]

The same congested tenements and sweatshops that exposed immigrant children to harmful germs often excluded the healthy rays of sunlight and fresh air that growing children especially needed. In the Lower East Side's Allen Street school, Jacob Riis heard students respond in rhyme to their teacher's daily inquiry, "What must I do to be healthy?":

> I must keep my skin clean,
> Wear clean clothes,
> Breathe pure air,
> And live in the sunlight.[82]

As Riis observed with a mixture of sadness and bitterness, "It seems little less than biting sarcasm to hear them [the children] say it, for to not a few of them all these things are known only by name. In their everyday life there is nothing even to suggest any of them."[83]

Pure air and sunlight were almost as rare commodities as bathtubs, and opportunity for proper exercise even rarer. Riis's photographs of children, his street arabs, surviving in the dark alleys and basements of New York City's immigrant neighborhoods spoke volumes about the need for clean air, sunlight, and organized play.

Public school health reformers Luther Gulick and Leonard Ayres observed that as the United States increasingly became an urban nation, children who worked no longer engaged in "muscular work which strengthened the muscles, enlarged the chest, and aided in giving the power to live." Even more important, though, play—the other "great source of muscular exercise and physical development" and the "heritage of all of the children of all of the world"—seemed under siege from three sources. All too frequently and especially among the urban poor, there seemed to be no time for play, "no space, for play and no traditions for play."[84] The reasons were many. Some children spent their days at work in factories or fields, or dodging traffic in city streets selling newspapers or making deliveries. However, even those fortunate enough to be in the classroom were too busy for healthy play. Gulick and Ayres complained that by 1910 many children were in school six hours a day for ten months of the year, with still another one to four hours of homework to keep them indoors. They lamented how little time was left for "play with dolls, wandering through the woods, or corresponding activities in which unconscious growth occurs."[85]

Prior to the 1890s, most public schools did not have playgrounds alongside them or space inside for recreation. However, by 1895, Jacob Riis and other reformers succeeded in persuading the New York State legislature to mandate that every new school constructed would have a playground. Before a substantial number of new schools were built with the outdoor facility, though, there was little place for urban children to play except in the streets. Gulick and Ayres shared the regret of many other reformers who deplored the building of cities without adequate parks and playgrounds, a sentiment that had led Frederick Law Olmsted, renowned urban planner and landscape architect, to place a vast park in the middle of Manhattan and to advocate green spaces for rest and recreation in other cities as well. However, too often such parks were designed to inspire tranquility: mowed green meadows were a priority, playgrounds were not. In the area of lower Manhattan south of 14th Street, "a scant space of three yards square for each child, only one child in ten can be given play room." The highly moralistic Gulick and Ayres, proponents of a "muscular Christianity," did not fail to mention that increased opportunity for vigorous exercise would not only make young bodies healthier but souls, too, eliminating "the prevalence of such games as craps." After all, that game of chance thrived because "It takes but little space, is quiet, can be played with varying numbers of players, is interesting, etc."[86]

As one historian has exclaimed, "By 1905 almost everyone concerned with social reform was concerned with play and almost every reform organization was involved in assisting the rise of organized play in one form or another."[87]

The Outdoor Recreation League (1897), Public School Athletic League (1903), the Playground Association of America (1906), New York City's vacation schools and playground schools (1898), and many other programs private and public represent efforts to foster middle-class American values and healthy bodies through play. Jewish philanthropists had much the same thought in mind with the founding of the Lower East Side's Educational Alliance in 1893, while settlement houses throughout the city sponsored programs of outdoor physical recreation, especially in summertime.

By 1899 the New York City Board of Education had hired a supervisor of physical culture. Three years later, the first citywide director of health education was appointed. The citywide program included physical fitness and hygiene. Games were taught to encourage physical activity, but in a bow to the immigrant origins of many students, ethnic folk dances were also included.[88]

Such programs seemed to be having a marked and favorable impact on the bodies of eastern European Jewish children, according to Dr. Maurice Fishberg. In eastern Europe young children were sent to schools, or *cheder.* According to Fishberg, "The unsanitary and unhygienic surroundings of these schools are well known to everybody. . . . During the period of most active growth the Jewish child there has no outdoor recreation and games, but remains indoors the greater part of the day, engaged in the study of the Bible and Talmud." Although Fishberg did not criticize this emphasis on scholarship and learning for the young, he did deplore the lack of physical activity that was entailed. It appeared to the doctor that the healthy development of young Jewish immigrant bodies was being "greatly retarded as a result of this lack of physical exercise during childhood." How much healthier was the child's life in American streets and public schools, according to Fishberg. "Here the child spends the greater part of the day outdoors, as can be witnessed in the streets of the lower East Side. . . . At six or seven it is sent to a public school which is a model of the hygienic and sanitary achievements of the century. . . . even the children of the poor, are of a better physical development as a result of this favorable environment. They are taller, their chests and muscular system are in better condition than those of their less happy brothers and sisters in Eastern Europe."[89] As in so much of what Fishberg said and wrote, the subtext appears to be that Jews are not inherently inferior and that healthy conditions allowed their young to become physical specimens every bit as healthy and robust as Gentiles.

Dr. Antonio Stella perceived the matter of fresh air and exercise for the Italian young from an entirely different perspective. After all, many Italian immigrants lived in the country and worked outdoors, in the warm sun. Life in the congested tenements and streets of urban America hardly seemed healthy for the children. Offering New York City data, Stella noted that while the general mortality figure for 1905–06 for all ages was 18.35 per 1,000 and under five years, 51.5, it was much higher in Italian neighborhoods, on some blocks as high as 25.0 for all ages and over 90.0 for those under five.[90]

Immigrant children and parents did not always embrace physical activity, even organized play, with open arms. Children objected to the adult supervision that went with the playground and equipment. Immigrant parents, always fearing the unknown, preferred their children close to home, even in streets clogged with traffic and choked by garbage, where they were within eyeshot of relatives and neighbors. Moreover, immigrants were often aware that the generosity of reformers was inspired by a desire to change the newcomers, Americanize them.[91]

Immigrant parents loved their children and wanted them to be healthy no less than American parents, but the value of play in conformity to American practice often escaped them. The contribution of sports to healthier bodies seemed especially farfetched. In 1903, a father wrote to the "*Bintel Brief*" (Bundle of Letters) column of the *Jewish Daily Forward* to ask the highly respected editor, Abraham Cahan, whether or not he should allow his son to play baseball. The distressed father wrote, "It makes sense to teach a child to play dominoes or chess. But what is the point of a crazy game like baseball? . . . I want my boy to grow up to be a *mensh* [sic], not a wild American runner." Cahan advised the parent to allow his child to play the game, not to lock him up in the house, provided that it did not interfere with his studies or get him into the company of ruffians. A key reason: health. "Baseball develops the arms, legs, and eyesight," wrote Cahan. "It is played in the fresh air."[92]

Educators and public health officials were aware that sick children in the schools often had a special need of "fresh air," and their conditions made it unlikely they would get any in a playground or baseball field. The result was the open-air classroom. In 1902 five schools received open-air playgrounds on their roofs. Six years later, the desire to combat tuberculosis among the young inspired the establishment of the first open-air school—on the water. A ferryboat, the *Southfield*, was declared an annex to P.S. 14. Three other such boats were later assigned to this duty, and an open-air school was built on the roof of the Vanderbilt Clinic. Public School 21 on Mott Street had the first open-air classroom in 1910 for children with respiratory diseases.

The demand for space in these classes by 1917 required the Bureau of Child Hygiene to establish those to receive priority eligibility. The list included children with arrested or cured tuberculosis, malnutrition, nervous diseases, cardiac disabilities, frequent colds and bronchitis, and those who showed signs of unusual fatigue. By the following year there were 109 open-air classes in operation and demand was soaring. Although not all children could be in open-air classes, many health professionals advocated open-window classes. In 1914, the New York Academy of Medicine's Public Health Committee concluded that open windows were as effective as open-air classes in providing a healthful atmosphere for those with special health problems.[93]

Even as public school teachers and public health officials were slowly prying open closed classroom windows to allow fresh air and sunlight into the lives of urban children, many of whom were foreign-born or the children of immigrants, American nativists were rushing to close doors. The doors they were trying to close were the ones that had allowed so many immigrants into the country in the first place. Contrary to the belief of educators and reformers that nurture, in the form of proper instruction in the rudiments of health, hygiene, and democratic citizenship, could transform the immigrant menace, nativists argued that most newcomers were by nature hopelessly inadequate, far below the standards of the native stock and incapable of change.[94] Prolific nativist writers such as E. A. Ross thought immigrants' unattractive appearance a clue to their physical inferiority to the American pioneering breed and argued for restriction. Ross, writing on one of his favorite themes—the danger of "race suicide"—explained to his readers in 1914 why immigration posed a danger:

> To the practised eye, the physiognomy of certain groups unmistakably proclaims inferiority of type. I have seen gatherings of the foreign-born in which narrow and sloping foreheads were the rule. The shortness and smallness of the crania were very noticeable. There was much facial asymmetry. Among the women, beauty, aside from the fleeting, epidermal bloom of girl-hood, was quite lacking. In every face there was something wrong—lips thick, mouth coarse, upper lip too long, cheekbones too high, chin poorly formed, the bridge of the nose tilted, or else the whole face prognathous. There were so many sugar-loaf heads, moon-faces, slit mouths, lantern-jaws, and goose-bill noses that one might imagine a malicious jinn had amused himself by casting human beings in a set of skew molds discarded by the creator.[95]

From Ross's perspective, public school health programs could not undo what the Creator had ordained.

With the passage of restrictive legislation in 1921 and 1924, America's golden door, if not wedged shut, remained only slightly ajar. Still, millions of immigrants had already arrived in the United States during the previous forty years. What would their impact be upon the sturdy native American stock that E. A. Ross, Lothrop Stoddard, and other racial nativists celebrated in their books and essays? The word *eugenics,* coined as a scientific term by Francis Galton in 1883, had become a popular buzzword in the United States by the 1920s, as well as a precise course of systematic population studies promoted by serious scientists as the means of perfecting the human race through genetic management.[96]

Local eugenics groups sprouted throughout the United States. Baby health competitions, or better-baby contests as they were often called, were held so that mothers could demonstrate the high quality of their offspring. Fairs and expositions across the country featured exhibits on Mendelian genetics, and there were fitter family contests, often sponsored by the American Eugenics Society at state and county fairs. Contestants provided the judges with the family's eugenic history. Had any close relatives been crippled or chronically ill or mentally unsound? All family members received medical examinations, including a Wassermann test, a psychiatric examination, and an intelligence test. Some fairs divided the winning contestants into small, medium, and large categories, depending on family size. Medals and trophies were awarded, accompanied by mellifluous praise for these model Americans.[97]

The public schools were making newcomers and their children healthy and clean by American standards. What the schools could not do, though, was to produce a new generation of Americans with altered pedigrees. Data are far from complete, so it may be that an immigrant family from eastern Europe or Italy or China entered a fitter family contest during the 1920s. Certainly there is no record of one winning.

The eugenics building at the Kansas Free Fair, site of Better-Baby contests, 1929. (Courtesy of American Philosophical Society, Philadelphia)

10

"Viruses and Bacteria Don't Ask for a Green Card": New Immigrants and Old Fears

Like rocks beneath a waterfall, every aspect of American society and culture has been slowly sculpted over time by the cascade of immigrants who have continually replenished the American population and altered its character across centuries. Public health in the United States and the culture of American medicine are no exceptions. They, too, have been fashioned and refashioned, pounded into one shape and then another across the centuries by the continual flow of immigrants down the gangplanks or over the borders and across the nation's threshold. Meanwhile, the newcomers themselves are buffeted by nativists' suspicions and accusations that they are accompanied across sea and land by silent travelers—germs and genes of an inferior sort—well capable of polluting the sturdy American population mainstream.

A glance at the period after Congress restricted immigration with a national origins quota system in the 1920s, and especially at the most recent wave of immigration pounding our shores, suggests that the connection between public health and immigration has remained a significant item on the agenda of both public health experts and those most concerned with the fate of immigrants and refugees—themselves. Natives and newcomers collided over public health matters even after the Johnson-Reed Act of 1924. Restriction may have slowed immigration's rush to a trickle, but previously arrived newcomers and their children continued to engage in a tug-of-war with American culture in matters of health and hygiene. More recent arrivals, too, especially the Latin Americans and Asians prominent in the surge of immigration during the past three decades, have echoed with remarkable resonance the voices of those who came before as they dicker with native-born Americans over the standards of health and hygiene they must meet and the varieties of health care accessible to them. Thus, the four themes presented to explain the linkage between immigration and public health are recycled in the present, as revealed in newspapers and the minutes and reports of government and private agencies concerned with the public health or the newcomers' welfare or both.

The first of these themes, the double helix of health and fear of the foreign-born, ever present throughout American history and especially evident during earlier peak periods of immigration, may well be the most resilient pattern. Certainly it has been reasserted time and again in the last two-thirds of the twentieth century. Just as U.S. Commissioner General of Immigration Terence Powderly feared that unrestricted immigration might yield an unsightly nation of blind, bald defectives in 1902, even as E. A. Ross warned of the "hirsute, low-browed, big-faced persons of obviously low mentality in 1914," critics of generous admission policies after 1929 when the Johnson-Reed Act became fully operable also focused upon the foreign-borns' unsightliness as an indicator of potential physical degeneracy to be avoided.

In the 1930s, popular opposition to the admission of Jewish refugees in flight from Hitler was rooted in an anti-Semitism that at times echoed European anti-Semites' hysterical disgust for the Jew as biologically inferior or simply repulsive. In 1939, with 933 refugee passengers of the *St. Louis*—most of them Jewish—anxiously waiting off the Cuban coast, Congress was debating the ill-fated Wagner-Rogers bill, a liberalization of the 1924 legislation that would have brought 20,000 German refugee children to the United States over two years. Undoubtedly, many of them would have been Jewish, a fact that contributed to the bill's defeat. Laura Delano, the wife of Immigration Commissioner James Houghteling and President Franklin Delano Roosevelt's cousin, was overheard to have framed succinctly the consequences of saving the children: "Her principal reserve on the Bill was that 20,000 charming children would all too soon grow into 20,000 ugly adults."[1] How many shared her rationale for opposing the rescue can never be calculated, but public opinion polls indicate little support for a generous refugee policy. The April 1939 *Fortune* poll listed 83 percent of all Americans opposed to an increase in the quotas.[2]

To receive a visa under the 1924 act, immigrants were required to obtain certification of mental and physical health from a physician at the issuing consulate or from a physician acceptable to consular officials. Ascertaining the mental health of immigrants who were refugees in the 1930s and 1940s—individuals in flight reeling from the effects of persecution or wartime chaos—remained a priority for consuls abroad and immigration officials at home, as did the desire to exclude those born mentally deficient. No matter how sympathetic they might be to the desperation of European refugees, immigration officials were charged with protecting the health and social stability of Americans.

If at times the flight of brilliant intellectuals such as scientists Albert Einstein and Enrico Fermi, novelist Thomas Mann, and theologian Paul Tillich yielded an intelligence dividend in America's favor, the nation's approach to less illustrious refugees was cautious. In 1944, one thousand refugees—87 percent of them Jewish refugees—with temporary status were admitted to the United States and interned at Oswego, New York, as part of a federal experiment in refugee rescue undertaken after Franklin Roosevelt formed the War Refugee Board.

Physical health in the camp was well maintained, but refugee life in Oswego took its psychological toll. Historian Sharon R. Lowenstein has observed that, while the refugees were pleased to be in the United States, "Being in America, but not *of* America put the refugees on an emotional roller coaster. The barbed wire-topped fence stunned them, customs confiscations angered them, mail censorship irritated them, and the loss of freedom hurt them."[3] As despair deepened, the camp was shaken when a young wife and mother of two committed suicide. Dr. Curt Bondy, chairman of the Psychology Department at Richmond Professional Institute of William and Mary College, who had been assistant director of the Kitchener Refugee Camp in England, was summoned and stayed at Oswego for four days. He recommended that the effects of persecution be taken into account in providing for the treatment of newcomers and that efforts be made to shore up mental health by increasing their feelings of control and by including refugees in the crafting of psychosocial therapies. Later, Rudolph Dreikurs, a Chicago psychiatrist who conducted individual evaluations of internees, concluded that camp residents presented an overall psychiatric profile no different than that of the rest of the population, except for an elevated number of cases of neurotic personality disorders. He found that personality disintegration, symptomized by a deterioration of reasoning skills and a high incidence of psychosomatic illnesses, although widespread, resulted from immediate circumstances and would be ameliorated by environmental improvements.[4]

The Oswego camp never did get a resident psychologist or psychiatrist, although the camp administrator repeatedly requested one from Washington.

The few refugees who exhibited severe antisocial behavior or became dysfunctional were hospitalized on Long Island or in Philadelphia, and a few received outpatient treatment in Syracuse in the form of electric shock treatments. Although they had been told to expect deportation after the war, all refugees were permitted to stay in the United States. Still, the internment, albeit benevolent, had exacted a price in mental well-being.

The health and well-being of foreign-born in the confinement of a camp was hardly a subject limited to the experience of refugees at Oswego. Japanese Americans, those born in Japan (Issei), and their American-born children (Nisei) found themselves in internment camps after 1942.[5] Responding to the fear that the Japanese would be disloyal during the war, the government evacuated West Coast Japanese from their homes and placed them in hastily established camps in California and other western states. Camps were built in stockyards, race tracks, and county fairgrounds. Eventually there were be ten permanent camps holding the 120,313 Japanese evacuees gathered and transported by the U.S. Army and held under the jurisdiction of the War Relocation Authority. A possibly apocryphal anecdote, much repeated, dealt with a child who entered a camp and noticed that everyone there looked to be Japanese except for a guard of armed soldiers. He is said to have inquired of an elder, "When can we go back to America?"[6] And indeed, the processing of internees upon their arrival was a kind of Ellis Island or Angel Island in reverse. Army physicians, usually assisted by evacuee physicians, examined throats and tongues and gave shots. Men were examined for venereal disease. In the camps themselves, medical personnel were in short supply and health problems abounded. Inadequate sanitation and the possibility of epidemics were persistent public health worries. Inoculations for smallpox and typhoid were given, but quarantine facilities were non-existent. At California's Turlock assembly center, where Japanese gathered before camp assignment, families of patients with communicable diseases continued to eat in common dining halls and wash in community sinks. At the Tanforan center, where Berkeley's 1,319 Japanese were gathered, there was a mild epidemic of measles.[7]

As in the Oswego camp, psychosocial adjustment was a major problem in internment camps. The very experience of evacuation was a major affront and expression of dishonor to the Japanese. The absence of privacy was acute. While many Nisei threw themselves into the work of the community, older internees, especially the Issei, often simply wandered the camp aimlessly. There were suicides. After the war, internees were released, but long-term physical and psychological effects of camp life remained.[8] In its camps, the United States government had managed to replicate some of the very public health problems that immigrants crowded into urban slums had confronted earlier in the century. Japanese immigrants and their American-born children were revisiting a past that many Issei thought they had escaped.

Postwar immigration policy toward refugees was as ungenerous as prewar policy. War brides, fiancees, and foreign-born children of U.S. servicemen were exempted from visa quotas. However, the Displaced Persons Acts of 1948 and 1950 admitted only 405,000 Europeans out of an estimated 5 million left homeless by the war.

In the displaced person (DP) camps of Europe, representatives of the United Nations Relief and Rehabilitation Administration (UNRRA) and voluntary organizations such as the American Friends Service Committee, the Jewish Joint Distribution Committee, and the Jewish Agency for Palestine focused on the problems of three distinct categories of Jews awaiting resettlement in a western nation or in Palestine—those who had survived the concentration camps, those who had hidden during the war, and those who had escaped to the Soviet Union during the war and now were fleeing west. All had undergone severe psychological trauma, raising questions in the minds of some Americans whether they were fit for America.

In the displaced persons camps, Jews without strong religious or political affiliations endured the most severe psychological crises. But psychiatrists and social workers remained in short supply until late 1945. Those who worked with the DPs observed that many were restless, sexually impotent, highly emotional, engaged in threats and tantrums, and, at times, completely lacked any emotional control. Some could neither concentrate nor work. A psychiatrist who lived in one DP center described the DPs as "asocial." In the Jewish camps, especially, observers described DPs who simply wandered around the compound, rolling their eyes and repeating their stories, almost without pause.

Often those who worked with the DPs experienced a cultural gap between themselves and those they were trying to assist. Non-Jewish workers became frustrated with ultraorthodox Jews who would not compromise the most minute detail of their rituals to accommodate more general camp needs. The workers failed to understand that these extremely observant Jews credited their survival to their piety. Others underestimated the profound psychological legacy of surviving concentration camp deprivations. When clothing was to be distributed, volunteers were shocked to see DPs wielding knives, clubs, and fists to assure themselves a place in line. Thousands of DPs also dreaded showers, delousing with DDT, and inoculations, all of which brought back the memories of how Jews were tortured and exterminated. An UNRRA worker complained bitterly of how hard it was to get "these people to take a steam bath voluntarily."[9] Women screamed and had to be pushed or carried under the shower nozzles. He was unaware of the Germans' method for gassing Jews. Non-Jewish DPs also showed the effects of their ordeals. Many had suffered the same torture as the Jews, but others had simply been in hiding. Observers noted that non-Jews seemed able to establish communities in the camps more readily than the Jewish DPs and did not seem to experience the emotional complexity—dramatic highs and lows—demonstrated by those who had survived the concentration camps.[10] It would take decades of research for physicians, psychiatrists and psychologists, to fully appreciate the damage that life in concentration camps exacted. Those bearing the psychological scars of persecution received care before and after relocation, but many bore the wounds for life.[11]

More recent refugees admitted under the 1980 act have also had to contend with being misunderstood and carelessly treated, with being stigmatized as a health menace and unfit for America. Responsible public health officials know that immigrants may at times be agents of infectious disease. However, careless language often discourages cautiousness, unintentionally fueling public fear, lending a "scientific" imprimatur to feelings grounded in racial or ethnic prejudice, a syndrome that spokespersons for refugee groups know only too well.

On a mild afternoon in April 1990, sixty-five thousand angry protesters tied up lower Manhattan traffic after pouring across the Brooklyn Bridge. Chanting and carrying placards, the marchers were protesting a new federal proposal on blood transfusions that demonstrators charged stigmatized Haitian and some African immigrants. Marchers—"college students, factory workers, and families with picnic baskets and umbrellas to protect them from the sun"—were angered by a United States Food and Drug Administration (FDA) recommendation that excluded all individuals from Haiti or the sub-Saharan region from donating blood on the grounds that large numbers of these groups tested positive for the HIV virus. The policy was not binding on blood banks, but because the FDA oversees the nation's blood supply, many local blood banks were complying.[12]

Protesters contended that Haitians in the United States, who numbered 350,000 in the New York Metropolitan area alone, were no more likely to carry the AIDS virus than any other group. Dr. Jean Claude Compas, chairman of the Haitian Coalition on AIDS, angrily told reporters, "This policy is on the basis that Haitian blood is dirty, that it is all infected with HIV virus. The decision is based . . . on nationality, ethnicity." Some in the crowd interpreted the recommendation as racist, describing such policies as evidence that AIDS is a weapon of "white folks." Others were less concerned by the motive than by the effect. Thirty-seven-year-old Herve La Guerre told the *New York Times*, "They're putting us in a position that we're not going to be able to survive here because uneducated people will discriminate against us if they think we have AIDS."[13] Several days later the FDA retreated from its recommendation in response to advice from its Blood Advisory Committee.

The FDA's clumsy attempt to curb donations of HIV-positive plasma to blood banks was reminiscent of the early 1980s when the Centers for Disease

Control in Atlanta (renamed the Centers for Disease Control and Prevention in 1992, but retaining the acronym CDC) sent a tremor of trepidation throughout New York, Miami, and other areas of Haitian concentration by announcing that Haitian newcomers had been classified in a special high-risk category for AIDS. By 1985, when the CDC finally dropped Haitians from its high-risk list, the psychological, financial, and social damage had been done. Tens of thousands of Haitians in the United States and in Haiti, most not infected with AIDS, had lost jobs, housing, and educational opportunities.[14]

What did AIDS have to do with Haitian nationality? Apparently very little. In April 1985, Dr. Walter Dowdle, director of the Center for Infectious Disease, acknowledged the spurious relationship, admitting, "The Haitians were the only risk group that were identified because of who they were rather than what they did."[15] Less than a decade later the Food and Drug Administration had only repeated the CDC's insensitivity toward these new immigrants.[16]

If scientific advances can demonstrate that being Haitian had nothing to do with being HIV-positive, at times medical discoveries demonstrate that the foreign-born do pose a threat to the public health, a threat that raises legal and ethical dilemmas. The second theme of this volume is the role of medical advances in binding immigration and public health. In the nineteenth and early twentieth centuries Californians faced with bubonic plague and New Yorkers confronting a first seemingly healthy typhoid carrier both turned to quarantine, though in very different ways and with equally different results. During the spike in immigration that began in the 1970s and has continued during the last two decades, tuberculosis, once thought conquered, has reappeared to an extent that has caused health professionals to refer to it as an epidemic, often among the foreign-born, and to again consider actions that would interfere with personal liberty on behalf of public health.[17]

There is little doubt in the minds of epidemiologists that the foreign-born are contributing disproportionately to high TB rates. A 1990 report from the CDC noted that "Foreign-born persons (as a group) residing in the United States have higher rates of tuberculosis than persons born in the United States. In 1989, the overall U.S. tuberculosis rate was 9.5 per 100,000 population; for foreign-born persons arriving in the United States, the estimated case rate was 124 per 100,000. In the period 1986-1989, 22% (20,316) of all reported cases of tuberculosis occurred in the foreign-born population. A majority of foreign-born persons who develop tuberculosis do so within the first five years after they enter the United States."[18]

Particular counties and cities with high concentrations of recently arrived immigrants echo at local levels what the CDC gleans from its national data. In 1991 New York City reported a 38 percent rise in the number of active TB cases (3,673 cases up from 3,520 in 1990 and 2,545 in 1989). The highest rates were in the African-American community, which accounted for 56 percent of the cases, with a rate of 112.2 per 100,000. However, "the rate for Hispanic residents was 52.3; for Asians 46.9, and for whites 13.2."[19]

Similarly, Montgomery County, bordering the District of Columbia, had the highest rate of tuberculosis in the state of Maryland in 1992, 12.9 per 100,00 as compared to a state rate of 8.9, a figure that county health officials attribute in part to the concentration of Asian and Hispanic resident immigrants. Lynn Frank, chief of communicable diseases and epidemiology for the Montgomery County Health Department, tied the spike in TB largely to "the number of immigrants who are from countries where the incidence of the disease is much higher."[20] Yvonne Richards, nurse manager for the county's refugee and migrant health program, observed, "Poor nutrition, poor access to health care, and inappropriate medical treatment in their countries may have made many people more susceptible to TB."[21] Clearly, newly arrived immigrants contributed to a public health problem by their very presence in the community. However, unlike with immigrants of an earlier era, newcomers from Latin America and Asia have not been widely castigated as health menaces. Health officials have been sympathetic to the victims of illness and sensitive to the potential for a nativist backlash nurtured by alarm over the public health. Undoubtedly physicians' ability to treat TB successfully has also prevented undue public hysteria.

As in earlier eras, the federal government has acted on TB as a "silent traveler" by trying to exclude it. Under the Immigration Act of 1990, a "dangerous contagious disease" is a medically excludable condition, as it was under previous legislation. Active TB has retained its Public Health Service definition as such a condition.[22] Not until treatment renders victims non-infectious are they eligible for admission. Screening consists of a chest radiograph for all applicants over fifteen years of age and a tuberculin skin test for those under fifteen who have had close contact with persons known to have or suspected of having tuberculosis.

The 1990 CDC report on TB in no way stigmatized the foreign-born as inherently diseased but strongly urged tightening screening procedures and encouraged state and local health departments to implement a program of preventive drug therapy for immigrants and refugees under thirty-five who had a positive tuberculin test. Those of any age with an abnormal chest radiograph were urged to start on preventive therapy within thirty days of arrival within the jurisdiction.

States and cities pursuing compliance with the CDC recommendations have not always succeeded. Immigrants and refugees are on the move, often do not stay in one jurisdiction long enough to complete treatment, and do not notify health officials in their new jurisdiction that they are receiving drug therapy. Discontinuation of treatment can have dire consequences for the individual newcomer who may find that the TB bacillus has mutated and become drug-resistant.[23] This, indeed, presents a public health danger to the surrounding community. However, because TB has hardly been a public health menace in recent decades, tuberculosis legislation is often archaic and inadequate. Only twenty-four states and the District of Columbia even grant health officials authority to impose treatment.

Because TB is not only increasingly common among immigrants but also among the homeless, internal migrants, and AIDS victims, jurisdictions have been embroiled in debates over the proper balance of patient rights and public health considerations. In reviewing TB legislation, Harvard health law professor Lawrence O. Gostin discovered that many states lacked guarantees that individuals forcibly confined while taking medication would have access to an attorney and a court hearing. Only 12 states plus the District of Columbia protect the confidentiality of patients.[24]

In March 1993, the New York City Board of Health adopted strict regulations designed to detain tuberculosis patients who do not complete treatment on their own. This measure could require confinement of patients for more than a year, some at a special 25-bed unit at Goldwater Memorial Hospital on Roosevelt Island. Advocates of the measure cite the desire to cure the specific patient, but even more so the need to stop production of deadly drug-resistant strains of the bacilli that develop in those who repeatedly stop and start medication.

Echoing the controversy over Mary Mallon earlier in the century, New York officials intentionally straddled the border between civil liberties and the public's health interests. Detained patients have a right to a lawyer paid for by the city and to a court hearing. Although patients must be guarded, it is to be in a style of security similar to that provided in psychiatric wards rather than a high-security prison. Moreover, New York City's health code revisions required the health commissioner to petition for a court order within two months of the detention—or within five days of a request for release from the patient—and to appear in court every three months to demonstrate with "clear and convincing evidence" the continued need for the order.[25] Given the data on new TB cases in New York City, many of those incarcerated are likely to be foreign-born. Those TB victims not in an active state and under treatment to avoid the infectious state are carefully monitored and offered incentives to continue the medication, including food vouchers, subway tokens, and small amounts of cash to help them meet clinical appointments.[26]

The study's third theme, then, the institutional response of national, state, and local governments to qualitative and quantitative dimensions of immigration as it affects and is affected by public health considerations, is echoed in the more contemporary response to TB. As in times past, a variety of nongovernmental organizations have also mounted institutional responses to the medical problems of immigrants and refugees. Because so much of a physician's ability to cure and a patient's ability to recover health or maintain it is grounded in communication, important institutional responses facilitate discussion between patient and physician. In his study of health services to immigrants in the 1920s, Michael Davis emphasized the crucial role of translators and the pitfalls of using close relatives of patients, who might interject themselves into the physician/patient dialogue. In December 1992, the *New York Times* reported a fresh solution to an old problem:

> In a cramped examination room of a Lower East Side health clinic recently, a distraught young mother, who spoke Chinese, stood holding a tiny baby while a nurse, who spoke only English, examined the child. The infant had been ill, the mother explained, and a relative had prescribed medicinal tea.
>
> The baby is too young for tea, the nurse said. The baby is too young for any medicine that the doctor or nurse does not order. The mother nodded.
>
> The exchange would have been impossible without the third person in the room—Lorraine Lau, 21, who is fluent in both English and Cantonese and who is earning credits at Hunter College for working as a translator in the pediatric clinic.[27]

The program that provided Ms. Lau is one of dozens of experiments going on around New York City to help the 28 percent of the population that is foreign-born cope with the perpetual problem in the health care system, their inability to communicate with doctors and nurses. As Beth Israel Medical Center President Dr. Robert G. Newman observed, "Language isn't an extra nicety to make someone feel more at home. It's essential to good medical care."[28] In response to that essential, Beth Israel, founded to cater to the needs of eastern European Jewish immigrants, opened a separate clinic for Japanese patients in 1990.

That same year, interested physicians, health care practitioners, community advocates and organizers, social workers, public health professionals, health care planners and administrators, and academicians formed the New York Task Force on Immigrant Health. With funding from New York City's United Hospital Fund and various foundations, the New York Task Force, under the directorship of Dr. Francesca Gany, was founded to "facilitate the effective delivery of culturally, linguistically, and epidemiologically appropriate health care to the city's growing numbers of immigrants throughout the five boroughs of New York."[29] The task force has sponsored roundtable forums on such topics as "Breast and Cervical Cancer, Birth Outcomes, Stressors Impacting on Immigrant Adolescents in New York City, Tuberculosis, and Family Planning and Contraceptive Use." It has developed an on-line comprehensive bibliography on immigrant health that had over 2,500 entries by the end of 1992 and is being continually updated. Activities in curriculum development and training are ongoing and there is much cross-fertilization of efforts and expertise.[30]

The minutes from the task force roundtable forum, "Tuberculosis in the Foreign-born and in Puerto Ricans," March 24, 1992, suggest the value of the task force in disseminating the vast array of data being amassed about the relationship of immigrants to the TB epidemic. In the pages of the minutes, too, are echoes of the past. Even as the International Ladies Garment Workers' Union, founded by southern and eastern Europeans, once struggled against tuberculosis contracted by members in shops and tenements, the ILGWU has more recently conducted a pilot study on tuberculosis among its Chinese-American union members.[31]

Dr. Pearl Korenblitt of the Mount Sinai School of Medicine's Division of Occupational Medicine reported to the New York Task Force roundtable forum that the ILGWU has 20,000 Chinese-American members, of whom one-third were recent immigrants. In the pilot, the union's health center nursing staff went out to eight garment shops in Chinatown to screen 199 members. Of these, 183 were at work on the day of the screening. Fifty-one percent tested positive on the tuberculin test, or PPD (purified protein derivative, the standard test material). Ninety-one percent of those who tested positive were women. The mean age was 41.6 years old. Only 2.7 percent had a history of being treated for active TB and only one person had a recent contact with someone known to be active.[32]

Offering preventive tuberculosis therapy to Chinese garment workers involves a difficult cultural negotiation, not unlike those of an earlier era. It suggests the fourth theme of this study, immigrant and refugee responses to differences between their own conceptions of therapy, health care, disease prevention and hygiene, as compared with those of native-born Americans. Only 20 percent of Chinese garment workers in the pilot were compliant with preventive therapy because TB is so stigmatizing in the Chinese community, even as it once was among eastern European Jews. Therefore, therapy must be presented with great discretion, much care being taken to avoid labeling. According to Korenblitt, workers in small shops often leave en masse if they discover that someone among them has tuberculosis. Although the Chinatown Health Clinic reports that one-third of tested clients have a positive tuberculin test, patients often refuse therapy if they feel healthy. The clinic's Susan Seto-yee observed that "patients feel that taking pills is not good for them and may even lie to the practitioner about their compliance."[33]

Commenting on the Haitians' encounter with TB, Dr. Jacques Mathieu of the Mount Sinai School of Medicine and the New York City Department of Health also defined it as a cultural negotiation. He finds that reliance on folk medicine and prayer to heal may play a role in the climbing morbidity rate. Hesitation to seek medical attention promptly in the absence of health insurance cannot be ignored. Some simply stop all medication when they feel well. Others are troubled by the appearance of the pills they are given. When the generic medication stocked by local clinics changes, so does the color of the pills, a change that causes some Haitians to become wary and stop all therapy.[34]

The last lines of the TB roundtable forum's minutes echo both earlier eras and that more recent Haitian experience that resulted in the traffic snarl in lower Manhattan. Participants agreed that "it is important to not blame immigrants for the current TB epidemic in New York City."[35] If fear of the foreign-born and fear of illness are locked in a timeless embrace, public awareness that the double helix persists suggests the possibility of breaking that bond.

Unlike immigrants, refugees are those who have suffered persecution or have a "well founded fear of persecution based on race, religion, nationality, membership in a particular social group, or political opinion."[36] Because refugees are often in flight and at times have suffered physical and emotional trauma prior to departure, their health and psychological problems are often distinct from those of legal or even undocumented aliens. During the past two decades millions of refugees have fled Southeast Asia and Latin America. In 1991, 100,229 refugees arrived in the United States. Most came from the former Soviet Union, 63,185, but second and third were the Southeast Asian countries of Vietnam, 25,412, and Laos, 8,726.

Refugee health, especially refugee mental health, has become a magnet for psychiatrists, psychologists, anthropologists, sociologists, social workers, and health care providers of all kinds.[37] The problem is how best to cross cultural lines to understand and treat ailments, which are as debilitating to one's ability to function in the United States as a contagious disease or a damaged limb, without shattering the immigrant's culture and emotional support system in the process.[38] The refugee dilemma too, then, echoes an aspect of this volume's fourth theme, the push and pull between traditional cultures and the pressures to acculturate.

Just as Russian Jewish immigrants explained to Dr. Maurice Fishberg that they wanted to return to Russia because the "air was too strong" in the United States, a way of expressing depression, Khmer refugees from war-torn Cambodia speak of being *bebotchit.*[39] In a culture that has no Western concept of depression, *bebotchit* is defined in Khmer as a "deep sadness inside oneself" caused by a specific set of circumstances. As explained by Dr. Richard Mollica of the Harvard Program in Refugee Trauma, "*bebotchit* is not a free floating feeling but a reactive state of depression related to experiencing an unfortunate event. The feeling of *bebotchit* is also so deep inside a person it can be hidden from being noticed by others."[40] Although precise cultural equivalents are difficult to establish with certainty, *bebotchit* is characterized by many of the same symptoms as what has been commonly called in Western psychiatry post-traumatic stress disorder (PTSD).[41] There are four categories of symptoms. One is recurrent memories related to the traumatic event—including, nightmares, daytime memories, or flashbacks. A second is persistent avoidance of anything associated with the trauma. A third is diminished emotional responsiveness to the external world, and a fourth, persistent symptoms of heightened nervousness not present before the trauma (e.g., sleep disturbances, being on guard, an exaggerated startle reaction).[42]

Mollica speculates that PTSD is "a form of cultural bereavement related to the loss of homeland, culture, tradition, and national identity."[43] Dissociation, too, is a recognized response in which victims begin to distance themselves from trauma by depersonalizing and detaching from the world. In dissociation, "specific internal mental contents (memories, ideas, feelings, perceptions) are lost to conscious awareness and become unavailable to recall."[44]

At times, dissociation can lead to psychosomatic illness. Chhean Im, a Khmer refugee who survived Cambodia's killing fields, cannot see through eyes that show no signs of physical damage, eyes that were forced to watch the Khmer Rouge kill some of her family and forcibly remove others from their homes. In prison camp, she saw individuals burned alive, thrown into fires by prison guards for minor infractions such as stealing a bit of rice. She told the *New York Times* about the last thing she remembers seeing, an especially brutal killing by the Angka, the Khmer Rouge's national police force. "They came and picked out one woman and one man and hit them over and over," she reported. "I started crying hard for a long time. It felt like there was a big needle pushing through my head." Some days later when she was no longer crying, Chhean Im could not see.[45] Having been blind for over a decade, she is suicidal.

Chhean Im's case is not unique. Over half of the 170,000 Cambodian refugees living in the United States live in Los Angeles County, most in a Long Beach enclave known as Little Phnom Penh. In 1984, local ophthalmologists began to notice an unusually high incidence of vision complaints among the neighborhood's female refugees. By 1991, 150 cases of individuals who had lost all or most of their sight had been reported in the community. Once physicians and psychologists became aware of the pattern, they began to plan therapy. Implementation has not been easy because signed consent forms are required to participate in a study of the condition, and many refugees are frightened of any contact with government officials because of their home country experiences. Dr. Gretchen Van Boemel, associate director of clinical electrophysiology at the Doheny Eye Institute, reported, "We needed their signed consent to participate, but the minute you say, 'Will you sign this?' they thought we were coming to take them away. It sounded to them like they were never coming back." Dr. Van Boemel and her associates learned that when the Khmer Rouge came during the night, they would knock on the door, saying "The Government needs you."[46] Other experts believe that the reluctance to seek medical treatment results less from fear or cultural differences but from the need to retain the blindness as a defense against the old memories and a strange new land.[47]

The need to comprehend and be sensitive to cultural and experiential differences between Americans and new arrivals is increasingly standard curriculum for health care givers. A medical anthropologist who teaches in a nursing program at California State University at Dominguez Hills uses case histories to achieve the kind of cultural sensitivity that Michael Davis advocated in his 1921 study of immigrant health. When a thirty-eight-year-old Cambodian male was hospitalized semicomatose with severe headache, nausea, vomiting, and lethargy, one of his relatives began to rub large welts on his skin with an oillike substance. Other Asian patients as well were found to have the red welts or were observed rubbing themselves with a quarter. If a nurse removed the coin, the patient became distraught. Nurses displayed "a combination of disbelief, disgust, curiosity, ridicule, intolerance, and finally attempts to understand." Most were distracted by the welts; some, treating children, thought it evidence of child abuse.

The nurses learned that patients were engaged in the Asian practice of coin rubbing. A heated or oiled coin is rubbed into the body to produce welts on the affected area in the belief that the illness must be drawn out. Red welts indicate to patients that the illness has been brought to the surface of the skin. They claim that such welts will appear only on the skin of an ill person, no matter how hard the skin is rubbed. Medical personnel note that even Asian immigrants who consult Western physicians for serious ailments continue coin rubbing for minor ills such as common colds. Nurses are taught that before becoming sidetracked by the welts or misreading their significance, "it would be a good idea to pull out a coin and mime rubbing the body with it. That, along with a questioning look, would probably convey the message. If the patient or family nods in agreement, the marks should be ignored."[48] In one tragic episode where such precaution was not taken, a Vietnamese man was arrested for child abuse when he brought his ill son to an emergency room and doctors did not recognize the marks as evidence of coining. Humiliated by his ordeal, the father later committed suicide.[49]

Increasingly, Western health care providers are willing to share turf with healers from non-Western medical traditions. Amos Deinard of the University of Minnesota Hospital community clinic and well versed in the problems of treating Laotian Hmong refugee patients has adopted an approach that would have pleased Dr. Hubert Dong of San Francisco's Chinese Hospital. According to pediatrician Deinard, "Our attitude is, you bring your shaman and we'll bring our surgeon and let's see if we can work on the problem together."[50] Deinard's view and that of many contemporary health care providers faced with treating the Hmong is that health care must be regarded as "a cooperative undertaking involving acculturation, a bi-directional process that requires the refugees (patients) to make all the changes."[51]

Dr. Philip Brickner of New York's St. Vincent's Hospital, which was founded to care for ailing Irish Catholic immigrants and protect their souls from proselytizing Protestants, supervises a Chinatown outreach program, "Living-At-Home," that brings medical care to the elderly. He is willing to tolerate alternate approaches to medicine as long as they do not interfere with the efforts of his teams, each of which consists of a physician, a nurse, a social worker, and an interpreter. The goal is to provide quality medical care at home so that the patients can avoid institutionalization. Often these elderly patients have never been treated by an American physician and have depended their entire lives on herbalists. The team accommodates the patient's cultural preferences as much as possible. "It sounds like birds' nests and penicillin but it has worked."[52]

Even as certain diseases such as tuberculosis make an encore among immigrants and refugees, the problem of medical quackery aimed at the foreign-born has also recurred. In the Adams-Morgan neighborhood of Washington, D.C., heavily populated with immigrants and refugees from Latin America, newcomers in need of medical care must be on their guard. Spiritualists and healers peddling herbal cures for which they charge twenty-five to seventy dollars have long been a benign presence in the community, but an influx of new arrivals in the 1990s has expanded the trade and the opportunity for high-priced fraud.

Maurico Zelaya from El Salvador, an expectant father worried about the future, was one of the victims. A spiritualist charged Zelaya $350 for tests that revealed the stiffness in his back as the work of "*algo malo*"—(something evil). The tests certainly seemed conclusive. A twenty-dollar bill, folded into a triangle, wrapped in aluminum foil and sealed with his own saliva, had blistered the palm of Zelaya's hand. A cotton swab rubbed across his chest and his "noble parts" had made water bubble and become the shade of blood. Seltzer, purchased at a nearby market, had turned to a coffee color after he gargled with it. For an additional $1,000, the healer persuaded Zelaya, who was a recovering alcoholic, to swallow a "cure" concocted from five eggs mixed in a bottle of red wine.[53]

Zelaya later told police that the cure resulted in a nine-day drinking binge during which he lost his job as a building superintendent and ended up in a detoxification center. "My mind was disturbed," Zelaya said, explaining why he spent the day after his release in Queens, New York, trying to track down the healer. He did not get the healer or his money back, but his physical pains were later relieved by purchasing "a wintergreen ointment made in Hong Kong" at a local supermarket that also sells herbal medications.[54]

The experience described by Zelaya, although as typical of medical quackery now as at the turn of the century, also raises another issue. Immigrants are attracted to healers for cultural reasons but also because most lack health insurance. Dr. Juan Romagoza, an immigrant physician from El Salvador who heads the Clinica del Pueblo in Washington, explains, "A lack of medical insurance frightens others who have learned that a visit to a doctor can result in expensive tests."[55] Romagoza clearly understands the significance of health insurance in his community. In 1990 the U.S. Census Bureau reported that sixty-three million Americans, or 28 percent of the population, lacked health insurance protection for substantial amounts of time during a twenty-eight month period. Only 48 percent of persons of Hispanic origin had full health coverage during the period, as compared with 72 percent of all Americans and 62 percent of the African-American community. In addition to African Americans and young adults of all backgrounds, Hispanics had one of the lowest rates of coverage.[56]

The absence of reliable health insurance is especially serious because immigrant workers get injured on the job more often and more seriously than nonimmigrant workers. Latino workers are a case in point. They tend to accept dangerous, low-paying jobs that other Americans will not. Studies conducted in Illinois, California, and New Jersey show that Latino workers—regardless of whether they are legal or undocumented—have injury rates two or three times higher than non-Latino, Caucasian Americans who hold the same jobs. African-American workers suffer higher rates than Caucasians, too, but lower rates than Latino workers.[57]

Language is part of the reason. Impartial health and safety experts interviewed by the *New York Times* describe Latino workers as young and inexperienced, most with barely a grade school education and a poor to nonexistent knowledge of English. Because employers rarely offer job training in Spanish, workers operating dangerous machines or handling toxic substances are at greater risk than non-Latinos. Being weak in English also makes it harder for such workers to quit jobs that they recognize as unsafe.[58]

Latino workers often do not understand their rights in the workplace, and undocumented workers are reluctant to complain, fearing deportation. Even though undocumented workers are as entitled as legal workers to medical care under workmen's compensation laws, illegals often fear that a doctor or a hospital will report them; so they turn to herbalists or other uncredentialed, nonallopathic healers, who are often quacks.

Although no organization collects national data on the ethnic origins of workers who get sick or sustain serious injury, several separate studies show similar patterns. A study examining 103 Latino patients and 110 non-Latinos at the Loyola University Medical Center near Chicago showed that 57 percent of injured Latinos had the most serious kinds of work-related injuries, including crushings and amputations, as compared with 22 percent of non-Latino workers. The California Department of Health, which requires that medical laboratories report blood tests showing elevated levels of lead, found that 49 percent of the workers in such reports had Latino surnames, a proportion almost double the state's 26 percent Latino population. The data consisted of records on 4,000 workers with excessive lead levels employed at 328 businesses between 1987 and 1990.[59]

Even as Alice Hamilton earlier in the century witnessed mortally ill Italian migrant laborers going home to die, foreign-born migrant labor continues to suffer acute health problems, especially those engaged in farm labor.[60] The federal Office of Migrant Health, established in 1962 estimated that in the late 1980s there were three million migrant and seasonal farm workers and dependents in the United States, of whom about a third, or one million, were migrants. Among the migrants Latinos predominate, making up as much as 90 percent of the migrant force, including Mexican-Americans, Mexicans, and Central Americans, although there are also substantial numbers of American Indians and Southeast Asians.

There are many health risks endemic to farm work, such as skin cancers from overexposure to the sun, but none that compare with the effects of chemical pesticides. Some forty-five thousand pesticide products are sold in the United States annually, most for use in agriculture. Exposure has been linked to leukemia, lymph node cancer, multiple myeloma (bone cancer) in adults, leukemia and brain cancer in children; birth defects, spontaneous abortion, sterility, and menstrual dysfunction; liver and kidney dysfunction, nervous system disorders such as poor motor coordination and thought processes, anxiety, and depression; and immunological abnormalities.[61]

Fearing job loss, migrant workers tend to avoid taking time off work to seek medical attention, especially undocumented migrant workers, who fear deportation should health care givers report them. Few migrants are fluent in English; a majority speak Spanish only. Language and cultural barriers, as well as the mobility of the migrant labor force, render health education and care problematical and make more appealing the ministration of *curanderos* (healers) and *brujas* (witches). Many migrant laborers in the Southwest have families in nearby Mexico and prefer to seek medical care there. Instead of finding care as they need it, migrants wait to return home, often at the expense of their own health and that of those around them.[62] Calling for more and improved health services for migrant laborers whatever their legal status, Surgeon General Antonia C. Novello in 1993 pointedly observed, "Viruses and bacteria don't ask for a green card."[63] Patterns common during the peak period of immigration between 1890 and 1920 are echoed in the data now being gathered on a more recent wave of newcomers.

Like water over a falls, immigration to America is ever changing but never ending. Different water cascades downward in the same direction, though at different volume and rate, depending on the season and weather. History moves similarly, escaping precise calculation. Or as American intellectual and cultural historian John Higham has reminded us, "History may move partly in cycles, but never in circles." Throughout American history, immigration and public health have shaped each other's progress. In a nation of nations, all things are affected by the continual arrival of newcomers from distant shores, their needs, and their cultural baggage. Most often the presence of the foreign-born has contributed to innovation, fueling medical discovery and institutional change as they alter the public health milieu in the United States. However, some things have remained the same. The double helix of health and fear that accompanies immigration continues to mutate, producing malignancies on the culture, neither fatal nor readily eradicated.

APPENDIX I

Classification of Excludable Medical Conditions According to the 1903 *Book of Instructions for the Medical Inspection of Immigrants*

CLASS A: "THOSE WHO ARE EXCLUDED FROM ADMISSION INTO THE COUNTRY BY REASON OF THE EXISTENCE OF A DISEASE OR ABNORMAL CONDITION OF A CHARACTER EXPRESSLY DECLARED BY THE LAW ITSELF TO CONSTITUTE A GROUND FOR SUCH EXCLUSION."

SUBDIVISION I: DANGEROUS CONTAGIOUS DISEASES

1. Trachoma
2. Pulmonary Tuberculosis

SUBDIVISION II: LOATHSOME DISEASES

1. Favus
2. Syphilis
3. Gonorrhea
4. Leprosy

SUBDIVISION III: INSANE PERSONS

"In the case of immigrants, particularly the ignorant representatives of emotional races, due allowance should be made for temporary demonstrations of excitement, fear, or grief, and reliance chiefly placed upon absolute assurance of the existence of delusions or persistent refusal to talk or continued abstinence from eating."

Subdivision IV: Idiots

Definition: "An idiot is a person exhibiting such a degree of mental defect, either inherited or developed during the early period of life, as incapacitates the individual for self-maintenance or ability to properly care for himself or his interests."

Class B: Aliens Excluded as Likely to Become Public Charges

"A few of the most common causes" listed are:

1. Hernia
2. Valvular heart disease
3. Pregnancy [unwed mothers were rejected]
4. Poor physique [such that they could not earn a living]
5. Chronic rheumatism
6. Nervous afflictions, including locomotive ataxia, spastic paraplegia, and "other incurable nervous diseases"
7. Malignant diseases—including carcinoma, sarcoma, etc.
8. Deformities—including kyphosis, lordosis scoliosis, mutilations of the extremities
9. Senility and debility
10. Varicose veins—"especially when affecting the lower extremities"
11. Eyesight—"serious defects of vision"
12. General considerations—"any disease or deformity which can not be placed in any of the above classes, but which will interfere with an immigrant's ability to earn a living, should be certified"—e.g. scar formation of the conjunctiva, entropion, infiltration of the cornea, etc.

APPENDIX II

Classification of Excludable Medical Conditions According to the 1917 *Book of Instructions for the Medical Inspection of Immigrants*

CLASS A: "ALIENS WHOSE EXCLUSION IS MANDATORY UNDER THE LAW BY REASON OF A CERTAIN SPECIFIED PHYSICAL OR MENTAL DEFECT OR DISEASE."

1. Idiots
2. Imbeciles
3. Feebleminded persons
4. Epileptics
5. Insane persons
6. Persons of constitutional psychopathic inferiority
7. Persons with chronic alcoholism
8. Persons certified as mentally defective
9. Persons afflicted with tuberculosis in any form
10. Persons afflicted with a loathsome contagious disease:
    Favus
    Ringworm of scalp and nails
    Sycosis barbae
    Actinomycosis
    Blastomycosis
    Frambesia (yaws)
    Mycetoma (Madura foot)

Leprosy
Oriental sore (cutaneous leishmaniasis)
Venereal diseases, "namely syphilis, gonorrhea and soft chancre"

11. Persons afflicted with a dangerous contagious disease:
Trachoma
Filariasis
Amoebiasis
Schistosomiasis
Other diseases caused by animal parasites, such as leishmaniasis, trypanosomiasis, paragonomiasis, clonorchiasis, etc.

CLASS B: "ALIENS NOT COMPREHENDED [*SIC*] UNDER CLASS A WHO ARE PHYSICALLY DEFECTIVE OR DISEASED, SUCH DEFECT OR DISEASE BEING OF A NATURE TO CAUSE DEPENDENCY OR TO AFFECT THE ABILITY OF THE ALIEN FOR SELF-MAINTENANCE"

Examples include but are not restricted to:
Hernia
Heart disease
States of permanently defective nutrition and of marked defective skeletal and muscular development
Chronic arthritis and myositis
Nervous affections
Malignant new growths
Deformities
Senility
Varicose veins
Eyesight—"defects of vision" where unaided vision is 20/70 or less
Chronic malaria
Uncinariasis
Pellagra
Beriberi
Cutaneous affections
Eruptive fevers
Anemia

CLASS C: "ALIENS WHO PRESENT DEFECTIVE OR DISEASED CONDITIONS OF A LESS SERIOUS CHARACTER, BUT WHO MUST BE CERTIFIED FOR THE INFORMATION OF THE IMMIGRATION OFFICERS AND BOARDS OF SPECIAL INQUIRY UNDER PROVISIONS OF THE LAW."

Pregnancy (Those accompanied by a male guardian or met by a male guardian were admitted.)

Notes

INTRODUCTION: THE DOUBLE HELIX OF HEALTH AND FEAR

1. Lynne Duke, "U.S. Camp for Haitians Described As Prison-Like," *Washington Post,* September 19, 1992.
2. In his June 8, 1993 ruling in *Haitian Centers Council v. McNary,* United States District Judge Sterling Johnson, Jr. admitted the 158 refugees still interned at the U.S. Naval Base at Guantánamo Bay. He noted that they had established their claim to political asylum and were being held only because they were HIV positive. Judge Johnson concluded his opinion with the observation that "the detained Haitians are neither criminals nor national security risks. Some are pregnant mothers and others are children. Simply put, they are merely the unfortunate victims of a fatal disease. . . . The Haitian camp at Guantánamo is the only known refugee camp in the world composed entirely of HIV positive refugees. The Haitian plight is a tragedy of immense proportion and their continued detainment is totally unacceptable to this Court." Opinion of Sterling Johnson, Jr., in *Haitian Centers Council, Inc. v. McNary,* June 8, 1993, United States District Court, Eastern District of New York, 39. The author is grateful to Lucas Guttentag, Director of the American Civil Liberties Union Foundation's Immigrants' Rights Project for providing a copy of Judge Johnson's opinion and his own insight into the case in a telephone conversation on June 11, 1993. See also *New York Times,* June 9, 1993, and *Washington Post,* June 9,10, 1993.
3. Interview with Fannie Kligerman quoted in David M. Brownstone, Irene M. Franck, and Douglass L. Brownstone, *Island of Hope, Island of Tears* (New York: Rawson, Wade, 1979), 169.
4. Howard F. Stein, *American Medicine as Culture* (Boulder: Westview Press,

1990). Stein observes, "Many members of ethnic cultures augment if not supersede their original medical cultural repertory with beliefs and practices common to national American society—for example, the high value on medical technology and purchase of over-the-counter drugs at pharmacies." Nor has this medical cultural exchange been one directional, according to Stein. "Acupuncture, regarded by the biomedical establishment in the early 1980s as an exotic folk practice of China, has by now been partially absorbed into biomedical professional culture and has become an alternative (popular) treatment modality." Even in the contemporary environment of international biomedical exchange, therapies are culturally shaped. See Lynn Payer, *Medicine and Culture, Varieties of Treatment in the United States, England, West Germany, and France* (New York: Henry Holt, 1988). The notion of the medical culture that is shaped by far more than science but economic priorities, power relationships, and other cultural values is also discussed by Charles Rosenberg in *The Care of Strangers: The Rise of America's Hospital System* (New York: Basic Books, 1987), 7. See also Lester S. King, *Transformations in American Medicine From Benjamin Rush to William Osler* (Baltimore: Johns Hopkins University Press, 1991).

5. Historian John Duffy's classic definition of public health in the United States was "community action to avoid disease and other threats to the health and welfare of individuals and the community at large." More recently, scholars have expanded that definition to include the notion that public health policy also actively promotes health, rather than merely maintains it. Public health historians who have acknowledged the crucial impact that peoples of diverse cultures have had on public health in the United States tend to treat the relationship as one directional, with medicine and hygiene being merely aspects of newcomers' acculturation. From this perspective, immigrants have historically played a passive role. Either they were turned away at America's door as physically or mentally unfit or, if admitted, made to accept prevailing American social constructions of health and disease. Beyond accepting their hosts' cultural assumptions, newcomers are depicted as willing to follow prescribed behavior—both therapeutic and preventive—for sustaining the public health. John Duffy, *The Sanitarians, A History of Public Health* (Urbana: University of Illinois Press, 1990), 1. Duffy's definition differs little from George Rosen's—"community action in the promotion of health and the prevention and treatment of disease"—in the author's classic volume, *A History of Public Health* (New York: MD Publications, Inc., 1958), 18.
6. I have consciously omitted discussion of involuntary immigrants such African slaves, because that subject is deserving of special treatment. Indeed, there are already a number of fine works in the field, including Kenneth Kiple and V.H. King, *Another Dimension to the Black Diaspora: Diet, Disease and Racism* (Cambridge, England: Cambridge University Press, 1981) and, more recently, Kiple, ed. *The African Exchange: Toward a Biological History of Black People* (Durham: Duke University Press, 1987). On

the slave trade, see the books and articles of Philip Curtin, including, "Epidemiology and the Slave Trade," *Political Science Quarterly* 83 (June, 1968): 191–216. On the antebellum slave community, see Todd L. Savitt, *Medicine and Slavery: The Diseases and Health Care of Blacks in Antebellum Virginia* (Urbana: University of Illinois Press, 1978).

7. The classic study of this genre of prejudice is John Higham, *Strangers in the Land, Patterns of American Nativism, 1860–1925* (New Brunswick, N.J.: Rutgers University Press, 1955). The classic work on prejudice remains Gordon W. Allport, *The Nature of Prejudice* (Reading, Mass.: Addison-Wesley, 1979; orig. 1954). See also George Eaton Simpson and J. Milton Yinger, *Racial and Cultural Minorities: An Analysis of Prejudice and Discrimination,* 5th edition (New York: Plenum Press, 1985).
8. Erving Goffman, *Stigma, Notes on the Management of Spoiled Identity* (Englewood Cliffs, N.J.: Prentice-Hall, 1963), 2–3.
9. The fear of contamination or pollution in human culture is treated by the anthropologist Mary Douglas, *Purity and Danger: An Analysis of the Concepts of Pollution and Taboo* (London: Routledge, 1989; orig. 1966). Pioneering work on the border between psychology and anthropology with respect to the concept of contagion and the emotion of disgust is being done by Paul Rozin. See Paul Rozin and Carol Nemeroff, "The Laws of Sympathetic Magic, A Psychological Analysis of Similarity and Contagion," in James W. Sigler, Richard A. Shweder, and Gilbert Herdt, eds., *Cultural Psychology, Essays on Comparative Human Development* (Cambridge: Cambridge University Press, 1990), 205–32. Mexican novelist and cultural critic Octavio Paz argues that North Americans are obsessed with purity that can be achieved best through separation. Because of the primacy of science as cultural authority, the obsession is expressed in a kind of "cult of hygiene" which "inspires certain attitudes toward sports, work, food, sex, and race." Octavio Paz, "Eroticism and Gastrosophy," *Daedalus* 101 (Fall, 1972): 67–85.
10. *Miami Herald,* April 11, 1985. The original classification appeared in the CDC's *Morbidity and Mortality Weekly Report,* March 4, 1983, 47.
11. In introducing her discussion of illness as a metaphor, Susan Sontag also uses migration and ethnicity as metaphors: "I want to describe, not what it is really like to emigrate to the kingdom of the ill and live there, but the punitive or sentimental fantasies concocted about the situation; not real geography, but stereotypes of national character." See Susan Sontag, *Illness As Metaphor* (New York: Random House, 1979) and later, *AIDS and Its Metaphors* (New York: Farrar, Straus & Giroux, 1989).
12. A volume that attributes the episode to preexisting patterns of American racism and ethnocentrism is Paul Farmer, *AIDS and Accusation, Haiti and the Geography of Blame* (Berkeley: University of California Press, 1992).
13. Lev. 19:33–34.
14. In answer to the question of what happens when peoples meet through migration, the broader history of disease's impact upon non-immune populations generally has been brilliantly analyzed in two works by Alfred J.

Crosby, Jr.: *Ecological Imperialism: The Biological Expansion of Europe, 900–1900* (New York: Cambridge University Press, 1986), and *The Columbian Exchange, Biological and Cultural Consequences of 1492* (Westport, Conn.: Greenwood Press, 1972). See also P.M. Ashburn, *The Ranks of Death: A Medical History of the Conquest of America* (New York: Coward-McCann, 1947) and William H. McNeill, *Plagues and People* (Garden City, N.Y.: Doubleday, 1976).

15. Foucault explains this linkage of a group's general behavior, especially their living conditions and their effect upon the public health as a pre-nineteenth century development in the North American and European communities. See Michel Foucault, "The Politics of Health in the Eighteenth Century," *Power/Knowledge* (New York: Harvester Press, 1980), 175–77.
16. Scholars influenced by the writings of Italian Marxist Antonio Gramsci have characterized the aspirations of well-off white Anglo-Saxon Protestants in the late nineteenth and early twentieth centuries as the perpetuation of a "cultural hegemony." There is no doubt that some nativists saw in scientific medicine an ideal instrument to employ in support of their designs. In his essay, "The Concept of Cultural Hegemony: Problems and Possibilities," *American Historical Review* 90(June, 1985): 567–93, T. J. Jackson Lears reminds readers of Gramsci's characterization of hegemony as "the 'spontaneous' consent given by the great masses of the population to the general direction imposed on social life by the dominant fundamental group; this consent is 'historically' caused by the prestige (and consequent confidence) which the dominant group enjoys because of its position and function in the world of production." Lears is not satisfied with that definition and I share that dissatisfaction for various reasons, not the least of which is the absence of "spontaneous consent" or passivity on the part of immigrants in the struggle of the native-born to retain domination over newer groups in the United States. As will be suggested in later chapters, different immigrant groups chose different methods to resist the use of health and hygiene by the native-born as instruments of cultural domination.
17. Edward Alsworth Ross, a harsh critic of immigration and a crusader against "the race suicide" he thought the inevitable result of unrestricted immigration, wrote the first substantial work on the notion of social control, *Social Control: A Survey of the Foundations of Social Order* (New York: Macmillan, 1901). However, in this volume Ross does not mention immigration. Instead, he was interested in commenting on a hot issue of late nineteenth- and early twentieth-century philosophical and sociological discourse, the breakdown of a "natural order" that had restrained unharnessed individualism and prevented it from becoming a destructive force. In light of this breakdown, Ross added his voice to the call for new types of social control to replace the rapid disappearance of concern for the common good that had been the traditional foundation of republican society. Another Progressive concerned by drift was Walter Lippmann, *Drift and Mastery, An Attempt to Diagnose the Current Unrest* (Madison: University of Wisconsin Press, 1985; orig. 1914).

18. The author heartily agrees with those critics who have denounced the frequent and careless use of the term *social control.* There is little new or worthy of note about the observation that throughout history, those who have held positions of power and authority have sought to use the material and cultural resources available to them to sustain and defend their position and to maintain quiescent the claims of those less wealthy or influential than themselves. Unfortunately, many of the new social historians both in the United States and elsewhere have all but defined every genre of social interaction among economic, racial, gender, and ethnic groups as the effort of one to assert social control over the other, and often in the cause of preserving a social order derived from the growth of industrial capitalism. See F. M. L. Thompson, "Social Control in Victorian Britain," *The Economic History Review* 34 (May, 1981): 189–209. I am indebted to my friend and colleague Dr. Janet Oppenheim for calling Thompson's article to my attention.
19. An excellent article on the potential for medicine's use by one group to assert social and cultural influence over another is Irving Kenneth Zola, "Medicine as an Institution of Social Control," in John Ehrenreich, ed., *The Cultural Crisis of Modern Medicine* (New York: Monthly Review Press, 1978): 80–100. Michel Foucault has also addressed the issue of how institutions can be used in "the subjugation of bodies and control of populations." See Foucault, *The History of Sexuality, Volume I: An Introduction,* trans. Robert Hurley (New York: Random House, 1990; orig. 1976), 137, 140.
20. The best single volume on the eugenicists remains Daniel J. Kevles, *In the Name of Eugenics, Genetics and the Uses of Human Heredity* (Berkeley: University of California Press, 1986; orig. 1985).
21. The argument is most clearly set forth in Morton Keller, *Affairs of State, Public Life in Late Nineteenth Century* (Cambridge: Belknap Press of Harvard University Press, 1977) and most broadly as well as clearly in Alan Dawley, *Struggles for Justice, Social Responsibility and the Liberal State* (Cambridge: Belknap Press of Harvard University Press, 1991).
22. C. S. Lewis, *The Abolition of Man or Reflections on Education With Special Reference to the Teaching of English in the Upper Forms of School* (New York: Macmillan Company, 1947), 35.
23. I have been especially influenced by Herbert Gutman's rejection of the notion that migration automatically resulted in a breakdown of social patterns with resulting "anomie and alienation." Immigrants' adaptation seems to me best characterized as a cultural negotiation that included all aspects of life, including that dealing with health and hygiene. See Gutman, "A Note on Immigration History," "Breakdown Models," and "Rewriting of the History of Immigrant Working-Class Peoples," in Herbert G. Gutman, *Power & Culture, Essays on the American Working Class,* ed. Ira Berlin (New York: Pantheon Books, 1987), 255–59. One of the first immigration scholars to reject the breakdown model was Rudolph J. Vecoli, "Contadini in Chicago: A Critique of *The Uprooted,*" *Journal of American History* 51 (Dec. 1964): 404–27.
24. *Washington Post,* November 6, 1988. I referred to this article in an earlier

publication, "Healers and Strangers: Immigrant Attitudes Toward the Physician in America—A Relationship in Historical Perspective," *Journal of the American Medical Association* 263 (April 4, 1990): 1807–11.

25. Amos S. Deinard and Timothy Dunnigan, "Hmong Health Care: Reflections of a Six Year Experience," *International Migration Review* 21(Fall, 1987): 857–65. For more on the Hmong and health care issues, see Glenn T. Hendricks, Bruce T. Downing, and Amos S. Deinard, eds., *The Hmong in Transition* (Staten Island, N.Y.: Center for Migration Studies, 1986), especially 331–446.
26. *New York Times,* September 29, 1990.
27. *New York Times*, December 13, 1990.

CHAPTER 1: "THE BREATH OF OTHER PEOPLE KILLED THEM"

1. Thomas Kamm, "Amazon Tragedy: White Man's Malaria and Pollution Imperil Remote Tribe in Brazil," *Wall Street Journal,* Mar. 21, 1990. The best work on the tribe is Napoleon A. Chagnon, *Yanomamo, The Last Days of Eden* (New York: Harcourt Brace Jovanovich, 1992; orig. 1983).
2. Ibid.
3. Ibid.
4. The author recognizes and respects the wishes of contemporary Native Americans to dispense with the term *Indian,* a misnomer that is a constant reminder of past oppression. However, to merely substitute one name for the other, especially when discussing the group's pretwentieth-century experience, would be ahistorical and a concession to a political correctness that is often artificial, trendy, and neglectful of a people's rich historical experiences, tragic though they may be. Therefore, I have opted to use Indians, Native Americans, Amerindians, and North American native-born interchangeably, intending to offend no one and in the spirit of William Shakespeare's comments on the significance of names.
5. The classic overview is William H. McNeill, *Plagues and People* (Garden City, N.Y.: Doubleday, 1976). McNeill demonstrates the role of disease in the flow of human history on a grand scale in a way that other works such as Hans Zinsser's *Rats, Lice and History* (Boston: Little, Brown, 1963; orig. 1934) do not.
6. John Duffy, "Smallpox and the Indians in the American Colonies," *Bulletin of the History of Medicine* 25 (July-Aug. 1951): 324.
7. Ibid., 325.
8. Donald J. Ortner, Noreen Tuross, and Agnes I. Stix, "New Approaches to the Study of Disease in Archeological New World Populations," *Human Biology* 64 (June 1992): 337–60.
9. Those who deny the presence of endemic syphilis in pre-Columbian America suggest that yaws, a disease prevalent in the humid tropics, and bejel, a similar disease found in arid warm climates, were mistaken for syphilis by earlier observers.

10. News in the unresolved mystery was the subject of an article in the *Washington Post*'s science supplement just prior to the Columbian Quincentenary. See Boyce Rensberger, "Did Syphilis Sail to Europe With Columbus and Crew?," *Washington Post*, July 27, 1992. In an interview after his return from examining the English skeletons, Smithsonian Institution anthropologist Dr. Donald Ortner, an expert on syphilis in ancient peoples, observed that the English findings were especially significant because they occurred in a climate where syphilis would have been a more likely form of treponemal disease than yaws or bejel. See Boyce Rensberger, "Findings Counter Theory on Syphilis," *Washington Post*, Nov. 19, 1992. Also, John Noble Wilford, "Clues Etched in Bone Debunk Theory of a Plague's Spread," *New York Times*, Nov. 17, 1992.

    The author wishes to thank Dr. Ortner for discussing the controversy with him at length after the examination of the English skeletons, Donald Ortner and Alan M. Kraut, telephone conversation, Nov. 30, 1992.
11. The best historical treatment of syphilis is Claude Quétel, *History of Syphilis*, trans. Judith Braddock and Brian Pike (Baltimore: Johns Hopkins University Press, 1992; orig. 1986). The dispute is discussed by Quétel, 39–44, 258–59.
12. Frank Livingston, "On the Origin of Syphilis: An Alternative Hypothesis," *Current Anthropology* 13 (Dec. 1991): 587–90.
13. Gerald Grob places the Amerindians' suffering into the larger context of how migration affects the disease environment in "Disease and Environment in American History," in *Handbook of Health, Health Care, and the Health Professions*, ed. David Mechanic (New York: Free Press, 1983), 3–22.
14. Francisco Guerra, "The Earliest American Epidemic: The Influenza of 1493," *Social Science History* 12 (Fall 1988): 305–25. For the significance of virgin soil epidemics of influenza in the decimation of Native Americans, see Alfred W. Crosby, "Virgin Soil Epidemics as a Factor in the Aboriginal Depopulation in America," *William and Mary Quarterly*, 3rd Series, 33 (1976): 293–94 and Henry F. Dobyns, *Their Number Become Thinned: Native American Population Dynamics in Eastern North America* (Knoxville: University of Tennessee Press, 1983), 18.
15. A fine discussion of zoonotic diseases from wildlife can be found in Calvin Martin, "Wildlife Diseases as a Factor in the Depopulation of the North American Indian," *Western Historical Quarterly* 7 (Jan. 1976): 47–62. See also Martin's *Keepers of the Game, Indian-Animal Relationships and the Fur Trade* (Berkeley: University of California Press, 1978). His innovative thesis that Indians' zealous pursuit of game animals was understood by them in the context of their cosmology as a war of retaliation that they must conduct against creatures who knowingly inflicted them with epidemic diseases—many of which were actually brought by Europeans—has provoked ample scholarly controversy. A fine volume of essays criticizing Martin's thesis is Shepard Krech III, ed., *Indians, Animals, and the Fur Trade, A Critique of*

"Keepers of the Game" (Athens: University of Georgia Press, 1981).

16. David Stannard raises questions about Guerra's thesis that the Indians on Hispaniola died of swine flu. He correctly observes that influenza has a short incubation period, "which makes it unlikely that the virus could have survived the lengthy ocean voyage (unless it was kept active by passing from host to host), and the difficulty of explaining how the virus was so well contained among the sows, even if they were stored below deck, and did not spread to the shipboard humans until the ships' arrival at the future site of Isabela." Recent research suggests that his own speculation is likely correct and Guerra's scenario is possible because in addition to direct host-to-host transferral, infection can come from "symptomless year-round carriers of the disease in whom contagion is triggered by an unknown mechanism during the so-called 'flu seasons.'" See Stannard, *American Holocaust, Columbus and the Conquest of the New World* (New York: Oxford University Press, 1992), chap. 3, n. 43, 300. See also R. E. Hope-Simpson and D. B. Golubev, "A New Concept of the Epidemic Process of Influenza A Virus," *Epidemiology and Infection* 99 (1987): 5–54. The role of animal infections in the decimation of Native American tribes in still speculative but increasingly intriguing as evidence mounts; see Martin, "Wildlife Diseases as a Factor in the Depopulation of the North American Indian."
17. P. M. Ashburn, *The Ranks of Death, A Medical History of the Conquest of America* (New York: Coward-McCann, 1947). The larger biological consequences of European exploration are treated in an especially provocative and engaging manner in Alfred W. Crosby, Jr., *The Columbian Exchange, Biological and Cultural Consequences of 1492* (Westport, Conn.: Greenwood Press, 1972). Several newer works were inspired by the Columbian Quincentennary and suggest some of the controversies that surfaced as writers sought to cut the cloth of the past to fit contemporary political consciousness. See especially Stannard, *American Holocaust*. A more balanced, less polemical approach can be found in two volumes of essays published by James Axtell, *After Columbus, Essays in the Ethnohistory of Colonial North America* (New York: Oxford University Press, 1988) and *Beyond 1492, Encounters in Colonial North America* (New York: Oxford University Press, 1992). In the latter work, see especially "Moral Reflections on the Columbian Legacy," 241–66.
18. This figure is a compromise estimate suggested by David J. Weber in *The Spanish Frontier in North America* (New Haven: Yale University Press, 1992), 28. Weber dismisses as too high Henry F. Dobyns's estimate of eighteen million for the entire Western Hemisphere in 1492. See Dobyns, *Their Number Become Thinned.* Seven million, a figure that Weber regards as too low, was calculated by Russell Thornton, *American Indian Holocaust and Survival, A Population History Since 1492* (Norman: University of Oklahoma Press, 1987). Another volume that treats the Indian depopulation of North America is Ann F. Ramenofsky, *Vectors of Death: The Archaeology of European Contact* (Albuquerque: University of New Mexico Press, 1987).

19. James Axtell, *The European and the Indian: Essays in the Ethnohistory of Colonial North America* (New York: Oxford University Press, 1981), 248.
20. Solorzano Pereira as quoted by Guerra, "Earliest Epidemic," 324.
21. Father Gabriel Sagard, *The Long Journey to the Country of the Hurons,* ed. George M. Wrong (Toronto: Champlain Society, 1939), 198. The passage is also cited in Virgil J. Vogel, *American Indian Medicine* (Norman: University of Oklahoma Press, 1970), 89.
22. James Adair, *The History of the American Indians* (London: printed for Edward and Charles Dilly, 1775), 124–25.
23. These practices are described in Axtell, *Beyond 1492,* 105.
24. John Lederer, *The Discoveries of John Lederer,* ed. William P. Cumming (Charlottesville: University of Virginia Press, 1958), 36–37.
25. See also Vogel, *American Indian Medicine.*
26. Ibid., 105–6.
27. James H. Merrell, *The Indians' New World, Catawbas and Their Neighbors from European Contact through the Era of Removal* (Chapel Hill: University of North Carolina Press for the Institute of Early American History and Culture, Williamsburg, Virginia, 1989), 19.
28. Francis Le Jau to the Secretary, June 13, 1710, in Frank J. Klingberg, ed., *The Carolina Chronicle of Dr. Francis Le Jau, 1706–1717,* in University of California Publications in History 53 (Berkeley: University of California Press, 1956), 78. Also cited in Merrell, *Indians' New World,* 21.
29. Merrell, *Indians' New World,* 24–25.
30. Ibid., 26–27.
31. Ibid., 237.
32. Patterns of human migration significantly affect the distribution of microorganisms among populations that do and do not have acquired immunities to particular microscopic creatures. If acquired immunities are present, the impact of contacts with new groups is inconsequential and often hardly detectable. However, the absence of such immunities, sometimes referred to as "virgin soil epidemics," can be devastating and often were for Native Americans. Francis Jennings, "Virgin Land and Savage People," *American Quarterly* 23 (1971): 319–41. Also, Jennings, *The Invasion of America: Indians, Colonialism, and the Cant of Conquest* (Chapel Hill: University of North Carolina Press, 1973). Jennings is quoted in James Axtell, *The European and the Indian,* 248.

    Epidemics have long attracted the attention of scholars, but as historian William H. McNeil has observed, "Failure to understand the profound differences between the outbreak of a familiar disease amid an experienced population and the ravages of the same infection when loosed upon a community lacking acquired immunities is, indeed, at the bottom of the failure of previous historians to give adequate attention to the whole subject." McNeil, *Plagues and People,* 3–4.
33. Thomas Morton, *New English Canaan* (1632) included in Peter Force, comp., *Tracts and Other Papers Relating Principally to the Origin, Settle-*

ment and Progress of the Colonies in North America, 4 vols. (Washington, D.C., printed by P. Force 1836–1847), 2, no.5, 18–19. Also cited in Axtell, *Beyond 1492,* 237.

34. Vogel, *American Indian Medicine,* 14. See also Anthony F. C. Wallace, "Dreams and Wishes of the Soul: A Type of Psychoanalytic Theory among the Seventeenth Century Iroquois," *American Anthropologist,* N.S., 60 (Apr. 1958): 234–48.
35. A brief but rich discussion of Indian interpretations of disease causation can be found in Vogel, *American Indian Medicine,* 14–15.
36. Martin, *Keepers of the Game,* passim, and for thoughtful critiques, Krech III, *Indians, Animals, and the Fur Trade.*
37. [James] *Adair's History of the American Indians* (1775), ed. Samuel Cole Williams (New York: Argonaut Press, 1966), 244–45.
38. Stannard, *American Holocaust,* 237.
39. Increase Mather, "An Historical Discourse Concerning the Prevalency of Prayer," in *A Relation of the Troubles Which Have Happened in New-England by Reason of the Indians There, from the Year 1614 to the Year 1675, etc.* (Boston: John Foster, 1677), 6.
40. Cotton Mather, *Magnalia Christi Americana,* vol. 1 (Hartford, Conn.: S. Andrus and Son, 1855), 55. Mather is quoted in C. D. O'Mally, ed., *The History of Medical Education* (Berkeley: University of California Press, 1970), 464. Also, Robert P. Hudson, *Disease and Its Control: The Shaping of Modern Thought* (Westport, Conn.: Greenwood Press, 1983), 27.
41. Harold M. Schmeck, "Last Samples of Smallpox Pose a Quandary," *New York Times,* Nov. 3, 1987.
42. William Bradford, *Of Plymouth Plantation, 1620–1647* (New York: Random House, 1981; orig.1856), 302–303.
43. Ibid.
44. The matter of inoculation was not without controversy in the Puritan community. While Cotton Mather was convinced of the practice's efficacy in preventing smallpox, others thought it a sin for healthy persons to bring sickness upon themselves, while others thought the practice a sinful means of interfering with God's will. See John R. Blake, "The Inoculation Controversy in Boston, 1721–1722," *New England Quarterly* 25 (1952): 489–506; also reprinted in Judith Walzer Leavitt and Ronald L. Numbers, eds., *Sickness and Health in America, Readings in the History of Medicine and Public Health,* 2d ed. (Madison: University of Wisconsin Press, 1985), 347–55.
45. The article perennially cited is Bernhard Knollenberg, "General Amherst and Germ Warfare," *Mississippi Valley Historical Review* (now *Journal of American History*) 41 (Dec. 1954): 489–94, 762–63. However, no historian has been able to verify the germ warfare that Knollenberg suggests.
46. Gary Nash, *Red, White and Black: The Peoples of Early America,* 3d ed. (Englewood Cliffs, N.J.: Prentice-Hall, 1982), 301.
47. Bernard Bailyn, *The Peopling of British North America, An Introduction* (New York: Knopf, 1986), 99–102. David Hackett Fischer, *Albion's Seed,*

Four British Folkways in America (New York: Oxford University Press, 1989), 251–52.

48. Robert P. Hudson, *Disease and Its Control, The Shaping of Modern Thought* (Westport, Conn.: Greenwood Press, 1983), 142. For the devastating effects of malaria upon early settlers, see St. Julien Ravenal Childs, *Malaria and Colonization in the Carolina Low Country, 1526–1696,* series 63 in the Johns Hopkins University Studies in Historical and Political Science (Baltimore: Johns Hopkins Press, 1940).
49. An excellent overview of the growth of the public health movement is John Duffy, *The Sanitarians, A History of American Public Health* (Urbana: University of Illinois Press, 1990).
50. George Rosen, *A History of Public Health* (New York: MD Publications, 1958), 288–89.
51. *Massachusetts Colony Acts and Resolves,* i, 452, as quoted in Emberson Edward Proper, *Colonial Immigration Laws, A Study of the Regulations of Immigration by the English Colonies in America* (New York: AMS Press, 1967), 29.
52. *Acts and Resolves,* ii, 337, as quoted in Proper, *Colonial Immigration Laws,* 30.
53. *Acts and Resolves,* iii, 982, as quoted in Ibid.
54. E. P. Hutchinson, *Legislative History of American Immigration Policy, 1798–1965* (Philadelphia: University of Pennsylvania Press, 1981), 396–404. Also, Roy L. Garis, *Immigration Restriction, A Study of the Opposition to and Regulation of Immigration to the United States* (New York: Macmillan, 1928), 22–27.
55. Rosen, *A History of Public Health,* 69. Scholars disagree over whether a forty-day period was first mentioned at Venice in the fourteenth century or even earlier, in 1127.
56. John B. Blake, *Public Health in the Town of Boston, 1630–1822* (Cambridge: Harvard University Press, 1959), 192–93.
57. Ernest Caulfield, *A True History of the Terrible Epidemic Vulgarly Called the Throat Distemper Which Occurred in His Majesty's New England Colonies Between the Years 1735 and 1740* (New Haven: *Yale Journal of Biology and Medicine* for the Beaumont Medical Club, 1939), 3.
58. John Duffy, *A History of Public Health in New York City, 1625–1866* (New York: Russell Sage Foundation, 1968), 86.
59. Friedrich Kapp, *Immigration and the Commissioners of Emigration* (New York: Arno Press, 1969; orig. 1870), 20.
60. Ibid.
61. Ibid., 21.
62. A third provision requiring specified quantities of food and water supplies for passengers applied only to departing ships. Rosebud T. Solis-Cohen, "The Exclusion of Aliens From the United States For Physical Defects," *Bulletin of the History of Medicine* 21 (1947): 33. Also E. P. Hutchinson, *Legislative History of American Immigration Policy, 1798–1965* (Philadelphia: University of Pennsylvania Press, 1981), 21–22.

63. The Act of February 22, 1847 (9 Stat. 127) stipulated "one passenger for every fourteen clear superficial feet of deck . . . on the lower deck or platform," but only "one passenger for every thirty such superficial feet" on the lowest or "orlop deck." The air and food requirements were legislated by the Act of May 17, 1848 (9 Stat. 220).
64. Solis-Cohen, "Exclusion of Aliens," 34. Hutchinson, *Legislative History of American Immigration*, 38–39.
65. Yellow fever's early symptoms are pains in the back, head, and limbs accompanied by chills. The victim suffers a high fever and constipation. Urination becomes infrequent and the emission albuminous. After several days the disease appears to go into remission, even the fever declining. However, relief is short-lived with the onset of symptoms even worse than those earlier. As the fever worsens, the victim's skin takes on a yellow cast, and he or she regurgitates stale blood that appears almost black. Hence yellow fever's other common name, black vomit. The intestinal mucous membrane hemorrhages. Patients then become lethargic, falling into a kind of stupor. Their tongues appear dry and brown, pulses speed up, becoming fainter. Patients become incontinent and appear to be wasting away as they decline toward death. Robert Berkow, ed., *The Merck Manual of Diagnosis and Therapy*, 15th ed. (Rahway, N.J.: Merck & Co., 1987), 192.
66. J. H. Powell, *Bring Out Your Dead, The Great Plague of Yellow Fever in Philadelphia in 1793* (Philadelphia: University of Pennsylvania Press, 1949), 13.
67. Erving Goffman, *Stigma: Notes on the Management of Spoiled Identity* (Englewood Cliffs, N.J.: Prentice-Hall, 1963).
68. Martin Pernick, "Politics, Parties, and Pestilence: Epidemic Yellow Fever in Philadelphia and the Rise of the First Party System," in Leavitt and Numbers, eds., *Sickness and Health in America*, 356–71. The essay is a revision of an earlier article that appeared in *The William and Mary Quarterly* 29 (1972): 559–86.
69. "Pestilence," a poem by Philip Freneau as quoted by Powell, *Bring Out Your Dead.*
70. Bache as quoted by Donald H. Stewart, *The Opposition Press of the Federalist Period* (Albany: State University of New York Press, 1969), 137. Also quoted in Pernick, "Politics, Parties and Pestilence," 359.
71. Thomas Jefferson to Benjamin Rush, Sept. 23, 1800, in Andrew A. Lipscomb and Albert E. Bergh, eds. *The Writings of Thomas Jefferson* (Washington, D.C.: Thomas Jefferson Memorial Association, 1904), v.10, 173. Also cited in Pernick, "Politics, Parties, and Pestilence," 360.
72. Benjamin Rush, *An Account of the Bilious Remitting Yellow Fever* (Philadelphia: T. Dobson, 1794), 147.
73. Pernick, "Politics, Parties, and Pestilence," 360.
74. Horace W. Smith, *Life and Correspondence of the Rev. William Smith, D.D. . . .* vol. I (Philadelphia: Sherman & Co., 1880), 395.
75. Deveze's background is discussed in Powell, *Bring Out Your Dead,* 159. For

the French physician's perspective on those who blamed refugees for the disease, see Jean Deveze, *An Enquiry into, and Observations upon; the Causes and Effects of the Epidemic Disease* (Philadelphia: Printed by Parent, 1794), 16.

76. The material on the Philadelphia Lazaretto is largely derived from Edward T. Morman, "Guarding Against Alien Impurities: The Philadelphia Lazaretto, 1854–1893," *The Pennsylvania Magazine of History* 107 (Apr. 1984): 131–52. A fuller discussion is in Morman, "Scientific Medicine Comes to Philadelphia: Public Health Transformed, 1854–1899" (Ph.D. diss., University of Pennsylvania, 1986).

CHAPTER 2: "A SCOURGE, A ROD IN THE HAND OF GOD"

1. The best study on the cholera epidemics of the nineteenth century, which also treats the issue of ethnic stigma, is Charles E. Rosenberg, *The Cholera Years, The United States in 1832, 1849 and 1866* (Chicago: University of Chicago Press, 1987; orig. 1962).
2. Kate H. Claghorn, "The Foreign Immigrant in New York City," *United States Industrial Commission Reports* (Washington, D.C.: Government Printing Office, 1901), vol. 15, 464.
3. *The Eighth Federal Census of the United States: 1860* (Washington, D.C.: R. Armstrong, 1863), 609.
4. Anti-Catholicism is analyzed in John Higham's classic study of nativism, *Strangers in the Land, Patterns of American Nativism 1860–1925* (New Brunswick, N.J.: Rutgers University Press, 1955). Antebellum anti-Catholicism is described most vividly in Ray Allen Billington, *The Protestant Crusade, 1800–1860, A Study of the Origins of American Nativism* (New York: Macmillan, 1938). Also, Dale T. Knobel, *Paddy and the Republic, Ethnicity and Nationality in Antebellum America* (Middletown, Conn.: Wesleyan University Press, 1986).
5. Richard J. Evans, *Death in Hamburg, Society and Politics in the Cholera Years, 1830–1910* (New York: Oxford University Press, 1987), 477. The stigmatization of medieval European Jewry for bubonic plague is described by Philip Ziegler, *The Black Death* (New York: Harper & Row, 1969), 96–109.
6. Although he does not discuss the issue of disease, one of the clearest discussions of fear of subversion in antebellum America remains David B. Davis, "Some Themes of Counter-Subversion: An Analysis of Anti-Masonic, Anti-Catholic, and Anti-Mormon Literature," *Mississippi Valley Historical Review* 47 (Sept. 1960): 205–24.
7. Both native and newcomer became bound up in a cultural discourse that historian Richard Hofstadter dubbed the "paranoid style," one marked by "heated exaggeration, suspiciousness and conspiratorial fantasy." Richard Hofstadter, *The Paranoid Style in American Politics and Other Essays* (New York: Random House, 1967), 3. The essay was a revised and expanded ver-

sion of the Herbert Spencer Lecture delivered by Hofstadter at Oxford in November 1963. An abridged version also appeared in *Harper's Magazine,* Nov. 1964.

8. Barbara Gutmann Rosenkrantz, *Public Health and the State, Changing Views in Massachusetts, 1842–1936* (Cambridge, Mass.: Harvard University Press, 1972), 1. An excellent discussion of exceptionalism is Dorothy Ross, *The Origins of American Social Science* (New York: Cambridge University Press, 1991), chap. 2.
9. *New York City Reports of Hospital Physicians and Other Documents in Relation to the Epidemic Cholera of 1832,* ed. Dudley Atkins (New York: G. C. and M. Carvill, 1832), 14–15.
10. *Niles' Register* 42 (July 21, 1832): 372.
11. Whitney R. Cross, *The Burned-Over District* (Ithaca, N.Y.: Cornell University Press, 1950) remains the most compelling account. Also, see Billington, *Protestant Crusade,* passim.
12. Daniel Walker Howe, *The Political Culture of the American Whigs* (Chicago: University of Chicago Press, 1979), 162–64.
13. Davis, "Themes of Counter-Subversion," 216–18.
14. Gardiner Spring, *A Sermon Preached August 3, 1832, a Day Set Apart in the City of New-York for Public Fasting, Humiliation and Prayer . . .* (New York: Jonathan Leavitt, 1832). Also cited by Rosenberg, *Cholera Years,* 44.
15. These varied religious responses are documented in Rosenberg, *Cholera Years,* 43–45.
16. Knobel, *Paddy and the Republic,* 56–57.
17. Samuel F. B. Morse, *Imminent Dangers to the Free Institutions of the United States Through Foreign Immigration* (New York: Arno Press, 1969; orig. 1835), 13.
18. *Newark Daily Advertiser*, July 29, 1854. Also quoted in Stuart Galishoff, *Newark, The Nation's Unhealthiest City 1832–1895* (New Brunswick, N.J.: Rutgers University Press, 1988), 60.
19. *New York Times,* Oct. 15, 1851. Also cited in Terry Coleman, *Going to America* (Garden City, N.Y.: Doubleday, 1973), 72.
20. Coleman, *Going to America,* 74.
21. Ibid., 75–76. René and Jean Dubos, *The White Plague, Tuberculosis; Man and Society* (New Brunswick, N.J.: Rutgers University Press, 1987; orig. 1953), 28–33, 94–104.
22. Stanley Joel Reiser, *Medicine and the Reign of Technology* (Cambridge: Cambridge University Press, 1978), 1–44.
23. Coleman, *Going to America,* 75.
24. Rosenberg, *The Cholera Years,* 135*n*5.
25. Friedrich Kapp, *Immigration and the Commissioners of Emigration* (New York: Arno Press, 1969; orig. 1870), 105–24.
26. Samuel Busey, *Immigration: Its Evils and Consequences* (New York: Arno Press, 1969; orig. 1856), 125.
27. Michael Les Benedict, "Contagion and the Constitution: Quarantine Cita-

tion from 1859 to 1866," *Journal of the History of Medicine and Allied Sciences* 25 (Apr. 1970): 177–93.

28. Rosenberg, *The Cholera Years,* 228.
29. Ibid., 229.
30. John Duffy, ed., *The Rudolph Matas History of Medicine in Louisiana,* 2 vols. (Baton Rouge: Louisiana State University Press, 1958–1962).
31. Rosenberg, *The Cholera Years,* 175–234, especially 192–212. Also, Rosenkrantz, *Public Health and the State,* 62–67 and John Duffy, *The Sanitarians, A History of American Public Health* (Urbana: University of Illinois Press, 1990), 138–54.
32. Robert Ernst, *Immigrant Life in New York City, 1825–1863* (New York: Ira J. Friedman, 1965; orig. 1949), 54.
33. Gerald N. Grob, *Mental Institutions in America, Social Policy to 1875* (New York: Free Press, 1973), 153–56; and Grob, *Mental Illness and American Society, 1875–1940* (Princeton: Princeton University Press, 1983), 37–38.
34. Ernst, *Immigrant Life in New York City,* 54.
35. Worcester Lunatic Hospital, *Twenty-Second Annual Report* (1855) as cited in David J. Rothman, *The Discovery of the Asylum, Social Order and Disorder in the New Republic* (Boston: Little Brown and Company, 1990; orig. 1971), 283.
36. Ernst, *Immigrant Life in New York City,* 54–55.
37. As historian Gerald Grob has observed, "When the institutionalized population was analyzed in terms of the age distribution of the entire native and foreign-born population, for example, the relative proportion of immigrants in hospitals declined precipitously." Additional corrections for sex distribution and urban/rural residence still further explained the discrepancy between newcomer and native-born rates of mental disease. Grob, *Mental Illness and American Society, 1875–1940,* 169–70.
38. Ernst, *Immigrant Life in New York City, 1825–1863,* 55.
39. Ibid.
40. Hasia Diner argues that Irish males in Ireland suffered the effects of mother-domination more than females. Emigrating males pined for home and mother and suffered higher rates of schizophrenia, while the females "made the decision to leave with a relatively light heart." Diner, *Erin's Daughters in America, Irish Immigrant Women in the Nineteenth Century* (Baltimore: Johns Hopkins University Press, 1983), 19.
41. Diner, *Erin's Daughters,* 110–11.
42. Grob, *Mental Institutions in America, Social Policy to 1875,* 230–35.
43. Rothman, *Discovery of the Asylum,* 285.
44. Grob, *Mental Institutions in America, Social Policy to 1875,* 238.
45. Ibid., 236–41.
46. Higham, *Strangers in the Land,* 11. Dale Knobel makes the distinction between race and ethnicity in antebellum culture, but he, too, argues for a protean stereotype of the Irish in *Paddy and the Republic,* xvii.

47. Knobel, *Paddy and the Republic,* 68–103.
48. "A Scene from Irish Life," *Harper's Monthly* 3 (1851): 833.
49. Orson S. Fowler and Lorenzo Fowler, *New Illustrated Self-Instructor in Phrenology and Physiology* (New York: Fowler and Wells, 1859), 43–45, 55–56.
50. Kerby A. Miller, "Class, Culture, and Immigrant Group Identity in the United States: The Case of Irish-American Ethnicity," in Virginia Yans-McLaughlin, ed., *Immigration Reconsidered, History, Sociology, and Politics* (New York: Oxford University Press, 1990), 110–11. See also Kerby A. Miller, *Emigrants and Exiles, Ireland and the Irish Exodus to North America* (New York: Oxford University Press, 1985).
51. Miller, "Class, Culture, and Immigrant Group Identity," 111–13. For a full discussion of Dr. William MacNeven as an Irish leader, see Victor R. Greene, *American Immigrant Leaders, 1800–1910, Marginality and Identity* (Baltimore: Johns Hopkins University Press, 1987), 25–29.
52. Thomas Addis Emmet, Jr., *Incidents of My Life* (New York: Putnam, 1911), 65.
53. George Potter, *To the Golden Door* (New York: Macmillan, 1960), 208–9.
54. Marvin R. O'Connell, "The Roman Catholic Tradition Since 1545," in Ronald L. Numbers and Darrell W. Amundsen, eds., *Caring and Curing, Health and Medicine in the Western Religious Traditions* (New York: Macmillan, 1986), 108–45.
55. Joseph and Helen McCadden, *Father Varela: Torch Bearer From Cuba* (New York, United States Catholic Historical Society, 1969), 165–67.
56. *New York Herald,* Aug. 8, 1849.
57. A letter from Right Rev. John Dubois, D.D., Bishop of New York, to Rev. ———, secretary of the Association for the Propagation of the Faith, Lyons, March 16, 1830, in United States Catholic Historical Society, *Historical Records and Studies* (New York: United States Catholic Historical Society, 1907), vol. 5, part 1, 228.
58. Pastoral letter of Bishop John Dubois, 1834, as quoted by Jay Dolan, *The Immigrant Church, New York's Irish and German Catholics, 1815–1865* (Baltimore: Johns Hopkins University Press, 1975), 130.
59. All studies of hospitals in the nineteenth century mention the difficulties that Roman Catholic clergy had serving their flock and protecting the ill from Protestant conversion. See Morris J. Vogel, *The Invention of the Modern Hospital, Boston, 1870–1930* (Chicago: University of Chicago Press, 1980), 127–28, and Charles E. Rosenberg, *The Care of Strangers, The Rise of America's Hospital System* (New York: Basic Books, 1987).
60 An excellent sociological discussion of this issue, which remains a matter of significance for religious groups in spiritually heterogeneous societies, can be found in William A. Glaser, *Social Settings and Medical Organization, A Cross-National Study of the Hospital* (New York: Atherton Press, 1970), 33–35.
61. *Freeman's Journal and Catholic Register,* Feb. 20, 1847.

62. John Gilmary Shea, *History of the Catholic Church in the United States,* 4 vols. (New York: John G. Shea, 1886–1892), v. 4, 154.
63. *Freeman's Journal and Catholic Register,* Feb. 5, 1848.
64. *First Annual Report of Saint Vincent's Hospital Under the Charge of The Sisters of Charity for the Year Ending January First, 1859,* 4. The best history of St. Vincent's Hospital is Sister Marie De Lourdes Walsh, *With A Great Heart, The Story of St. Vincent's Hospital and Medical Center of New York, 1849–1964* (New York: St. Vincent's Hospital, 1965).
65. On the role of female orders in the development of Catholic hospitals in New York City, see Bernadette McCauley, "'Who Shall Take Care of the Sick?' Roman Catholic Sisterhoods and Their Hospitals in New York City, 1850–1930," (Ph.D. diss., Columbia University, 1992).
66. Historian Marvin R. O'Connell claims that "the sisters and their institutions of healing clearly gave priority to cure of soul over cure of body. They acted out of no mere humanitarian impulse, but out of the conviction that performing the corporal works of mercy was a primary means of securing salvation for those who performed them, as well as an occasion of supernatural grace to those for whom they were performed." Thus, O'Connell views the bravery of the nuns in the face of disease as an aspect of their own religious odysseys. O'Connell, "The Roman Catholic Tradition Since 1545," 137.
67. A useful treatment of the creation of the first Roman Catholic hospital in St. Louis is Ann Kathryn Webster, "The Impact of Catholic Hospitals in St. Louis" (Ph.D. diss., Saint Louis University, 1968).
68. *Fifty-First Annual Report of the St. Vincent's Hospital of the City of New York for the Year 1900* (New York: The Meany Printing Co., 1901), 8.
69. Ibid., 22.
70. *Freeman's Journal and Catholic Register,* Jan. 26, 1850.
71. *First Annual Report of St. Vincent's Hospital (1859),* 4.
72. Ibid., 5.
73. Ibid.
74. *Fifty-First Annual Report of the St. Vincent's Hospital* (1900), 22.
75. The material on these two important Catholic hospitals in Philadelphia has been largely drawn from two fine articles: Gail Farr Casterline, "St. Joseph's and St. Mary's: The Origins of Catholic Hospitals in Philadelphia," *Pennsylvania Magazine of History and Biography* 108 (July 1984): 289–314 and Judith G. Cetina, "In Times of Immigration," in Ursula Stepsis, C.S.A., and Delores Liptak, R.S.M., *Pioneer Healers, The History of Women Religious in American Health Care* (New York: Crossroad, 1989), 86–117.
76. Casterline, "St Joseph's and St. Mary's," 298.
77. *Catholic Herald,* Feb. 10, 1848. Also cited in ibid., 299.
78. *Catholic Herald,* June 23, 1849. Also cited in ibid., 301.
79. Ibid.
80. Casterline, "St. Joseph's and St. Mary's," 303–4.
81. Rosenberg, *Care of Strangers,* 109–21.

CHAPTER 3: "PROPER PRECAUTIONS"

1. "Status of the Work," *Proceedings of the Association of Medical Officers of American Institutions for Idiotic and Feeble-Minded Persons, June 1888 at Orillia, Canada* (Philadelphia: Lippincott, 1889), 77.
2. The National Board of Health's original mandate was to advise state and local health boards, collect and disseminate health information, conduct investigations of specific public health menaces, and plan a permanent national health agency. It was soon empowered to offer financial aid to states and localities in support of their public health efforts and to federalize quarantine duties when the states did not meet their responsibilities. *National Board of Health Bulletin* 1 (1879): 1–3. A useful discussion of how the board actually functioned can be found in Margaret Humpreys, *Yellow Fever and the South* (New Brunswick, N.J.: Rutgers University Press, 1992), 63–76.
3. Fiorello H. LaGuardia, *The Making of An Insurgent, An Autobiography: 1882–1919* (New York: Capricorn, 1961; orig. 1948), 53–61.
4. The relationship of temporary emigration and remittances in the minds of Italian officials is amply demonstrated in Betty Boyd Caroli, *Italian Repatriation from the United States, 1900–1914* (New York: Center for Migration Studies, 1973), 57–64. See also, Dino Cinel, *The National Integration of Italian Return Migration, 1870–1929* (Cambridge: Cambridge University Press, 1991).
5. An overview of this peak era of immigration is Alan M. Kraut, *The Huddled Masses, The Immigrant in American Society, 1880–1921* (Arlington Heights, Ill.: Harlan Davidson, 1982). The economic significance of this migration for both host and donor nations as part of a larger capitalist system is explored in John Bodnar, *The Transplanted: A History of Immigrants in Urban America* (Bloomington: Indiana University Press, 1985).
6. This formulation of late-nineteenth-century American society's joust with the hydra-headed specter of modernization, industrialization, and immigration has been most articulately and concisely presented by Robert Wiebe, *The Search for Order, 1877–1920* (New York: Hill and Wang, 1967).
7. Even after the Civil War the states were still the primary source of regulatory power. Medical experts failed to persuade the federal government to assume quarantine responsibilities. See Michael Les Benedict, "Contagion and the Constitution: Quarantine Agitation from 1859 to 1866," *Journal of the History of Medicine and Allied Sciences* 25 (Apr. 1970): 177–93.
8. On modernization in American government, see William E. Nelson, *The Roots of American Bureaucracy, 1830–1900* (Cambridge, Mass.: Harvard University Press, 1982) and Stephen Skowronek, *Building a New American State: The Expansion of National Administrative Capacities, 1877–1920* (Cambridge: Cambridge University Press, 1982). The tensions between localism and the increasing need for greater federal involvement in public life is most fully described in Morton Keller, *Affairs of State, Public Life in Late*

Nineteenth Century America (Cambridge, Mass.: Belknap Press of Harvard University, 1977).

9. *New York Times,* Jan. 2, 1892.
10. The details of the line inspection at New York's Ellis Island have been well documented by historians. See Elizabeth Yew, "Medical Inspection of the Immigrant at Ellis Island, 1891–1924," *Bulletin of the New York Academy of Medicine* 56 (1980): 488–510; Rosebud T. Solis-Cohen, "The Exclusion of Aliens From the United States for Physical Defects," *Bulletin of the History of Medicine 21* (1947): 33–50; Thomas M. Pitkin, *Keepers of the Gate: A History of Ellis Island* (New York: New York University Press, 1975); and Harlan Unrau, *Ellis Island,* 3 vols. (Washington, D.C.: U.S. Department of the Interior/ National Park Service), vol.2, 575–732.
11. Catherine Bolinski, as quoted in David M. Brownstone, Irene M. Franck, and Douglass L. Brownstone, *Island of Hope, Island of Tears* (New York: Rawson, Wade, 1979), 177.
12. Fannie Kligerman as quoted in ibid., 168–69.
13. Euterpe Dukakis, transcript of unpublished interview, 1985, in Ellis Island Oral History Project. Euterpe Dukakis is the mother of former Massachusetts governor and 1988 Democratic presidential candidate, Michael Dukakis.
14. Mary T. Mernin, M.D., A.A. Surgeon, U.S.P.H.S., "Medical Inspection of Aliens at Ellis Island With Special Reference to the Examination of Women and Children," *Medical Women's Journal* (June 1924): 172–75. Mernin was the senior physician in the women's examining room on Ellis Island when this essay was published.
15. Brownstone et al., *Island of Hope, Island of Tears,* 208.
16. Ibid., 169. In fairness the PHS was not intentionally destroying immigrants' clothing. There were few chemical agents available for delousing and most were quite corrosive.
17. The *huanglian* was known to be a bitter herb that only a foolish person would knowingly bite into.
18. Anonymous poems scrawled on a wall by Chinese immigrants at the Angel Island immigration depot in San Francisco Bay (c. 1910). Him Mark Lai, Ginny Lim, and Judy Yung, *Island* (Seattle: University of Washington Press, 1991; orig. 1980), 100–102. The poem is reprinted in its entirety here with the permission of the authors and the University of Washington Press.
19. Paul Starr, *The Social Transformation of American Medicine* (New York: Basic Books, 1982), especially chap. 3.
20. William Williams's activities on Ellis Island are thoroughly documented in his personal papers in the Rare Book and Manuscript Division of the New York Public Library.
21. Terence Powderly, "The Immigrant Menace to National Health," *North American Review* 175 (1952): 53–60.
22. LaGuardia, *The Making of An Insurgent,* 64.
23. For a description of the change in American attitudes toward germ theory,

see Phyllis A. Richmond, "American Attitudes Toward the Germ Theory of Disease (1860–1880)," in Gert Brieger, ed., *Theory and Practice in American Medicine: Historical Studies From the Journal of the History of Medicine and Allied Sciences* (New York: Science History Publications, 1976), 58–84. For a discussion of one who resisted the germ theory, see James H. Cassedy, "The Flamboyant Colonel Waring: An Anticontagonist Holds the American Stage in the Age of Pasteur and Koch" (orig. 1962), in Judith Leavitt and Ronald Numbers, eds., *Sickness and Health in America: Readings in the History of Medicine and Public Health* (Madison: University of Wisconsin Press, 1985), 451–58.

24. Victoria A. Harden, *Inventing the NIH: Federal Biomedical Research Policy, 1887–1937* (Baltimore: Johns Hopkins University Press, 1986).
25. Daniel M. Fox, *Health Policies and Health Politics: The British and American Experience, 1911–1965* (Princeton: Princeton University Press, 1986). Gerald N. Grob, "New Wine in New Bottles: The History of Health Policy," *Reviews in American History* 15 (Sept. 1987): 365–73.
26. Circular entitled "Quarantine Restrictions upon immigration to aid in the prevention of the introduction of cholera into the United States," Sept. 1, 1892, included in *Abstract of Sanitary Reports,* Sept. 2, 1892, vol. 7, no. 336, 445. Also, *New York Times*, Sept. 2, 1892.
27. Some state officials thought the federal government was overstepping its bounds. In New York, Dr. William T. Jenkins, health officer of the port of New York, had responded testily to the circular order, promising to use "his own discretion and exercise his own authority under the State laws, in the matter of detaining immigrant vessels at Quarantine." The *New York Times* defended Jenkins against those who felt the state bureaucrat too anxious to put the public health at risk for the sake of states rights, expressing confidence that "[Jenkins] will not be guilty of the indiscretion of using his State authority to diminish the safeguards against cholera provided in the President's order." The influential newspaper concluded that "It would be inadvisable for any local authority to challenge the Federal power when so wisely and discreetly employed." Perhaps, but some feisty New Yorkers did not hesitate to pick up arms when they felt that state authorities were acting irresponsibly to carry out the federal mandate. *New York Times,* Sept. 3, 1892.

    State and local authorities were at each others' throats, as well. New York State officials rented several buildings on Fire Island for first- and second-class passengers' detention (steerage passengers were confined to the ship). However, when word got out, the local board of health deputized all Fire Island residents wishing to resist the entry into their midst of potential cholera carriers. When an armed mob of locals lined the pier, the board ordered a contingent of National Guard summoned by the governor to disperse them. *New York Herald,* Sept. 9, 12–14, 1892.
28. Act of Feb. 15, 1893 (27 Stat. 452).
29. E. P. Hutchinson, *Legislative History of American Immigration Policy,*

1798–1965 (Philadelphia: University of Pennsylvania Press, 1981), 106–7. Also, John Duffy, *The Sanitarians, A History of American Public Health* (Urbana: University of Illinois Press, 1990), 172.

30. The standard history of the United States Public Health Service remains the volume by Ralph C. Williams, *The United States Public Health Service* (Washington, D.C.: Government Printing Office, 1951). The PHS did not join the Immigration Bureau in the Department of Commerce and Labor but remained in the Treasury Department. PHS physicians were detailed for duty under the immigration law of 1891 and only in matters pertaining to administration were they under the general direction of the immigration commissioners in the various ports where their services were required.
31. Victor Heiser, M.D., *An American Doctor's Odyssey, Adventures in Forty-five Countries* (New York: Norton, 1936), 16. The Supreme Court had upheld New York's right to take precedence over the federal government in demanding detailed lists of passengers from ship captains, defining it as a police power belonging to the states in *New York* v. *Miln,* 11 Pet. 102 (1837). An excellent discussion of the conflict between state and federal jurisdiction in public health matters can be found in Humphreys, *Yellow Fever and the South.*
32. Heiser, *An American Doctor's Odyssey,* 17.
33. Ibid., 16.
34. Harlan D. Unrau, *Ellis Island,* vol. 2, 575–650.
35. Alfred C. Reed, "Going Through Ellis Island," *Popular Science Monthly* 82 (1913): 5–18. Also, Reed, "Immigration and the Public Health," *Popular Science Monthly* 83 (1913): 320–38.
36. Howard A. Knox, "A Diagnostic Study of the Face," pamphlet reprinted from the *New York Medical Journal,* June 14, 1913.
37. John C. Thill, transcript of unpublished interview, 1977. Office of the Statue of Liberty–Ellis Island Collaborative in New York City. Copy on file at the National Library of Medicine, Bethesda, Maryland.
38. Heiser, *An American Doctor's Odyssey,* 15.
39. Victor Safford, *Immigration Problems: Personal Experiences of An Official* (New York: Dodd, Mead, 1925), 244–45.
40. Ibid., 246.
41. Ibid., 247.
42. Williams, *United States Public Health Service,* 515. According to Ralph C. Williams, the PHS became a uniformed service after the Civil War: "With the Civil War recently ended and all of the physicians on duty in the marine hospitals having had military service, it was but a natural step that uniforms be developed for the use of medical officers and other personnel at the marine hospitals. This was particularly logical since most of the patients at these marine hospitals were merchant seamen and officers and enlisted men of the Coast Guard, persons accustomed to discipline and to dealing with uniformed personnel."
43. Heiser, *An American Doctor's Odyssey,* 14.

44. Ibid., 14–15.
45. Ibid.
46. Alfred C. Reed, M.D., "Immigration and the Public Health," 320–38.
47. H. M. Lai, "Island of Immortals, Chinese Immigrants and the Angel Island Immigration Station," *California History* 57 (Spring 1978): 94, 102*n*28.
48. Bernard Marinbach, *Galveston: Ellis Island of the West* (Albany: State University of New York Press, 1983), 59–61.
49. Israel Zangwill as quoted by Marinbach, *Galveston: Ellis Island of the West,* 158.
50. Of the additional 11,072 certified for a disease or defect that might affect their ability to earn a living, 5,573 were classified as senile and 1,286 as suffering from hernia, while another 883 were described by PHS physicians as having a "lack of physical or poor muscular development." U.S. Public Health Service, *Annual Report of the Chief Medical Officer of Ellis Island, June 30, 1911,* RG 90, File 53274. United States National Archives and Public Records Service.
51. These figures were calculated from data in the *Annual Report of the Surgeon General of the Public Health Service* (Washington, D.C.) for the years 1892–1924.
52. Joel B. Reisser, *Medicine and the Reign of Technology* (Cambridge: Cambridge University Press, 1978), 63–65, 85.
53. U.S. Public Health Service, *Regulations Governing the Medical Inspection of Aliens* (Washington, D.C.: Government Printing Office, 1917).
54. U.S. Bureau of Immigration, *Annual Report of the Commissioner General of Immigration* (Washington, D.C.: Government Printing Office, 1899), 40.
55. Ibid.
56. U.S. Treasury Department, Bureau of Public Health and Marine-Hospital Service, *Book of Instructions for the Medical Inspection of Immigrants, Prepared by Direction of the Surgeon-General* (Washington, D.C.: Government Printing Office, 1903). See also, U.S. Treasury Department, Public Health Service, *Regulations Governing the Medical Inspection of Aliens* (Washington, D.C.: Government Printing Office, 1917).
57. William Williams, "Remarks on Immigration," unpublished address to the Senior Class of Princeton University, 1904, in William Williams Papers, New York Public Library.
58. In addition to a common concern over an invasion of harmful and contagious bacteria, American policymakers were concerned over the social implications of illness or deformity, especially the possible dependence upon public assistance of a disabled individual. A "likely to become a public charge clause" (lpc) clause had been grounds for exclusion since the immigration law of 1882. What had changed was the commitment of government officials, including many progressives, to making medical grounds the primary reason for declaring an immigrant to be lpc. The most thorough treatment of the "likely to become a public charge" clause is a doctoral dissertation by Patricia R. Evans, "'Likely to Become a Public Charge': Immigration

in the Backwaters of Administrative Law," (Ph.D. diss., George Washington University, 1987). For an excellent, concise discussion of the lpc clause and health issues, see Hutchinson, *Legislative History of American Immigration Policy, 1798–1965,* 410–19. The medicalization of social decisions began well before the twentieth century. The massive problems posed by industrialization, urbanization, and immigration accelerated the process, at most. There are several provocative essays on the involvement of medicine in the management of society and as an instrument of social control; see I. K. Zola, "Medicine as an Institution of Social Control," in J. Ehrenreich, ed., *The Cultural Crisis of Modern Medicine* (New York: Monthly Review Press, 1978), 80–100 and Charles Rosenberg, "Disease and Social Order in America," *Milbank Quarterly* 64 (1986, supplement 1): 34–55.

59. Nathan Bijur, unpublished letter to William P. Dillingham, Feb. 9, 1907. Oscar Strauss Papers, Manuscript Division of the Library of Congress. Bijur was an attorney and immigration law expert who later became a justice of the New York State Supreme Court. He was also an activist in many Jewish causes.
60. By 1917, the PHS specified that medical diagnosis was to be "for the medical officer to determine." The "question of the effect of such a disease or defect on the alien's earning capacity is a practical one, and therefore for the immigration officer to determine." However, the regulations did leave the door open a crack, allowing that an immigration officer "may desire, and having obtained, may consider an expression of opinion by the medical officer on the practical phase of the matter." U.S. Public Health Service, *Regulations Governing the Medical Inspection of Aliens* (1917), 8.
61. Philip Cowen, *Memories of An American Jew* (New York: Arno Press, 1975; orig. 1932), 148.
62. Ibid., 148–49.
63. Immigration Act of 1882 (22 Stat. 214).
64. *Journal of the American Medical Association* 14 (1890): 688.
65. U.S. Public Health Service, *Book of Instructions for the Medical Inspection of Aliens* (1903).
66. Thomas W. Salmon, "The Diagnosis of Insanity in Immigrants," *Annual Report of the Surgeon General of the Public Health and Marine-Hospital Service* (1905): 271.
67. Robert Watchorn to [Commissioner General of Immigration Frank P. Sargent], June 11, 1906, RG 85, General Immigration Files, U.S. National Archives.
68. Grover A. Kempf, transcript of unpublished interview, 1977. Office of the Statue of Liberty–Ellis Island Collaborative in New York City. Copy on file at the National Library of Medicine, Bethesda, Maryland.
69. E. K. Sprague, "Mental Examination of Immigrants," *Survey* 31 (1914): 466–68.
70. Fiorello H. LaGuardia, *The Making of An Insurgent,* 65.
71. Ibid., 65–66.

72. Howard A. Knox, *Alien Mental Defectives: A Collection of Papers* (Chicago: C.H. Stoetling, 1914), 1–5 and Knox, "A Diagnostic Study of the Face." Also, E. H. Mullan, *Mentality of the Arriving Immigrant* (New York: Arno Press, 1970; orig. 1917) and Bertha Boody, *A Psychological Study of Immigrant Children at Ellis Island* (New York: Arno Press, 1970; orig. 1926).
73. Henry H. Goddard, "Feeble-mindedness and Immigration," *Training School Bulletin* 9 (Oct. 1912): 91–94.
74. Henry H. Goddard, "The Feeble-minded Immigrant," *Training School Bulletin* 9 (Nov. 1912): 109–113.
75. Ibid., 110–11.
76. Howard A. Knox, "The Moron and the Study of Alien Defectives," *Journal of the American Medical Association* 60 (1913): 105.
77. E. H. Mullan to Surgeon General Rupert Blue, Sept. 9, 1913, RG 90, Box 36, File 219.
78. Henry H. Goddard, "Mental Tests and the Immigrant," *Journal of Delinquency* 2 (Sept. 1917): 243–77.
79. Cowen, *Memories of an American Jew*, 174–75.
80. *New York Times*, Oct. 14, 1913.
81. Howard A. Knox, "A Scale Based On The Work At Ellis Island For Estimating Mental Defect," *Journal of the American Medical Association* 62 (Mar. 7, 1914): 741–45.
82. Anthony Stella, *Some Aspects of Italian Immigration to the United States* (New York: Arno Press, 1975; orig. 1924), 40–53.
83. Morris J. Karpas to Irving Lipsitch, Nov. 5, 1913, RG 90, Box 36, File 219, U.S. National Archives.
84. L. L. Williams, "Examination of Mentally Defective Aliens," *American Journal of Insanity* 70 (1914): 26.
85. Goddard, "Mental Tests and the Immigrant," 243. A virtual industry in Goddard bashing was founded with the publication of Leon Kamin, *The Science and Politics of I.Q.* (Hillsdale, N.J.: Lawrence Erlbaum Associates, 1974). Kamin and others argued that early mental testers such as Robert Yerkes, Lewis Terman, Carl Brigham, and Goddard were motivated in their research by nativist assumptions and that their research was invaluable to those arguing for immigration restriction such as that finally achieved in the Johnson-Reed Immigration Act of 1924. The argument was extended and brought to the attention of nonspecialists by Stephen Jay Gould, *The Mismeasure of Man* (New York: Norton, 1981). Also, Gould, "Science and Jewish Immigration," in *Hen's Teeth and Horses Toes* (New York: Norton, 1984), 291–303. A more recent trend has been to examine Goddard's decision to publish his 1913 results in the larger context of his own career and the racial nativism that was pervasive in the intellectual climate of his time. The latter is argued persuasively by Steven A. Gelb, "Henry H. Goddard and the Immigrants, 1910–1917: The Studies and Their Social Context," *Journal of the History of the Behavioral Sciences* 22 (Oct. 1986): 324–32. A

thorough and balanced perspective on Goddard's career, including the testing program on Ellis Island, is available in Leila C. Zenderland, "Henry Herbert Goddard and the Origins of American Intelligence Testing" (Ph.D. diss., University of Pennsylvania, 1986), especially 561–76. The extent to which Goddard's test results from the Ellis Island studies proved valuable to eugenicists plumping for exclusionary legislation on the grounds of the foreign-born's innate inferiority remains an open historiographical question. A fine essay claiming that restrictionists in favor of the Johnson-Reed Immigration Act of 1924 drew little political ammunition from Goddard's work is Mark Snyderman and R. J. Herrnstein, "Intelligence Tests and the Immigration Act of 1924,"*American Psychologist* 38 (Sept. 1983): 324–32.

86. Goddard, "Mental Tests and the Immigrant," 265. Also quoted by Zenderland, "Henry Herbert Goddard and the Origins of American Intelligence Testing," 576.
87. Some monkeys were inoculated with material from mumps and scarlet fever patients and sent to the Hygienic Laboratory. United States Public Health Service, *Annual Report of the Surgeon General* (1915), 204.
88. Ibid.
89. Edward Morman has examined the changing attitudes toward port quarantine in relation to increases in immigration in his study of quarantine in Philadelphia. Edward Mormon, "Guarding Against Alien Impurities: The Philadelphia Lazaretto, 1854–1893," *Pennsylvania Magazine of History* 107 (Apr. 1984): 131–52. Also, Morman, "Scientific Medicine Comes to Philadelphia: Public Health Transformed, 1854–1899" (Ph.D. diss., University of Pennsylvania, 1986).
90. Rosenberg, "Disease and Social Order in America: Perceptions and Expectations," 34–55 and in the "Afterward 1987" of *The Cholera Years* (Chicago: University of Chicago Press, 1987; orig. 1962) and "Deconstructing Disease," *Reviews in American History* 14 (Mar. 1986): 110–15. Also, Peter Wright and Andrew Treacher, eds., *The Problem of Medical Knowledge: Examining the Social Construction of Medicine* (Edinburgh: Edinburgh University Press, 1982).
91. Rosenberg, "Disease and Social Order in America," 34–35.
92. Taliaferro Clark and J. W. Schereschewsky, "Trachoma: Its Character and Effects," *Public Health Service Bulletin* 19 (Washington, D.C.: Government Printing Office, 1910).
93. Reed, "Going Through Ellis Island," 9.

CHAPTER 4: A PLAGUE OF NATIVISM

1. Mary Roberts Coolidge, *Chinese Immigration* (New York: Arno Press, 1969; orig 1909), 401.
2. An intriguing study of the Chinese laundry originally written as a doctoral dissertation in 1953 at the University of Chicago is Paul C. P. Siu, *The Chi-*

nese Laundryman, A Study of Social Isolation, ed. John Kuo Wei Tchen (New York: New York University Press, 1987).

3. Stuart Creighton Miller, *The Unwelcome Immigrant, The American Image of the Chinese, 1785–1882* (Berkeley: University of California Press, 1969). Other excellent volumes on opposition to the Chinese in California, specifically, are Alexander Plaisted Saxton, *The Indispensable Enemy, Labor and the Anti-Chinese Movement in California* (Berkeley: University of California Press, 1971) and the indispensable Elmer Clarence Sandmeyer, *The Anti-Chinese Movement in California* (Urbana: University of Illinois Press, 1973; orig. 1939). The best recent monograph on the Chinese experience in California is Sucheng Chan, *The Bittersweet Soil: The Chinese in California Agriculture, 1860–1910* (Berkeley: University of California Press, 1986). Abundant examples of anti-Chinese stereotypes are available in Cheng-Tsu Wu, ed., "*Chink!*" (New York: World Publishing, 1972).
4. Hinton Rowan Helper, *The Land of Gold: Reality versus Fiction* (Baltimore: H. Taylor, 1855), 94–96.
5. Arthur B. Stout, "Report on Chinese Immigration," *First Biennial Report, State Board of Health in California* (Sacramento: State Board of Health, 1870–1871), 63.
6. *San Francisco Municipal Reports for the Fiscal Year, 1876–77, Ending June 30, 1877* (San Francisco: Board of Supervisors, 1877), 394.
7. Ibid., 397.
8. Ibid.
9. Leprosy is more properly known as Hansen's disease, for Gerhard Armauer Hansen of Norway, who isolated the *Mycobacterium leprae* in the mid-1870s. The bacillus has never been cultivated in vitro and even the mode of transmission remains puzzling. Infection is usually thought to be through the skin or by way of the respiratory tract. Although some researchers admit the possibility of insect vectors, others believe that a human carrier is necessary. Whether lengthy and intimate contact with an active case is required or merely brief exposure also remains undetermined, as does the length of incubation. Researchers are quite confident that susceptibility is genetically linked and that patterns of immunity vary regionally across the globe. Individuals thought to be genetically susceptible can contract the lepromatous leprosy with dramatic disfiguration that has come to characterize the disease in the media and the popular mind. However, more benign forms do occur among those without the genetic predisposition. An excellent study of the historic relationship of leprosy to social stigma is Zachary Gussow, *Leprosy, Racism, and Public Health* (Boulder, Colo.: Westview Press, 1989).
10. *Congressional Globe,* 41st Cong., 2d sess., 1870, 756, pt. 7.
11. U.S. Congress, *Report of the Joint Special Committee to Investigate Chinese Immigration,* Rept. 689, 44th Cong., 2d sess., 1877.
12. Ibid., also cited in Coolidge, *Chinese Immigration,* 106.
13. *North American Review,* 126 (1878): 516, 524–25.

14. *New York Times,* Sept. 28, 1878. This and other episodes are also mentioned by Gussow, *Leprosy, Racism and Public Health,* 122–29.
15. *Congressional Record,* 47th Cong., 1st sess., 1882, 2973.
16. Joan B. Trauner, "The Chinese as Medical Scapegoats in San Francisco, 1870–1905," *California History* 57 (Spring 1978): 70–87. An earlier version of this essay appeared in *The Bulletin of the Chinese Historical Society of America* 9 (Apr. 1974): 1–18.
17. *San Francisco Chronicle,* Mar. 9, 1900. The episode is also described in Silvio J. Onesti, Jr., "Plague, Press, and Politics," *Stanford Medical Bulletin* 13 (Feb. 1955): 1; Philip A. Kalisch, "The Black Death in Chinatown, Plague and Politics in San Francisco," *Arizona and the West* 14 (1972): 113–36; Loren G. Lipson, "Plague in San Francisco in 1900," *Annals of Internal Medicine* 77 (1972): 303–10; and Guenter B. Risse, "Politics, Commerce and Public Health: The Plague Outbreak in San Francisco, 1900," in *The History of Public Health and Prevention,* published conference proceedings, Stockholm, Sept. 6–8, 1991. Although the victim's name has been spelled in different ways, the author has chosen to follow that used by the *San Francisco Chronicle.*
18. Except where otherwise indicated, the chronicle of events in the crisis is taken from Charles McClain's superb article on the plague crisis in San Francisco, "Of Medicine, Race, and American Law: The Bubonic Plague Outbreak of 1900," *Law and Social Inquiry* 13 (Summer 1988): 447–513. As of yet there is no single-volume study of the episode. The best single piece on the larger issue of the Chinese as the targets of stigma for reasons related to disease and hygiene is Trauner, "The Chinese As Medical Scapegoats in San Francisco, 1870–1905," 70–87. Also, Victor H. Haas, "When Bubonic Plague Came to Chinatown," *American Journal of Tropical Medicine* 8 (1959): 141–47, and Kalisch, "The Black Death in Chinatown, Plague and Politics in San Francisco, 1900–1904."
19. Robert Berkow et al., eds., *The Merck Manual of Diagnosis and Therapy* (Rahway, N.J.: Merck, Sharp & Dohme Research Laboratories, 1987), 95–98. Also, James B. Wyngaarden and Lloyd H. Smith, *Cecil Textbook of Medicine,* 18th ed. (Philadelphia: W.B. Saunders, 1988), vol. 2, 1661–63. L. F. Hirst, *The Conquest of Plague* (Oxford: Clarendon Press, 1953), 29.
20. The notion that the Chinese diet diminished their resistance to plague can be found in Walter Wyman, *Annual Report of the Surgeon-General of the Marine Hospital Service of the United States for the Fiscal Year 1897* (Washington, D.C.: Government Printing Office, 1898), 411–19. Also cited and explained in Guenter B. Risse, "'A Long Pull, A Strong Pull, and All Together': San Francisco and Bubonic Plague, 1907–1908," *Bulletin of the History of Medicine* 66 (Summer 1992): 264.
21. *San Francisco Municipal Reports* (1890) 316.
22. *Chung Sai Yat Po,* Mar. 8, 1900, as quoted by McClain, "Of Medicine, Race and American Law," 455.

23. *San Francisco Examiner,* Mar. 7, 1900. The poem is also reprinted in a footnote in McClain, "Of Medicine, Race and American Law," 454–55, n. 24.
24. *San Francisco Chronicle,* Mar. 8, 1900.
25. *San Francisco Bulletin,* Mar. 9, 1900.
26. *San Francisco Examiner,* Mar. 13, 1900.
27. Ibid.
28. Mc Clain, "Of Medicine, Race, and American Law," 466.
29. Ralph C. Williams, *The United States Public Health Service, 1798–1950* (Washington, D.C.: Government Printing Office, 1951), 249–50.
30. The ambiguities in the language and circumstances are clearly set forth in McClain, "Of Medicine, Race, and American Law," 468–69.
31. Quoted in ibid., 470.
32. *Department Circular 93* (1900), Marine Hospital Service, May 22, 1900, RG 90, U.S. National Archives. Also, ibid., 471.
33. *Chung Sai Yat Po,* May 18, 1900. A paraphrase of the translation is offered by McClain, "Of Medicine, Race, and American Law," 472.
34. Sacramento *Record-Union,* May 20, 1900.
35. Paraphrase of "A Notice Given By the Wicked Health Officers," *Chung Sai Yat Po,* see McClain, "Of Medicine, Race, and American Law," 474.
36. The Chinese consul was sufficiently incensed over the racial affront and the implicit insult to China to use his influence to ensure that first-rate legal counsel was retained. *Wong Wai* v. *Williamson,* Civil Case No. 12, 937, RG 21, U.S. National Archives, Pacific-Sierra Regional Branch, San Bruno, California. A case for Japanese plaintiffs was filed separately, see Obata, *Negoro et al.* v. *Williamson,* Civil Case No. 12, 938, RG 21, also located at the National Archives Branch at San Bruno.
37. *Wai Wong* v. *Williamson,* 103 *Fed. Rep.,* 1900, 7.
38. *San Francisco Call,* May 29, 1900. Account also cited in McClain, "Of Medicine, Race, and American Law," 484.
39. *San Francisco Examiner,* May 31, 1900.
40. Ibid.
41. *San Francisco Call,* May 31, 1900.
42. "Tents Are to Be Put Up," *Chung Sai Yat Po* translation as quoted by McClain, "Of Medicine, Race, and American Law," 492.
43. *San Francisco Call,* June 6, 1900.
44. *Jew Ho* v. *Williamson,* Case File No. 12, 940, RG 21, National Archives at San Bruno.
45. Full arguments are summarized in McClain, "Of Medicine, Race, and American Law," 496–504.
46. Ibid.
47. *San Francisco Examiner,* June 16, 1900.
48. McClain, "Of Medicine, Race, and American Law," 505–6.
49. "The Report of the Government Commission on the Existence of Plague in San Francisco," reprinted in *Occidental Medical Times* 15 (Apr. 1901): 102–3.

50. W. M. Dickie, *Plague in California, 1900–1924,* a pamphlet reprint of the *Proceedings of the Conference of State and Provincial Health Authorities of North America,* 1926. Also, Trauner, "The Chinese As Medical Scapegoats," 78–81.
51. Frank Morton Todd, *Eradicating Plague From San Francisco, Report of the Citizen's Health Committee and An Account of Its Work,* March 31, 1909. Also, Trauner, "The Chinese as Medical Scapegoats," 80–81 and Risse, "'A Long Pull, A Strong Pull, and All Together': San Francisco and Bubonic Plague, 1907–1908," 260–86.
52. *Montgomery Journal,* Sept. 24, 1986.
53. According to the epidemiologist Dr. George A. Soper, after his investigation of the Typhoid Mary case Simon Flexner called his attention to an address that Robert Koch had delivered in 1902, mentioning the results of Conradi and Drigalski in using a new culture medium that allowed detection of typhoid bacteria in the stools of visibly well subjects. On the occasion of Robert Koch's sixtieth birthday, the 1903 *Festschrift Zum Sechzigsten Geburtstag* published in his honor contained several papers on the healthy carrier. Other Europeans, too, were publishing essays on the phenomenon.
54. Walter Reed, Victor C. Vaughn, and Edward O. Shakespeare, *Abstract of Report on the Origin and Spread of Typhoid Fever in U.S. Military Camps During the Spanish War of 1898* (Washington: Government Printing Office, 1900), 200.
55. Maj. George A. Soper, "Typhoid Mary," *The Military Surgeon* 45 (July 1919): 1–15 and "The Curious Career of Typhoid Mary," *Bulletin of the New York Academy of Medicine* 15 (Oct. 1939):698–712. There has been no single-volume history of the case and not many articles. See Mary C. McLaughlin, M.D., "Mary Mallon: Alias Typhoid Mary," *The Recorder of the American Irish Historical Society* 40 (1979): 44–57; and an entertaining popular essay by John F. Wukovits, "Destroying Angel," *American History Illustrated* (Mar./Apr. 1990): 68–72. That the case continues to inspire popular interest abroad as well as in the United States is also suggested by a recent German novel, J. F. Federspiel, *The Ballad of Typhoid Mary,* trans. Joel Agee (New York: Dutton, 1983). Renewed scholarly interest in the scientific implications of the case is suggested by Judith Walzer Leavitt, "'Typhoid Mary' Strikes Back, Bacteriological Theory and Practice in Early Twentieth Century Public Health," *ISIS* 83 (Dec. 1992): 608–29.
56. In his senior thesis at Harvard University, John Andrew Mendelson argues that Mary Mallon's case was a cause celebre because the concept of the healthy carrier was both unfamiliar and seemed unbelievable to those who had come to accept the germ theory of disease. If specific bacilli caused specific diseases, how could an individual harbor those bacilli and not be ill? See John Andrew Mendelson, "Typhoid Mary, Medical Science, the State, and the 'Germ Carrier'," (Senior Thesis, Harvard University, 1988). I am indebted to him for sending me a copy of the thesis and for sharing his insights with me.

57. Berkow et al., eds., *The Merck Manual of Diagnosis and Therapy* 85–88.
58. Soper, "Typhoid Mary," 8 and "The Curious Career of Typhoid Mary," 702.
59. Soper, "The Curious Career of Typhoid Mary," 698.
60. Ibid., 704.
61. S. Josephine Baker, M.D., *Fighting for Life* (New York: Macmillan, 1939), 73.
62. Ibid.
63. Ibid., 75.
64. Soper suspected that in cases where the test proved negative, the samples taken for analysis were too small or some other technical error was made in the analysis. The author has chosen to use the nineteenth century term *bacillus typhosa.* Today it is known that many of these organisms were *salmonella typhi.* However, it is impossible to say with certainty which organism was involved because of changes in bacterial taxonomy in the intervening century.
65. Mallon may have done well to be skeptical of the surgical procedure. Testing in the 1920s by the New York Department of Health included following five carriers who had agreed to removal of their gallbladders. In none of the cases was the victim's infectiveness eliminated by the surgery. A discussion of the broader scientific implications of Mary Mallon's case is Leavitt, "'Typhoid Mary' Strikes Back, Bacteriological Theory and Practice in Early Twentieth Century Public Health," 608–29.
66. On June 30, 1909, the *New York American* ran the headline, "'Typoid [*sic*] Mary' Never Ill, Begs Freedom." It seems that the nickname had been informally used by some public health officials, including PHS physician Milton Rosenau, in a critique of William Park's paper on typhoid carriers at the 1908 annual meeting of the American Medical Association. The Rosenau commentary is reproduced in the *Journal of the American Medical Association* (1908), 982. The nickname appears there in quotation marks and with a lowercase "t", which as one investigator suggests may mean that it had not been used before in print. Certainly usage after the *New York American* story was to capitalize the "t" and often to omit the quotation marks. See also, Mendelsohn, "Typhoid Mary," 92–93.
67. *New York American,* June 30, 1909.
68. *New York American,* Feb. 21, 1910.
69. Ibid.
70. Ibid.
71. McLaughlin, "Mary Mallon: Alias Typhoid Mary," 52.
72. Stanley Walker, "Profiles: Typhoid Carrier No. 36," *The New Yorker* 10 (Jan. 26, 1935): 21–25.
73. Soper, "The Curious Career of Typhoid Mary," 712.
74. The observation was that of the physician Mary C. McLaughlin, "Mary Mallon: Alias Typhoid Mary," 56.
75. Mendelson, "Typhoid Mary: Medical Science, the State, and the 'Germ Carrier,'" passim.

76. Walker, "Profiles: Typhoid Carrier No. 36," 21–25.
77. Soper, "The Curious Career of Typhoid Mary," 698.
78. The notion of a cult of true womanhood was most lucidly explicated by Barbara Welter, "The Cult of True Womanhood," in Welter, *Dimity Convictions: The American Woman in the Nineteenth Century* (Athens: Ohio University Press, 1976). An excellent synthesis of the history of women in America is Sara M. Evans, *Born for Liberty, A History of Women in America* (New York: Free Press, 1989).
79. An especially fine study of Irish women and mobility is Hasia Diner, *Erin's Daughters in America, Irish Immigrant Women in the Nineteenth Century* (Baltimore: Johns Hopkins University Press, 1983).
80. Kerby A. Miller, *Emigrants and Exiles, Ireland and the Irish Exodus to North America* (New York: Oxford University Press, 1985), 495–96.

CHAPTER 5: "THAT IS THE AMERICAN WAY. AND IN AMERICA YOU SHOULD DO AS AMERICANS DO"

1. *Annual Report of the Bureau of Health of the City of Philadelphia for the Year Ending December 31, 1915* (Philadelphia: City of Philadelphia, 1916), 156–57. For a full discussion of the episode in the light of Philadelphia's public health politics, see Michael P. McCarthy, *Typhoid and the Politics of Public Health in Nineteenth Century Philadelphia* (Philadelphia: American Philosophical Society, 1987), 92–93.
2. Antonio Stella, *Some Aspects of Italian Immigration to the United States, Statistical Data and General Consideration Based Chiefly Upon the United States Censuses and Other Official Publications* (New York: Arno Press, 1975; orig. 1924), 66. The second quotation is from Stella, "Tuberculosis and the Italians in the United States," *Charities* 12 (1904): 486–89.
3. Infection from this highly contagious virus occurs only among humans and only through direct contact. In developing countries where sanitation and hygiene are poor, especially in tropical countries, the virus circulates year-round and children acquire infection and immunity in the first few years of life. However, with improved sanitation and hygiene, infection is delayed. Older children and adults remain susceptible; summer epidemics affecting these older groups were quite common prior to the introduction of the Salk and Sabin vaccines in the 1950s. The best single volume on polio remains John R. Paul, *A History of Poliomyelitis* (New Haven: Yale University Press, 1971). For a brief description of the disease, see Robert Berkow et al., eds., *The Merck Manual of Diagnosis and Therapy,* 15th ed. (Rahway, N.J.: Merck Sharp & Dohme Research Laboratories, 1987), 2037.
4. For a discussion of polio epidemics in the United States prior to the 1930s, see Naomi Rogers, *Dirt and Disease, Polio Before FDR* (New Brunswick, N.J.: Rutgers University Press, 1992). It is based upon her dissertation on the subject, "Screen the Baby, Swat the Fly: Polio in the Northeastern

United States, 1916" (Ph.D diss., University of Pennsylvania, 1986). Also, Saul Benison, "The History of Polio Research in the United States: Appraisal and Lessons," in Gerald Holton, ed., *The Twentieth Century Sciences: Studies in the Biography of Ideas* (New York: Norton, 1972), 308–43.

5. The 1916 epidemic as it affected New York City was chronicled by New York Health Commissioner Haven Emerson, *A Monograph on the Epidemic of Poliomyelitis (Infantile Paralysis) in New York City in 1916 Based on the Official Reports of the Bureau of the Department of Health* (1917; reprinted New York: Arno Press, 1977). See also Emerson's presentation in "Transactions of a Special Conference of State and Territorial Health Officers with the United States Public Health Service, for the Consideration of the Prevention of the Spread of Poliomyelitis: Held at Washington, D.C., August 17 and 18, 1916," *Public Health Bulletin* 83 (1917) and C. H. Lavinder, A. W. Freeman, and W. H. Frost, "Epidemiological Studies of Poliomyelitis in New York City and the North-Eastern United States During 1916, *Public Health Bulletin* 91 (1918).
6. Emerson, *A Monograph on the Epidemic of Poliomyelitis,* 108–9.
7. Ibid.
8. Charles Rosenberg discusses the meaning of the epidemic as an event and its "dramaturgic form" in "What is an Epidemic? AIDS in Historical Perspective," *Daedalus* 18 (Spring 1989): 1–17.
9. This was not the first time that immigrants had been linked to polio. The historian Saul Benison describes how in 1910 Dr. Simon Flexner of the Rockefeller Institute received a letter from Mrs. A. O. Langmuir claiming that of the thirty-six polio cases in her town of Chouteau, Montana, thirty-five of the victims were Norwegian by birth. Engaging in the not uncommon practice of blaming the victim, she speculated that what she regarded as Norwegian ignorance of "sanitation and ventilation" was responsible for the pattern. Saul Benison, "Poliomyelitis and the Rockefeller Institute: Social Effects and Institutional Response," *Journal of the History of Medicine and Allied Sciences* 29 (1974): 80–81.

   It is important to note that in 1916 this element of nativism was absent from the social construction of the disease in cities with few immigrants. There is no hint of it in Washington, D.C., for example, where preparations were underway should the nation's capital begin to experience the epidemic on a scale similar to New York's. By the end of 1916, there had been only thirty-nine cases and eight deaths from polio without any pattern suggesting linkage to a particular group. Cheri McKinless, "A Firm Hand in a Confusing Time: Dr. William C. Woodward and the Control of Poliomyelitis, Washington, D.C., 1910–1916" (unpublished graduate paper, The American University, 1990).
10. *New York Times,* July 1, 1916.
11. *New York Times,* July 11, 1916. New York City Health Commissioner Dr. Haven Emerson's inquiry was documented in a cover letter and memoran-

dum from C. H. Lavinder, the physician on Ellis Island who responded to the inquiry, to Surgeon General Rupert Blue in Washington, D.C., July 5, 1916, RG 90, File 1712, United States National Archives, Washington, D.C. Final assurance that polio was not being detected at the port of embarkation was requested and received. Memo from Acting Surgeon General to Eager at the American Consulate, Naples, Italy, July 6, 1916, RG 90, File 1712.

12. Richmond Mayo-Smith, *Emigration and Immigration: A Study in Social Science* (New York: Scribner's, 1890), 133.
13. Edward Alsworth Ross, *The Old World in the New, The Significance of Past and Present Immigration to the American People* (New York: Century, 1914), 113.
14. Louis I. Dublin and Gladden W. Baker, "The Mortality of Race Stocks in Pennsylvania and New York," reprinted from the *Quarterly Publication of the American Statistical Association* (Mar. 1920) located in Papers and Speeches file of the Louis Dublin Papers at the Metropolitan Life Insurance Company Archives.
15. All data in table were excerpted from Dublin and Baker, "Mortality of Race Stocks in Pennsylvania and New York."
16. *New York Times,* July 9, 1916. Also, Rogers, *Dirt and Disease,* 47.
17. *New York Times,* July 12, 1916. Also, Rogers, "Screen the Baby, Swat the Fly," 39.
18. Steeped in a culture centered on the domus and believing that illness resulted from a jealous stranger's *mal'occhio* (evil eye) Italians often preferred to seek health care from family members and cures from the intercession of holy figures, such as Harlem's Madonna of Mount Carmel, than from doctors and visiting nurses. See Robert Anthony Orsi, *The Madonna of 115th Street, Faith and Community in Italian Harlem, 1880–1950* (New Haven: Yale University Press, 1985).
19. E. Holmes, Report of Nurse, August 24, 1916, Department of Health, City of New York, Special Investigation of Infantile Paralysis, Rockefeller Institute Papers, A Category, American Philosophical Society, Philadelphia, Pennsylvania.
20. Kissing the dead is mentioned as part of the "ritualized outpouring of grief" by Rudolph J. Vecoli, "Cult and Occult in Italian-American Culture, The Persistence of a Religious Heritage," in Randall M. Miller and Thomas D. Marzik, eds., *Immigrants and Religion in Urban America* (Philadelphia: Temple University Press, 1977), 33.
21. Ernest J. C. to Mayor John P. Mitchel, July 10, 1916, Mayor John P. Mitchel Papers, General Correspondence files, Municipal Archives, New York City.
22. *New York Times,* July 23, 1916.
23. This overview is based on many studies of the Italian immigrant experience, including Dino Cinel, *From Italy to San Francisco: The Immigrant Experience* (Stanford, Calif.: Stanford University Press, 1982); Virginia Yans-McLaughlin, *Family and Community, Italian Immigrants in Buffalo,*

1880–1930 (Ithaca: Cornell University Press, 1977); Donna Gabaccia, *From Sicily to Elizabeth Street, Housing and Social Change Among Italian Immigrants, 1880–1930* (Albany: State University of New York Press, 1983) and *Militants and Migrants: Rural Sicilians Become American Workers* (New Brunswick, N.J: Rutgers University Press, 1988); and Rudolph Vecoli's seminal essay, "*Contadini* in Chicago, A Critique of *The Uprooted*," *Journal of American History* 51 (Dec. 1964): 404–17.

24. Leonard Covello, *The Social Background of the Italian-American School Child* (Leiden, Holland: E. J. Brill, 1967; orig.1944).
25. Much of the material to follow is from Phyllis H. Williams, *South Italian Folkways in Europe and America, a Handbook for Social Workers, Visiting Nurses, School Teachers, and Physicians* (New Haven: Yale University Press, 1938). The volume is one in a series published by the Institute of Human Relations at Yale. The book, designed to have practical application for those who would serve the Italian immigrant community, is a valuable study conducted in the best social scientific tradition of its era. In her preface, Williams explains how profoundly she has drawn upon work done in Italy compiling cultural traditions, including Dr. Giuseppe Pitre's twenty-five-volume study of popular Sicilian traditions, *Biblioteca delle Tradizioni Populari Siciliane* (Turin-Palermo: Carlo Clausen, 1871–1913). Williams is also careful to note that the people whose folkways and mores she describes do not represent a cross section of the Italians either in Italy or in the United States but chiefly "peasants and fishing folk" from the six southern provinces, including Sicily.
26. Ibid., 135. Williams observes that "the word pagan is derived from *paganus*, meaning peasant or country dweller."
27. Antoinette T. Ragucci, "Generational Continuity and Change in Concepts of Health, Curing Practices, and Ritual Expressions of the Women of An Italian-American Enclave" (Ph.D. diss., Boston University, 1971), 9–10.
28. Williams, *South Italian Folkways*, 135.
29. Ibid., 136–37. Although local saints were often represented iconographically as the Virgin Mary, each was an embodiment of some saint or historical figure whose past was associated with a particular village or town. This local saint veneration was derived from the polytheism of the Roman departmental deities.
30. Giuseppe Pitre, "The Jettatura and the Evil Eye," translated from *Biblioteca delle tradizioni Popolari Siciliane*, vol. 17, *Usi e Custumi, Credenze e Pregiudizi del Popolo Siciliano vol. 4* (Palermo, 1889), 235–49, in Alan Dundes, ed., *The Evil Eye* (Madison: University of Wisconsin Press, 1992), 130–42.
31. Ibid., 142.
32. Ibid., 143–44.
33. Joseph Collins, *My Italian Year* (New York: Scribner's, 1910), 10. Also in Williams, *South Italian Folkways*, 145.
34. Williams, *South Italian Folkways*, 160.

35. Harriette Matteini, *Letters from Florence, Italy, in 1866* (New Haven: Fanny Winchester Hotchkiss, 1893), 52–53. Also cited in ibid., 160–61.
36. Williams, *South Italian Folkways,* 162.
37. Ibid., 162.
38. Ibid.
39. Antonio Stella, "Tuberculosis and the Italians in the United States," *Charities* 12 (1904): 486–89.
40. Williams, *South Italian Folkways,* 164.
41. Ibid., 165.
42. Ibid. On this transformation to modern therapeutics in the U.S., see John Harley Warner, *The Therapeutic Perspsective, Medical Practice, Knowledge, and Identity in America, 1820–1885* (Cambridge, Mass.: Harvard University Press, 1986).
43. Ibid., 166.
44. Ibid., 169.
45. Carlo Levi, *Christ Stopped at Eboli* (New York: Farrar, Straus and Giroux, 1963; orig. 1947), 237.
46. Ibid., 238.
47. Ibid., 240.
48. Williams, *South Italian Folkways,* 174.
49. Yans-McLaughlin, *Family and Community,* 69–71.
50. Antonio Stella, "The Effect of Urban Congestion on Italian Women and Children," *Medical Record* 74 (May 2, 1908): 732.
51. John Foster Carr, *Guide for the Immigrant Italian in the United States of America,* published under the auspices of the Connecticut Daughters of the American Revolution (New York: Arno Press, 1975; orig. 1911), 3.
52. Ibid., 21.
53. Ibid., 38.
54. Ibid.
55. Ibid., 46.
56. Ibid.
57. Ibid.
58. Ibid., 48.
59. Ibid., 50.
60. Workers of the Federal Writers' Project, Works Progress Administration in the City of New York, *The Italians of New York: A Survey* (New York: Random House, 1938), 112–13.
61. Carr, *Guide for the Immigrant Italian,* 50.
62. Ibid., 51.
63. Dino Cinel, *The National Integration of Italian Return Migration, 1870–1929* (New York: Cambridge University Press, 1991), 105.
64. Most of the biographical data on Antonio Stella is drawn from *The National Cyclopaedia of American Biography* (New York: James T. White and Company, 1929), 20, 80–81.
65. Stella, *Some Aspects of Italian Immigration,* 66.

66. Ibid., 70.
67. Ibid., 70–71.
68. Ibid., 66–67.
69. Ibid., 68.
70. Ibid.
71. Stella, "Tuberculosis and the Italians in the United States," 486–89, also reprinted in Lydio F. Tomasi, *The Italian in America: The Progressive View, 1891–1914* (New York: Center for Migration Studies, 1978), 169–73.
72. Ibid., 170.
73. Ibid., 170–71.
74. Stella, "The Effects of Urban Congestion on Italian Women and Children," 724.
75. Stella, "Tuberculosis and the Italians in the United States," 171.
76. Commissariato dell' Emigrazione, *Bolletino dell' emigrazione,* no. 20 (1905): 28, also quoted by Betty Boyd Caroli, *Italian Repatriation from the United States, 1900–1914* (New York : Center for Migration Studies, 1973), 66–68.
77. Caroli, *Italian Repatriation,* 67.
78. Dr. T. Rosati as quoted in ibid., 67–68; Luigi Rossi, ibid., 68.
79. John Shaw Billings, M.D., *Vital Statistics of New York City and Brooklyn Covering a Period of Six Years Ending May 31, 1890* (Washington, D.C.: Government Printing Office, 1894), 15.
80. Ibid., 19.
81. When mortality per 100,000 was measured with nativity, the 1906 data on Italians for all causes of death showed a mortality rate of 1,816. However, when parents' nativity was used, the rate skyrocketed to 3,643.
82. William H. Guilfoy, "The Deathrate of the City of New York As Affected by the Cosmopolitan Character of Its Population," *Medical Record* 73 (Jan. 4, 1908–June 27, 1908): 133–34.
83. Ibid., 135.
84. Ibid.
85. Ibid.
86. Robert Morse Woodbury, *Infant Mortality and Its Causes* (Baltimore: Williams and Wilkins Company, 1926), 108–9. The eight cities were Johnstown, Pennsylvania; Manchester, New Hampshire; Saginaw, Michigan; Brockton and New Bedford, Massachusetts; Waterbury, Connecticut; Akron, Ohio; and Baltimore, Maryland.
87. Ibid.; Samuel H. Preston and Michael R. Haines, *Fatal Years: Child Mortality in Late Nineteenth Century America* (Princeton: Princeton University Press, 1991), 101.
88. Orsi, *The Madonna of 115th Street,* xix–xx. The term used by anthropologist Orsi was borrowed from Emmanuel LeRoy Ladurie, who used it in his study of Montaillou, see Ladurie, *Montaillou: The Promised Land of Error* (New York: Vintage Books, 1979).
89. Williams, *South Italian Folkways,* 175.

90. Orsi, *The Madonna of 115th Street,* 181–82, 193–95, 209–10.
91. Ibid., 181.
92. Williams, *South Italian Folkways,* 180.
93. Michael M. Davis, Jr., *Immigrant Health and Community* (Montclair, N. J.: Patterson Smith, 1971; orig. 1921), 138–39.
94. Williams, *South Italian Folkways,* 178.
95. Ibid., 180.
96. Ibid., 182.
97. Antoinette T. Ragucci, "Generational Continuity and Change in Concepts of Health, Curing Practices and Ritual Expressions of the Women of an Italian-American Enclave," 268.
98. Ibid., 268.
99. Ibid., 272.
100. Ibid.
101. Ibid., 273.
102. The best overview of the history of urban settlement houses is Allen F. Davis, *Spearheads for Reform: The Social Settlements and the Progressive Movement, 1890–1914* (New York: Oxford University Press, 1967). Well known are Jane Addams's *Twenty Years at Hull House* (New York: Macmillan, 1910) and Lillian D. Wald, *The House on Henry Street* (New York: Henry Holt, 1915). On Hull House and its founder, also see Davis, *American Heroine, The Life and Legend of Jane Addams* (New York: Oxford University Press, 1973); Mary Lynn McCree Bryan and Allen F. Davis, eds., *100 Years at Hull House* (Bloomington: Indiana University Press, 1990; orig. 1969); and Rivka Shpak Lissak, *Pluralism and Progressives: Hull House and the New Immigrants, 1890–1919* (Chicago: University of Chicago Press, 1989). On Lillian Wald, see Beatrice Siegel, *Lillian Wald of Henry Street* (New York: Macmillan, 1983).
103. This perspective is epitomized by Davis in *Spearheads for Reform.*
104. A critical view of settlement workers can be found in John F. McClymer, *War and Welfare: Social Engineering in America, 1890–1925* (Westport, Conn.: Greenwood Press, 1980).
105. An especially intelligent discussion of settlement house reform and its relationship to the desire for preserving the moral order by socially controlling new arrivals and other potentially dissident elements of the population is Paul Boyer, *Urban Masses and Moral Order in America, 1820–1920* (Cambridge, Mass.: Harvard University Press, 1978).
106. Starr Center Annual Report, 1908, 5. I wish to thank Dr. Fred Miller of the National Endowment for the Humanities, formerly head of the Urban Archives at Temple University, for calling this material to my attention.
107. Ibid., 2
108. Starr Center *Annual Report,* 1911, 7.
109. Rocco Brindisi, "The Italian and Public Health," *Charities* 12 (1904): 486.
110. Thomas J. Riley, "Poverty and Poliomyelitis," *Survey* (July 29, 1916): 447–48.

CHAPTER 6: *GEZUNTHAYT IZ BESSER VI KRANKHAYT*

1. The discussion of the conditions of the Russian Jews in this chapter is based upon S. N. Dubnow, *History of the Jews in Russia and Poland from the Earliest Times Until the Present Day*, trans. I. Friedlander (Philadelphia: Jewish Publication Society of America, 1916), 3 vols., and Salo Baron, *The Russian Jews Under the Tsars and Soviets*, 2d ed. (New York: Macmillan, 1976).
2. Bertha Pearl, "Those Days," *The Montefiore Echo* 2 (Aug. 1916): 1.
3. Alan Dundes, ed., *The Blood Libel Legend, A Casebook in Anti-Semitic Folklore* (Madison: University of Wisconsin Press, 1991).
4. Sander L. Gilman's research has been pathbreaking in its analysis of this obsession. Sander L. Gilman, "The Jewish Body: A 'Footnote,'" *Bulletin of the History of Medicine* 64 (Winter 1990): 588–602. See Gilman's larger study, *The Jew's Body* (New York: Routledge, 1991).
5. An exquisitely drawn portrait of western European Jewish intellectuals struggling to define categories to explain the differences between Jews, especially less assimilated eastern European Jews, and Gentiles can be found in Sander L. Gilman, *Jewish Self-Hatred, Anti-Semitism and the Hidden Language of the Jews* (Baltimore: Johns Hopkins University Press, 1986).
6. Entry for Maurice Fishberg, *Encyclopedia Judaica, Decennial Book* (Jerusalem: Macmillan Company, 1982), 1328–29.
7. Maurice Fishberg, "Health and Sanitation of the Immigrant Jewish Population of New York," (July 1902): 40–41.
8. Ibid., 41.
9. Jeanne S. as quoted by Sydney Stahl Weinberg, *The World of Our Mothers, The Lives of Jewish Immigrant Women* (Chapel Hill: University of North Carolina Press, 1988), 4.
10. Fishberg, "Health and Sanitation of the Immigrant Jewish Population of New York," (July 1902): 41.
11. Ibid., 42.
12. Arthur Ruppin, *The Jews of Today* (New York: Macmillan, 1913), 76, 79.
13. Fishberg, "Health and Sanitation of the Immigrant Jewish Population of New York," (July 1902): 42.
14. Lillian Mermin Feinsilver, *The Taste of Yiddish* (South Brunswick, N.J.: A.S. Barnes, 1970), 36, 210.
15. Mark Zborowski and Elizabeth Herzog, *Life Is with People, The Culture of the Shtetl* (New York: International Universities Press, 1952; reprinted 1976), 355.
16. Ibid.
17. Interview with Clare Kaplan as quoted by Elizabeth Ewen, *Immigrant Women in the Land of Dollars, Life and Culture on the Lower East Side, 1890–1925* (New York: Monthly Review Press, 1985), 141.
18. Interview with Sophie Abrams, ibid.
19. Zborowski and Herzog, *Life Is with People,* 355. For a discussion of feldsh-

ers and rural health care in pre-Revolutionary Russia, see Samuel C. Ramer, "Feldshers and Rural Health Care in the Early Soviet Period," in Susan Gross Solomon and John F. Hutchinson, eds., *Health and Society in Revolutionary Russia* (Bloomington: Indiana University Press, 1990), 121–45.

20. The interpretation belongs to Maimonides, the twelfth-century Spanish physician who moved to Egypt, where he healed the sick and wrote commentary. Moses Maimonides, *Mishneh Torah, Hilcot Rotze'ach* 4:1 as cited by Fred Rosner, "Communicable Diseases and the Physician's Obligation to Heal," in Rosner, ed., *Medicine and Jewish Law* (Northvale, N.J.: Jason Aaronson, 1990), 73.
21. For a discussion of the modernization of European Jewry, see Calvin Goldscheider and Alan S. Zuckerman, *The Transformation of the Jews* (Chicago: University of Chicago Press, 1984).
22. Zborowski and Herzog, *Life Is with People,* 355.
23. The Pale was the region of tsarist Russia that extended about 386,000 square miles from the Baltic Sea to the Black Sea, including Congress Poland, Lithuania, Byelorussia, and the Ukraine (excluding Kiev).
24. Dr Maurice Fishberg's essays describe general health conditions in pre-Revolutionary Russia and especially the health conditions of Russian Jews. Fishberg published a series, "Health and Sanitation of the Immigrant Jewish Population," in a journal, *Menorah,* between July and September 1902. Most of these articles were later republished verbatim in Maurice Fishberg, "Health and Sanitation: New York," in Charles S. Bernheimer, *The Russian Jew in the United States, Studies of Social Conditions in New York, Philadelphia, and Chicago, with a Description of Rural Settlements* (Philadelphia: The John Winston Co., 1905). See Fishberg in Bernheimer, ed., 303. See also, Nancy M. Frieden, *Russian Physicians in an Era of Reform and Revolution, 1856–1905* (Princeton: Princeton University Press, 1981) and John F. Hutchinson, "Tsarist Russia and the Bacteriological Revolution," *Journal of the History of Medicine and Allied Sciences* 40 (1985): 420–39.
25. Nancy M. Frieden, "The Russian Cholera Epidemic, 1892–1893, and Medical Professionalization" *Journal of Social History* 10 (1977), 538–59. Richard J. Evans, *Death in Hamburg, Society and Politics in the Cholera Years, 1830–1910* (New York: Oxford University Press, 1987), 280–81.
26. *New York Times,* Sept. 9, 1892.
27. The term "promised city" is from the title of Moses Rischin's excellent study of Jews in New York, *The Promised City, New York's Jews, 1870–1914* (Cambridge, Mass.: Harvard University Press, 1962).
28. Rischin, *Promised City,* 94.
29. The most comprehensive and descriptive contemporary study of tenements is Robert W. DeForest and Lawrence Veiller, *The Tenement House Problem, Including the Report of the New York State Tenement House Commission of 1900,* 2 vols. (New York: Macmillan, 1903).
30. *Fifth Report, Tenement House Department, City of New York, 1909,* (New

York: J. J. Little & Ives, Co., 1910), 106–7, 112–13. Also cited by Rischin, the *Promised City,* 84.

31. Fishberg, "Health and Sanitation of the Immigrant Jewish Population of New York," (July 1902): 45.
32. Tenement house reform is covered in Roy Lubove, *The Progressives and the Slums, Tenement House Reform in New York City, 1890–1917* (Pittsburgh: University of Pittsburgh Press, 1962). An anthology on reform in New York City is Ronald Lawson, ed., with the assistance of Mark Naison, *The Tenement Movement in New York City, 1904–1984* (New Brunswick, N.J.: Rutgers University Press, 1986).
33. Fishberg, "Health and Sanitation of the Immigrant Jewish Population of New York," 44.
34. William Dean Howells, *Impressions and Experiences* (New York: Harper, 1896), 253.
35. Ibid.
36. Jacob A. Riis, *How the Other Half Lives, Studies Among the Tenements of New York* (New York: Hill and Wang, 1957; orig., 1901), 83–84.
37. Manly H. Simons, "The Origin and Condition of the Peoples Who Make Up the Bulk of Our Immigrants at the Present Time and the Probable Effect of the Absorption Upon Our Population," *The Military Surgeon* 23 (Dec. 1908): 433.
38. Ibid.
39. Edward Alsworth Ross, *The Old World in the New, The Significance of Past and Present Immigration to the American People* (New York: Century, 1914), 289–90.
40. Madison Grant, *The Passing of the Great Race* (New York: Scribner's, 1916), 14–16.
41. William H. Guilfoy, "The Death Rate of the City of New York as Affected by the Cosmopolitan Character of Its Population," *Medical Record* 73 (Jan. 4, 1908–June 27, 1908): 132–35.
42. Franz Boas, *Changes in Bodily Form of Descendants of Immigrants* in *Reports of the Immigration Commission,* 61st Cong., 2d Sess., Senate Document No. 208 (Washington, D.C.: Government Printing Office, 1911), vol. 38.
43. Guilfoy, "The Death Rate of the City of New York," 132–135. This conclusion is also drawn by Jacob Jay Lindenthal, "*Abi Gezunt*: Health and the Eastern European Jewish Immigrant," *American Jewish History* 70 (June 1981): 420–41. Another excellent article on the subject is Deborah Dwork, "Health Conditions of Immigrant Jews on the Lower East Side of New York, 1880–1914," *Medical History* 25 (1981): 1–40.
44. Fishberg, "Health and Sanitation of the Immigrant Jewish Population of New York," *Menorah* 33 (Aug. 1902), 73.
45. Ibid., 74.
46. Ibid., 74.
47. Ibid.
48. Leroy-Beaulieu as quoted by Fishberg, "Health and Sanitation of the Immi-

grant Jewish Population of New York," *Menorah* 33 (Aug. 1902): 75. Anatole Leroy-Beaulieu published *Les Juifs et l'antisémitisme* in France in 1893. Two years later it was translated into English as *Israel Among the Nations*. The words of praise for the Jews that Fishberg cites are a sharp contrast to the volume's main argument, which concerns the physical degeneration manifest in eastern European Jewry. Leroy-Beaulieu sounds somewhat like American E. A. Ross, observing, "Their [the Jews'] physical strength, their muscular power, has diminished in each generation; their blood has become poorer, their stature smaller, their shoulders and chests narrower. Many Jews of the larger Jewries have an emaciated, pallid look. Many of them show signs of racial decline and degeneracy." Leroy-Beaulieu saw part of the problem as heredity, "marriages between near relations," but also environment: "their age-long confinement, their lack of exercise, of pure air and wholesome nourishment."

49. Fishberg, "Health and Sanitation of the Immigrant Jewish Population of New York," (Aug. 1902): 75.
50. Ibid., 76.
51. Seymour Siegel, "A Guide to Observance," in Samuel H. Dresner, ed., *The Jewish Dietary Laws* (New York: Burning Bush Press, 1959), 55.
52. Dresner, "Their Meaning in Our Times," in ibid., 21.
53. Ibid., 29.
54. An anthropological discussion of the dietary laws suggests that the laws of *kashrut* are "like signs which at every turn inspired meditation on the oneness, purity, and completeness of God." Those species considered unclean are those that are imperfect members of their class. Thus, the pig is forbidden because it falls short of being a cloven-hoofed, cud-chewing ungulate, the model of the proper kind of food for pastoralists such as the ancient Israelites. Mary Douglas, *Purity and Danger, An Analysis of Pollution and Taboo* (Hammondsworth, England: Pelican Books, 1970), 54–72.
55. Siegel, "A Guide to Observance," 70, n.9.
56. *New York Times*, July 30, 1893. See also, Alan M. Kraut, "The Butcher, the Baker, the Pushcart Peddler: Jewish Foodways and Entrepreneurial Opportunity in the East European Immigrant Community, 1880-1940," *Journal of American Culture* 6 (Winter 1983): 71–83.
57. Rischin, *Promised City*, 56–57.
58. Paula E. Hyman, "Immigrant Women and Consumer Protest: The New York City Kosher Meat Boycott of 1902," *American Jewish History* 70 (Sept. 1980): 91–105.
59. New York City Bureau of Standards and the New York City Bureau of Municipal Research, *Report on the Cost of Living for an Unskilled Laborer's Family in New York City* (New York, 1915), 15. The report mentions Louis Dublin's dissent. A social worker's study of New York City wage earners as consumers offers some valuable insight into the kind of health expenses families actually met. See Louise Bolard More, *Wage-Earners' Budgets: Study of Standards and Cost of Living in New York*

(New York: Arno Press and The New York Times, 1971; orig. 1907).

60. M. A. Lipkind, "Some East Side Physicians at the Close of the Nineteenth Century," *Medical Leaves* 1 (1937): 103–9. Also, A. J. Rongy, "Half A Century of Jewish Medical Activities in New York City," *Medical Leaves* 4 (1941): 151–63.
61. There is an increasingly rich literature on childbearing and the role of midwives. See Judith Walzer Leavitt, *Brought to Bed, Child-Bearing in America, 1750–1950* (New York: Oxford University Press, 1986). Also, Charlotte G. Borst, "Wisconsin's Midwives as Working Women: Immigrant Midwives and the Limits of a Traditional Occupation, 1870–1920," *Journal of American Ethnic History* 8 (Spring 1989): 24–59; Nancy Schrom Dye, "Review Essay: The History of Childbirth in America," *Signs* 6 (Autumn 1980): 97–108; Janet Bogdan, "Care or Cure? Childbirth Practices in Nineteenth-Century America," *Feminist Studies* 4 (June 1978): 92–99; and Richard W. Wertz and Dorothy C. Wertz, *Lying-In* (New York: Free Press, 1977).
62. F. Elizabeth Croswell, "The Midwives of New York," *Charities and Commons* 17 (1907): 667. Also cited in Ewen, *Immigrant Women in the Land of Dollars,* 131–32.
63. The great degree to which poor women as well as rich were active agents in moving childbirth from "the traditional family to the scientific hospital" is emphasized in Virginia Anne Metaxas Quiroga, "Poor Mothers and Babies: A Social History of Childbirth and Child Care Institutions in Nineteenth Century New York City" (Ph.D. diss., State University of New York at Stony Brook, 1984).
64. Mary Boyd, "The Store of Dammerschalf," *Survey* 32 (1914): 129.
65. Lillian Wald, *The House on Henry Street* (New York: Henry Holt and Company, 1915), 57–58.
66. Ibid.
67. Lipkind, "Some East Side Physicians," 104. Rischin, *Promised City,* 72.
68. Lipkind, "Some East Side Physicians," 104.
69. The material on Jewish immigrant attitudes toward physicians first appeared in Alan M. Kraut, "Healers and Strangers, Immigrant Attitudes Toward the Physician in America—A Relationship in Historical Perspective," *Journal of the American Medical Association* 263 (Apr. 4, 1990): 1807–11.
70. Maurice Fishberg, *The Jews: A Study of Race and Environment* (New York: Walter Scott Publishing Co., Ltd., 1911). This volume summarized and modified many of the conclusions in earlier articles so that Fishberg could make clear his opposition to the perspective that the Jews were a separate race whose nature rather than nurture accounted for the group's rates of morbidity and mortality.
71. John Shaw Billings, "Vital Statistics of the Jews in the United States," *Census Bulletin No. 19* (Washington, D.C.: United States Bureau of the Census, 1890), 3. The best biography of Billings remains Fielding H. Garrison, M.D., *John Shaw Billings, A Memoir* (New York: Putnam's, 1915).

72. The rate of response was remarkable; 10,618 families returned completed forms, accounting for 60,630 individuals, or 70.8 percent of those who received inquiries. Native-born families in the survey numbered 1,332; 8,263 of the families surveyed were resident in the country for a period of fifteen years or more, and 522 had been here between five and ten years.
73. Billings, "Vital Statistics of the Jews in the United States," 19.
74. Billings conducted a second study, *Vital Statistics of New York City and Brooklyn* (1894), utilizing municipal health data for a period of six years ending May 31, 1890. According to Billings's statistics, Russian and Polish Jews had the lowest rates of mortality in the city for the period covered by the study: only 14.85 per 1,000 as compared with 43.57 for Bohemians, 35.29 for Italians, and 32.51 for Irish. The mortality of children was the lowest among the Russian Jews, 28.67 per 1,000, as opposed to such high figures as 82.57 for the Bohemians and 76.41 for the Italians.
75. Louis I. Dublin and Gladden W. Baker, "The Mortality of Race Stocks in Pennsylvania and New York," reprinted from the *Quarterly Publication of the American Statistical Association* (Mar. 1920), located in Papers and Speeches file of the Louis Dublin Papers of the Metropolitan Life Insurance Company Archives. The authors, collecting data by nationality, observed, "In New York State, there were in 1910, 558,952 Russian born persons, constituting 20.5 per cent of the foreign born and 6.2 per cent of the total white population, . . . a heterogeneous group which includes a large proportion of Jews as well as Poles and true Russians. Of all the Russian immigrants to the United States in the period 1899–1910, 43.8 per cent were Jews, and it is very probable that a much larger proportion of the Russians in New York State in 1910 were Jews." They assumed, then, that their "figures for Russians are really for a group composed mainly of Jews."
76. Ibid., 10–11.
77. Ibid., 11.
78. Robert Morse Woodbury, *Causal Factors in Infant Mortality: A Statistical Study Based on Investigations in Eight Cities* (Washington, D.C.: Government Printing Office, 1925): 104–7.
79. Gretchen A. Condran and Ellen A. Kramarow, "Low Child Mortality in the United States in the Early Twentieth Century: An Examination of a Jewish Immigrant Population," unpublished paper, Population Studies Center, University of Pennsylvania. I am indebted to Gretchen Condran for generously sharing her paper and her insights with me.
80. Fishberg, *The Jews,* 176.
81. Infant mortality rate is calculated by the number of deaths under one year of age per 1,000 live births. A. Hamilton, "Excessive Child-bearing as a Factor in Infant Mortality," proceedings of Conference on the Prevention of Infant Mortality, *Prevention of Infant Mortality* (New Haven: American Academy of Medicine, 1909), 74–80. In the same volume, see E. T. Devine, "The Waste of Infant Life," 95–112.
82. Samuel H. Preston and Michael R. Haines, *Fatal Years, Child Mortality in*

Late Nineteenth Century America (Princeton: Princeton University Press, 1991), 28.

83. U. O. Schmelz, *Infant Mortality Among the Jews of the Diaspora* (Jerusalem: Institute of Contemporary Jewry, The Hebrew University of Jerusalem, 1971), 77–83. Gretchen A. Condran and Ellen A. Kramarow, "Child Mortality among Jewish Immigrants to the United States," *Journal of Interdisciplinary History* 22 (1991): 223–54.
84. Irving Howe, *World of Our Fathers, The Journey of the East European Jews to America and the Life They Found and Made* (New York: Harcourt Brace Jovanovich, 1976), 149–50.
85. Edward Alsworth Ross, *The Old World in the New,* 289–90.
86. Simons, "The Origin and Condition of the Peoples Who Make Up the Bulk of Our Immigrants," 433.
87. Valuable historical studies of tuberculosis are Rene and Jean Dubos, *The White Plague, Tuberculosis, Man, and Society* (1952; reprinted New Brunswick, N.J.: Rutgers University Press, 1987); J. Arthur Myers, *Captain of All These Men of Death, Tuberculosis Historical Highlights* (St. Louis: Warren H. Green, 1977); Guy P. Youmans, *Tuberculosis* (Philadelphia: Saunders, 1979); Barbara Bates, *Bargaining for Life, A Social History of Tuberculosis, 1876–1938* (Philadelphia: University of Pennsylvania Press, 1992). A useful study of the public health campaign against TB is Michael E. Teller, *The Tuberculosis Movement, A Public Health Campaign in the Progressive Era* (Westport, Conn.: Greenwood Press, 1988). An intriguing dissertation on the transformation of the tubercular patient's image from the romantic view of the eighteenth and nineteenth centuries to the impoverished, wasted image of the late nineteenth- and twentieth-century victim is Nan Marie McMurry, "'And I? I Am in a Consumption': The Tuberculosis Patient, 1780–1930" (Ph.D. diss., Duke University, 1985).
88. The precise origins of the term *white plague* are unclear. J. Arthur Myers claims that Oliver Wendell Holmes coined the term in 1891. See Meyers, *Captain of All These Men of Death* (St. Louis: Warren H. Green, 1977), xi.
89. The eminent demographer of disease, Thomas McKeown, in his influential volume, *The Modern Rise of Population* (New York: Academic Press, 1976), concluded that nonmedical factors, particularly improvements in nutrition and other general aspects of the standard of living, contributed to the decline of infectious diseases, including tuberculosis. However, historian of medicine Leonard G. Wilson has argued that McKeown was mistaken and that declines in TB mortality were a direct result of removing pulmonary tuberculosis, or phthisis, patients from their families and co-workers to the confines of institutions, increasingly in the United States to TB sanataria. See Leonard G. Wilson, "The Historical Decline of Tuberculosis in Europe and America: Its Causes and Significance," *Journal of the History of Medicine and Allied Sciences* 45 (July 1990): 366–96 and Wilson, "The Rise and Fall of Tuberculosis in Minnesota: The Role of In-

fection," *Bulletin of the History of Medicine* 66 (Spring 1992): 16–52. Most recently, Barbara Bates has challenged contentions that separating TB patients from those not yet infected sufficiently accounts for declining rates of the disease, especially in northeastern states with their highly congested urban areas. See Bates, *Bargaining for Life,* 313–27. For more on the issue, see the heated correspondence between Lind Bryder and Leonard Wilson in the *Journal of the History of Medicine and Allied Sciences* 46 (1991): 358–68.

90. For the years 1901–1903, for example, the distribution of pulmonary tuberculosis among various religious groups per 10,000 of the population were Catholics, 38.3 percent; Protestants, 24.6 percent; and Jews, 13.1 percent. For all forms of TB, the figures were Catholics 49.6; Protestants, 32.8; and Jews, 17.8. Death rates computed for various cities with high concentrations of Jews suggest a similar mortality pattern. Lemberg (now Lvov): Jews, 30.64, Christians, 63.51; Cracow: Jews, 20.49, Christians, 66.41. Compiled from European sources by Maurice Fishberg, "The Comparative Pathology of the Jews," *New York Medical Journal* 73 (1901): 540–41; also, Fishberg, "Tuberculosis Among the Jews," *Transactions of the Sixth International Congress on Tuberculosis* 3 (New York, 1908): 415–28, also reprinted in *Medical Record,* Dec. 26, 1908.

91. Data collected by contemporary physicians show that, in wards comprising mostly Jews, the mortality rate from tuberculosis for the years 1897, 1898, and 1899 was less than that found in the twenty other wards of New York City. The seventh, tenth, eleventh, and thirteenth wards, located on the Lower East Side and consisting mostly of Jews, experienced mortality rates of 213.74, 172.44, 155.27, and 110.54 per 100,000 inhabitants, respectively, while the fourth ward, which comprised largely non-Jewish laborers, had a death rate of 565.06. Maurice Fishberg, "The Relative Infrequency of Tuberculosis Among Jews," *American Medicine* 2 (1901): 695–98.

92. Simons, "The Origin and Condition of the Peoples Who Make Up the Bulk of Our Immigrants," 433.

93. Fishberg, "The Relative Infrequency of Tuberculosis Among Jews," 698.

94. Kate Levy, M.D., [Health and Sanitation in] "Chicago" in Bernheimer, ed., *The Russian Jew in the United States,* 327.

95. Fishberg, *The Jews,* 291. Fishberg's doubts about infection via ingesta and his data hardly closed the question. See J. Arthur Myers, "Development of Knowledge of the Unity of Tuberculosis and the Portals of Entry of Tubercle Bacilli," *Journal of the History of Medicine and Allied Sciences* 29 (1974): 213–28.

96. Ibid., 290.

97. Ibid., 291.

98. Ibid., 290.

99. Fishberg may well have been concerned about the genetic component of European racial discourse finding an increasingly large space in American

anti-Semitic rhetoric. See Robert Singerman, "The Jew as Racial Alien: The Genetic Component of American Anti-Semitism," in David A. Gerber, ed., *Anti-Semitism in American History* (Urbana: University of Illinois Press, 1986), 103–28.

100. Fishberg, *The Jews*, 293.
101. Ibid.
102. Ibid., 292.
103. Ibid., 295.
104. Fishberg, "Tuberculosis Among the Jews," 14.
105. Edward Livingston Trudeau, *An Autobiography* (Garden City, N.Y.: Doubleday Page, 1916). For a recent and popular treatment of Trudeau and the Saranac Lake sanatorium, see Mark Caldwell, *The Last Crusade, The War on Consumption, 1862–1954* (New York: Atheneum, 1988).
106. Isaac Metzker tells the anecdote in his commentary on a letter to the editor of the *Jewish Daily Forward*'s *"Bintel Brief"* ("Bundle of Letters") column in Metzker, ed., *A Bintel Brief* (Garden City, N.Y.: Doubleday, 1971), 83. Also quoted by Stefan Kanfer, *A Summer World, The Attempt to Build a Jewish Eden in the Catskills, From the Days of the Ghetto to the Rise and Decline of the Borscht Belt* (New York: Farrar, Straus and Giroux, 1989), 52.
107. Dorothy Levenson, *Montefiore, The Hospital as Social Instrument, 1884–1984* (New York: Farrar, Straus & Giroux, 1984), 3.
108. Ibid., 63–86.
109. Ibid., 65.
110. Patient publications from tuberculosis sanatoria offer a unique perspective on immigrants' feelings about their lives, the institutions caring for them, and the disease ravaging their bodies. *Montefiore Echo* 2 (Aug. 1916): 1.
111. The historian of the West, Patricia Nelson Limerick, observes that many places in the West were refuges for those who had trouble breathing, including victims of asthma and bronchitis, as well as TB, in *The Legacy of Conquest, The Unbroken Past of the American West* (New York: Norton, 1987), 89.
112. *Hatikvah* 1 (Apr. 1923): 6.
113. I am indebted to my mother-in-law, ethnomusicologist Irene Heskes, for calling the score to my attention and translating the lyrics from the Yiddish. *Menschen-Fresser,* words and music by Solomon Small (Smulewitz), arranged by H. A. Russoto (New York: Hebrew Publishing Co., 1912).
114. *Sanatorium* 3 (Jan. 1909): 26–27.
115. *Hatikvah* 2 (Sept. 1924): 61.
116. "A Letter From A Patient To A Friend," signed Nat in *Hatikvah* 1 (Mar. 1923): 13.
117. *Sanatorium* 1 (May 1907): 49.
118. Maurice Fishberg, "Health Problems of the Jewish Poor," a paper read before the Jewish Chautauqua Assembly on Monday, July 27, 1903, at Atlantic City, New Jersey, later a pamphlet reprinted from *The American Hebrew* (New York: Press of Philip Cowen, 1903), 5.

119. Ibid., 4–5.
120. Ibid.
121. Ibid., 16.

CHAPTER 7: "THE OLD INQUISITION HAD ITS RACK AND THUMBSCREWS"

1. United States Immigration Commission, *Immigrants in Industries* (Washington, D.C.: Government Printing Office, 1910), vol. 2, 297. Full data on all industries in the survey are in volumes 17 and 18 of the study.
2. *New York Times,* Dec. 27, 1908.
3. Crystal Eastman, *Work-Accidents and the Law,* vol. 2 of Paul Underwood Kellogg, ed., *The Pittsburgh Survey, Findings in Six Volumes* (New York: Arno Press, 1969; orig. 1910), 14.
4. Alice Hamilton, *Exploring the Dangerous Trades: The Autobiography of Alice Hamilton, M.D.* (Boston: Little, Brown, 1943), 123–24.
5. Arno Dosch, "Our Expensive Cheap Labor," *The World's Work* 26 (Oct. 1913): 699–703.
6. Hamilton, *Exploring the Dangerous Trades,* 115.
7. Paul S. Peirce, "Industrial Diseases," *North American Review* (Oct. 1911): 537.
8. Dosch, "Our Expensive Cheap Labor."
9. Jeremiah W. Jenks and W. Jett Lauck (5th ed. revised by Rufus D. Smith), *The Immigration Problem, A Study of American Immigration Conditions and Needs* (New York: Funk and Wagnalls, 1922), 34.
10. Fitzhugh Mullan, *Plagues and Politics: The Story of the United States Public Health Service* (New York: Basic Books, 1989), 61.
11. J. W. Schereschewsky, "The Health of Garment Workers," in U.S. Public Health Bulletin 71, *Studies in Vocational Diseases* (Washington, D.C.: Government Printing Office, 1915), 13.
12. Peirce, "Industrial Diseases," 530.
13. Ibid., 534. See also "Phosphor-Necrosis," in "Service of J. E. Garretson, M.D.," reported by Dr. DeF. Willard in *Medical Times,* May 15, 1872: 305–7; J. Ewing Mears, "Phosporus-Necrosis of the Jaws," extracted from the *Transactions of the American Surgical Association,* vol. 3, 1885 (Philadelphia: Collins, Printer, 1885); Alice Hamilton, *Industrial Toxicology* (New York: Harper & Brothers, 1934), 86–91.
14. Ibid., 534–35.
15. Hamilton, *Exploring the Dangerous Trades,* 117–18.
16. *Report of the Illinois Commission on Occupational Diseases to His Excellency Governor Charles S. Deneen* (Chicago: Warner Printing Co., 1911). Some of Hamilton's correspondence concerning the Illinois study is available in Barbara Sicherman, *Alice Hamilton, A Life in Letters* (Cambridge, Mass.: Harvard University Press, 1984), especially 153–83.
17. *Report of Illinois Commission,* 24.

18. Ibid.
19. Ibid.
20. Pierce, "Industrial Diseases," 535.
21. Michael Gold, *Jews Without Money* (New York: Avon Books, 1965; orig. 1930), 78.
22. Philip Zausner, *Unvarnished: The Autobiography of a Union Leader* (New York: Brotherhood Publishers, 1941), 31–32.
23. Alice Hamilton, "The White Lead Industry in the United States, With An Appendix on the Lead-Oxide Trade," in U.S. Department of Commerce and Labor, *Bulletin of the Bureau of Labor* 95 (July 1911): 189–282.
24. Ibid., 215.
25. Ibid., 219.
26. Ibid., 220.
27. Ibid., 221.
28. Ibid., 226.
29. Pierce, "Industrial Diseases," 535.
30. Ibid., 536. The notion that bacilli, not dust particles, caused tuberculosis took a long while to achieve popular understanding and acceptance.
31. Ibid.
32. David Rosner and Gerald Markowitz, *Deadly Dust, Silicosis and the Politics of Occupational Disease in Twentieth-Century America* (Princeton: Princeton University Press, 1991). See also, Markowitz and Rosner, "'The Street of Walking Death': Silicosis, Health and Labor in the Tri-State Region, 1900–1950," *Journal of American History* 77 (Sept. 1990): 525–52. On the broader issue of health in the workplace, see the fine essays in Rosner and Markowitz, eds., *Dying for Work: Workers' Safety and Health in Twentieth-Century America* (Bloomington: Indiana University Press, 1987). According to Rosner and Markowitz, recent research suggests that respiratory diseases were often misdiagnosed. Tuberculosis, recognized as a germ disease by the turn of the century, was the catchall diagnosis when in fact silicosis, a respiratory impairment caused by industrial dust, was often the ailment crippling and sometimes killing workers.
33. David Rosner and Gerald Markowitz, "Consumption, Silicosis, and the Social Construction of Industrial Disease," *Yale Journal of Biology and Medicine* 64 (1991): 481–89. Also, in *Deadly Dust,* 38–41, 143. The condition of silico-tuberculosis is mentioned in A. D. Hosey, V. M. Trasko, and H. B. Ashe, "Control of Silicosis in the Vermont Granite Industry: Progress Report," *U.S. Public Health Service Publication No. 557* (Washington, D.C.: Government Printing Office, 1957): 57. Also, H. E. Ayer, J. M. Dement, K. A. Busch, H. B. Ashe, B. T. H. Levadie, W. A. Burgess, and L. Di Berardinis, "A Monumental Study—Reconstruction of a 1920 Granite Shed," *American Industrial Hygiene Journal* (May 1973): 206–11.
34. Kenneth Durr, "'For the Ones That Are to Follow': Silico-Tuberculosis and the Barre, Vermont, Granite Trade, 1910–1930," unpublished paper, a revised version of an essay written for a class in "Health, Disease, and

Medicine in American History" at The American University, Washington, D.C., 1992.

35. Rosner and Markowitz, *Deadly Dust,* 3.
36. The author's father, Harry Kraut, and uncle, Louis Kraut, worked in the jewelry trade as plater and polisher, respectively. Long after the passage of legislation by the New York State legislature, small family-owned nonunion factories, many employing immigrants and their children, managed to evade the laws by bribing and cajoling inspectors. Hacking coughs and ill health, including premature death from emphysema, was the price that some workers paid so that the factories employing them could evade the regulations, keeping profit margins high and their lofts fully staffed by workers who were paid well below union scale and enjoyed none of the security of pension plans.
37. Pierce, "Industrial Diseases," 532.
38. A helpful anthology is Eileen Boris and Cynthia R. Daniels, eds., *Homework: Historical and Contemporary Perspectives on Paid Labor at Home* (Urbana: University of Illinois Press, 1989). See especially Daniels's essay, "Between Home and Factory: Homeworkers and the State," 13–32.
39. Jacob A. Riis, *How the Other Half Lives: Studies Among the Tenements of New York* (New York: Hill & Wang, 1957; orig. 1901), 103–4.
40. Charles Dickerman, "Cigar Makers' Neurosis," *National Eclectic Medical Association Quarterly* 10 (Sept. 1918): 62–63. See also, Patricia A. Cooper, *Once a Cigar Maker: Men, Women, and Work Culture in American Cigar Factories, 1900–1919* (Urbana: University of Illinois Press, 1987), 101. On the broader issue of health in cigar shops and mutual assistance among Italian and Cuban cigar workers in Tampa, see Gary R. Mormino and George E. Pozzetta, *The Immigrant World of Ybor City, Italians and Their Latin Neighbors in Tampa, 1885–1895* (Urbana: University of Illinois Press, 1987), 175–205.
41. Frieda S. Miller, "Industrial Homework in the United States," *International Labor Review* 43 (Jan. 1941): 1–50, especially 13. Also, Daniels, "Between Home and Factory," 20.
42. Data from New York State, *Second Report of the Factory Investigating Commission* (1913) vol. 2: 677, 691, 729, was used to derive this estimate by Cynthia Daniels, "Between Home and Factory," 15.
43. Ibid.
44. Riis, *How the Other Half Lives,* 80–81.
45. Mary Van Kleeck, *Artificial Flower Makers* (New York: Russell Sage Foundation, 1913).
46. Ibid., 92.
47. Ibid., 116.
48. Ibid., 93.
49. Ibid.
50. Ibid., 99.
51. Ibid., 98–99.

52. Henry White, "Perils of the Home Factory," *Harper's Weekly* (Feb. 11, 1911): 10.
53. New York State Department of Labor, *Annual Report of the Commissioner of Labor, 1911* (New York: State Printing Office, 1911), 214–19.
54. Van Kleeck, *Artificial Flower Makers,* 138–39.
55. *Report on Women and Child Wage-earners in the United States,* as quoted by Van Kleeck, *Artificial Flower Makers,* 141.
56. Ibid., 143.
57. Rheta Childe Dorr, "The Child Who Toils at Home," *Hampton Magazine* (Apr. 1912): 187.
58. "For Healthy Bread, Buy Union Made Bread," *Women's Label League Journal* 2 (June 1913): 13.
59. Riis, *How the Other Half Lives*, 177.
60. Marie Ganz, *Rebels* (New York: Dodd, Mead, 1919), quoted in Irving Howe and Kenneth Libo, *How We Lived, A Documentary History of Immigrant Jews in America, 1880–1930* (New York: Richard Marek, 1979), 136.
61. *The Outlook,* 81, (Nov. 21, 1903).
62. Sydelle Kramer and Jenny Masur, eds., *Jewish Grandmothers* (Boston: Beacon Press, 1976), 95.
63. Riis, *How the Other Half Lives,* 180.
64. Louise C. Odencrantz, *Italian Women in Industry, A Study of Conditions in New York City* (New York: Russell Sage Foundation, 1919), 80.
65. Ibid.
66. *New York World,* Mar. 26, 1911.
67. *Milwaukee Journal*, Mar. 27, 1911.
68. Louis Waldman, *Labor Lawyer* (New York: Dutton, 1944), 33–34.
69. Elizabeth Hasanovitz, *One of Them: Chapters from a Passionate Autobiography* (Boston: Houghton Mifflin, 1918), 214.
70. *The Survey,* 26 (Apr. 8, 1911) 81–87. Also cited in Irving Howe, *World of Our Fathers, The Journey of the East European Jews to America and The Life They Found and Made.* (New York: Simon and Schuster, 1976), 305.
71. There is no single volume on the Triangle Shirtwaist Fire by an historian. Journalist and labor activist Leon Stein was for many years the editor of the ILGWU journal, *Justice.* His re-creation is grounded on reliable research, though it cannot be relied upon in all details and matters of interpretation as might a thoroughly annotated work. See Stein, *The Triangle Fire* (Philadelphia: J.B. Lippincott, 1962).
72. David Dubinsky and A. H. Raskin, *David Dubinsky: A Life with Labor* (New York: Simon and Shuster, 1977), 40.
73. Joseph Schereschewsky was born in Peking, China, in 1873, the son of an eastern European Jewish immigrant from Lithuania who converted to Christianity in the United States and became an Episcopal priest, a missionary to China, and, in 1877, a bishop. Joseph received an A.B. from Harvard in 1895 and a medical degree from Dartmouth in 1899, joining the Public Health Service that year. He was in charge of the Office of Field Investiga-

tion of Occupational Diseases from 1913 to 1918. He died in 1940. *Who Was Who in America* 1 (Chicago: A. N. Marquis Co., 1943): 1087.

74. Schereschewsky, "The Health of Garment Workers," 98.
75. Ibid.
76. Ibid., 101.
77. Ibid., 102.
78. I benefited enormously from examining the Workers' Health Bureau files in the Charlotte Stern Collection at the Tamiment Institute Library at New York University. Two fine articles on the Workers' Health Bureau are Angela Nugent, "Organizing Trade Unions to Combat Disease: The Workers' Health Bureau, 1921–1928," *Labor History* 26 (Summer 1985): 423–46 and David Rosner and Gerald Markowitz, "Safety and Health as a Class Issue: The Workers' Health Bureau of America during the 1920s," in Rosner and Markowitz, *Dying for Work,* 53–64.
79. Robert L. Brandfon, *Cotton Kingdom of the New South, A History of the Yazoo Mississippi Delta from Reconstruction to the Twentieth Century* (Cambridge, Mass.: Harvard University Press, 1967), 144–45, 147–48.
80. "Recent Immigrants in Agriculture," in *Immigrants in Industries,* vol. 24 of *Reports of the Immigration Commission* (Washington, D.C.: Government Printing Office, 1911), 44.
81. The Italians at Sunnyside were usually northern Italians who had been imported by New York businessman Austin Corbin to help work the 4,700 acres under cultivation on the 10,000-acre Arkansas plantation fifteen miles below Greenville, Mississippi. In 1909, there were 127 Italian families, or 600 Italians working land they rented for $6.00 to $7.50 per acre. Later, after Corbin's death, when another lessee took over management of the property, southern Italians were recruited in even greater numbers.
82. "Recent Immigrants in Agriculture," 310.
83. Ibid., 333.
84. The extent of Southern manipulation of the Immigration Commission's report is suggested by Pete Daniel, *The Shadow of Slavery, Peonage in the South, 1901–1969* (New York: Oxford University Press, 1972), 102, 107.
85. I am indebted to both Randolph H. Boehm of the University Press of America and Dr. Pete Daniel of the Museum of American History, Smithsonian Institution, for their generosity in sharing their thoughts and photocopies of documents on peonage with me.
86. Randolph H. Boehm, "Mary Grace Quackenbos and the Federal Campaign Against Peonage: The Case of Sunnyside Plantation," Ernesto Milani, "Peonage at Sunnyside and the Reaction of the Italian Government," and Bertram Wyatt-Brown, "Leroy Percy and Sunnyside: Planter Mentality and Italian Peonage in the Mississippi Delta," delivered earlier versions of these essays at the annual meeting of the Organization of American Historians held in Washington, D.C., in March 1990. Revised versions, an additional paper, William B. Gatewood, Jr., "Sunnyside: The Evolution of an Arkansas Plantation, 1848–1945," and a commentary by Pete Daniel were

published in *The Arkansas Historical Quarterly* 50 (Spring 1991).

87. Michele Berardinelli, "Report on Italian peonage matters in Mississippi," Department of Justice Records, National Archives, R.G. 60, Box 10803, 21–22.
88. "Recent Immigrants in Agriculture," 333.
89. Mary Grace Quackenbos, *Report on Sunnyside Plantation, Arkansas,* Sept. 28, 1907, Department of Justice records, National Archives, RG 60, Straight Numerical file, 100937.
90. Ibid., 60.
91. Ibid. Before including her recommendations in her September report, Quackenbos made them directly to the owners of Sunnyside, Leroy Percy and O. B. Crittenden, in Mary Grace Quackenbos to LeRoy [*sic*] Percy, Esq., Aug. 17, 1907, RG 60, Straight Numerical File, 100937.
92. Ibid.
93. Quackenbos to Percy, Aug. 17, 1907.
94. LeRoy [sic] Percy to Mrs. Mary Grace Quackenbos, Aug. 17, 1907.
95. Quackenbos to Percy, Aug. 17, 1907. Also, Quackenbos, *Report on Sunnyside Plantation, Arkansas,* Sept. 28, 1907.
96. Percy to Quackenbos, Aug. 17, 1907.
97. Joseph E. Rocconi, *My Memories,* typescript in Bolivar County Library, Cleveland, Mississippi. This document was found and generously shared with me by Randolph Boehm.
98. Commissariato dell' Emigrazione, *Bolletino dell' emigrazione* 20 (1905): 28; also quoted by Betty Boyd Caroli, *Italian Repatriation from the United States, 1900–1914* (New York: Center for Migration Studies, 1973), 63–68.
99. Senate Documents, "Report on Peonage," *Abstracts of Reports of the Immigration Commission,* 42 vols. 61st Cong., 3d sess., 1910–1911, no. 747, 2: 444–46.
100. Hamilton, *Exploring the Dangerous Trades,* 139.

CHAPTER 8: "THERE COULD ALSO BE MAGIC IN BARBARIAN MEDICINE"

1. Oral history with Dr. Hubert Dong, M.D., transcribed in Collen P. Quock, ed., *Chinese Hospital Medical Staff, San Francisco* (1982): 15, on file in the Californiana Collection of the San Francisco Public Library.
2. There is a discussion of these social service departments in Charles Rosenberg, *The Care of Strangers, The Rise of America's Hospital System* (New York: Basic Books, 1987), 312–16.
3. Michael M. Davis, Jr., *Immigrant Health and the Community* (Montclair, N.J.: Patterson Smith, 1971; orig. 1921), 307.
4. Ibid.
5. This brief profile of Michael Davis is based upon Ralph E. Pumphrey's sketch in Walter I. Trattner, ed., *Biographical Dictionary of Social Welfare in America* (Westport, Conn.: Greenwood Press, 1986), 211–13. See also by

Pumphrey, "Michael M. Davis and the Development of the Health Care Movement, 1900–1928," *Societas* 2 (1972): 27–41 and "Michael M. Davis and the Transformation of the Boston Dispensary," *Bulletin of the History of Medicine* 49 (1975): 451–65.

6. Davis and his colleagues conducted interviews with approximately 150 doctors and an equal number of individual immigrants on the subject of hospitals. Davis, *Immigrant Health and the Community,* 309.
7. Ibid., 320–21.
8. Davis procured the names of twenty-six dieticians from the American Dietetic Association and asked them whether they accommodated immigrant patients with special diets of familiar foods, whether they perceived foreign-born patients to be dietary problems, and whether they thought it "desirable and practicable" to adjust hospital diets to suit the cultural preferences of patients. Thirteen of those polled responded. Nine were doing nothing about matching diet to preference of immigrant patients. Several had maintained a separate kosher kitchen for observant Jewish patients but abandoned it when there were too few requests to justify the expense. Many hospitals served fish on Fridays routinely to accommodate any patients who might be observant Catholics. Some said they would adjust the diet if they had a majority of patients from one particular ethnic group. One complained that in cases where diet was an aspect of therapy, immigrant patients either resisted or returned to their preferred diet as soon as they left the hospital, regardless of any health risk at stake. Ibid., 311.
9. A sample of thirty hospitals, each located in a city with a large foreign-born population and known to have a high proportion of immigrant patients, was used by Davis and his team to study the language issue. Twenty-one responded. Only one employed a professional interpreter, a large municipal institution with approximately two thousand beds. The institution's size and the diversity of its patients caused Davis to characterize the single interpreter as "humorously inadequate." Fourteen called upon employees or patients to interpret for physicians and staff, as needed, while five hospitals did nothing about providing interpreters for non-English-speaking patients except on occasion. Ibid., 314.
10. Rosenberg, *Care of Strangers,* 112.
11. "Il Dottor Giuseppe Fabiani e il suo Ospedale," in *La Colonia Italiana di Filadelfia* (1906), located in the files of the Urban Archives at Temple University.
12. Thomas Neville Bonner, *Medicine in Chicago, 1850–1950: A Chapter in the Social and Scientific Development of the City,* 2d ed. (Urbana: University of Illinois Press, 1991), 152.
13. "A Hospital for Poles," *Dziennik Chicagoski,* Nov. 13, 1893, translation in WPA Foreign Language Press Survey for Chicago, II D3, located at Immigration History Research Center, University of Minnesota.
14. "Polish Hospital Proposed for Chicago, Play to be Staged to Raise Funds," ibid., Apr. 6, 1894.

15. Joan B. Trauner, "The Chinese As Medical Scapegoats in San Francisco, 1870–1905," *California History* 57 (Spring 1978): 70–87. An earlier version of the essay appeared in *The Bulletin of the Chinese Historical Society of America* 9 (Apr. 1974): 1–18.
16. Ibid, 83–84.
17. In 1887, the Denver Chamber of Commerce reported that "Colorado is the Mecca of Consumptives, and rightfully; for dry air, equable temperature and continuous sunshine are as yet the most reliable factors in the cure of this [tuberculosis] disease." *Fourth Annual Report of the Denver Chamber of Commerce,* Denver, 1887. The arrival of impoverished Jewish consumptives from eastern cities soon caused apprehension in the non-Jewish community and among the established German-Jewish elite. See Lyle Dorsett and Michael McCarthy, *The Queen City, A History of Denver,* 2d ed. (Boulder, Colo.: Pruitt Publishing Company, 1986), 176–80 and Robert H. Shikes, *Rocky Mountain Medicine: Doctors, Drugs, and Disease in Early Colorado* (Boulder, Colo.: Johnson Books, 1986), 156–70. A biography of Israeli Prime Minister Golda Meir mentions the political conflict in Denver occasioned by impoverished tubercular Jews. See Ralph G. Martin, *Golda, Golda Meir: The Romantic Years* (New York: Scribner's, 1988), 31. An anecdotal but useful study of Denver's Jewish community and its efforts to cope with its unusual burdens is Ida Libert Uchill, *Pioneers, Peddlers and "Tsadikim"* (Boulder, Colo.: Quality Line Printing Co., 1957).
18. Charles Denison, *Rocky Mountain Health Resorts: An Analytical Study of High Altitudes in Relation to the Arrest of Chronic Pulmonary Disease* (Boston: Houghton, Osgood and Company, 1880).
19. Financial problems, some of them occasioned by the depression of 1893, slowed progress. However, when B'nai B'rith undertook to make the hospital a national institution for the care of tuberculosis patients, the National Jewish Hospital for Consumptives was truly on its way to becoming a reality.
20. *First Annual Report of National Jewish Hospital at Denver, Colorado,* (1900). See Mary Ann Fitzharris, *A Place to Heal, The History of National Jewish Center for Immunology and Respiratory Medicine* (Boulder, Colo.: Johnson Publishing Company, 1989), 1.
21. Jeanne Lichtman Abrams is the single most knowledgeable scholar on the history of the JCRS. Jeanne Lichtman Abrams, "Chasing the Cure: A History of the Jewish Consumptives' Relief Society" (Ph.D. diss., University of Colorado, 1983). She has also written on Anna Hilkowitz, one of the female volunteers for the JCRS, see Abrams, "*Unsere Leit* ('Our People'): Anna Hilkowitz and the Development of the East European Jewish Woman Professional in America," *American Jewish Archives* 37 (Nov. 1985): 275–89. Jewish consumptives in Denver often also suffered family dislocation. See Abrams, "'For a Child's Sake': The Denver Sheltering Home for Jewish Children in the Progressive Era," *American Jewish History* 79 (Winter 1989–90): 181–202.

22. Charles Spivak, "The Genesis and Growth of the Jewish Consumptives' Relief Society," Part 1, *The Sanatorium* 1 (Jan. 1907): 6–7.
23. Ibid., Part 2 (Mar. 1907): 26.
24. Letter to Charles Spivak from Workman's Circle of New York, February 28, 1908, Workman's Circle Folder, Jewish Consumptives' Relief Society Archives, held by the Rocky Mountain Jewish Historical Society, University of Denver.
25. Letter to Charles Spivak from Rochester Workman's Circle, June 6, 1908, ibid.
26. Irving Howe, *World of Our Fathers* (New York: Harcourt Brace Jovanovich, 1976), 149.
27. Spivak's life has been pieced together by Abrams, "Chasing the Cure," 11–15.
28. *The Sanatorium* 12 (May–Aug. 1918): 54–55.
29. German Jews and eastern European Jews had long experienced cultural tensions in America. In the pages of the New York *Yiddishe Gazetten,* an eastern European Jew contrasted the efficient charity of assimilated German Jews with their "beautiful offices, desks, all decorated, but strict and angry faces," to the *Zedakah,* charity delivered with loving kindness as a religious obligation, that one experienced among his own eastern European brethren, "who speak his tongue, understand his thoughts and feel his heart." *Yiddishe Gazetten,* Apr. 1894, as quoted by Moses Rischin, *The Promised City, New York's Jews, 1870–1914* (New York: Harper & Row, 1970; orig. 1962), 104.
30. *Jewish Outlook* 1 (Apr. 15, 1904): 7, 8.
31. Dr. Philip Hilkowitz, "President's Report," *The Sanatorium Sixth Annual Report, 1910,* 38.
32. Rabbi William S. Friedman, "Modern Methods of Fighting Tuberculosis, What the National Jewish Hospital for Consumptives Is Doing," address delivered before the National Jewish Chautauqua at Atlantic City, New Jersey, July 28, 1905, National Jewish Hospital Archive, Denver, Colorado.
33. *Jewish Outlook* 1 (Oct. 7, 1904): 6.
34. Dr. Adolph Zederbaum, "Kosher Meat in Jewish Hospitals and Sanatoria," *The Sanatorium* 2 (Nov. 1908): 275, JCRS Archives. In 1925, National Jewish finally did establish a kosher kitchen. However, the cultural tensions embodied in the debate over *Kashrut* took much longer to abate.
35. A study of Catholic hospitals that goes well beyond the founding of the first institutions to serve Irish and German Catholics is Bernadette McCauley, "'Who Shall Take Care of Our Sick?': Roman Catholic Sisterhoods and Their Hospitals, New York City, 1850–1930," (Ph.D. diss., Columbia University, 1992). Much useful information on the formation of Columbus Hospital in New York as an institution dedicated to the treatment of Italian immigrants can be found in Mary Louise Sullivan, M. S. C., *Mother Cabrini, "Italian Immigrant of the Century"* (New York: Center for Migration Studies, 1992), esp. chap. 22.
36. Data are offered by McCauley, "'Who Shall Take Care of Our Sick?'" 2.

37. Gail Farr Casterline, "St. Joseph's and St. Mary's: The Origins of Catholic Hospitals in Philadelphia," *Pennsylvania Magazine of History and Biography* 108 (July 1984): 289–314. See also, Jon M. Kingsdale, *The Growth of Hospitals, 1850–1939* (New York: Garland, 1989). However, a more suggestive approach is Joseph Ryan, O.S.A., "The Evolution of St. Agnes Hospital of Baltimore, Maryland, From an Asylum to a Modern Hospital" (unpublished graduate seminar paper, The American University, 1991).
38. Corrigan's letter is quoted by Sullivan, *Mother Cabrini,* 40.
39. *First Annual Report, Columbus Hospital, 1896,* 10.
40. Mother Cabrini to Leone Reynaudi, July 20, 1907, Appendix B, ibid., 255.
41. The count and names of the hospitals can be found in Daniel Ethan Bridge, "The Rise and Development of the Jewish Hospital in America, 1850–1984" (Rabbinic Ordination thesis, Hebrew Union College, 1985), 1, 70–129.
42. See the excellent study of Boston's struggle to establish a lasting Jewish hospital in Arthur J. Linenthal, *First a Dream: The History of Boston's Jewish Hospitals, 1896 to 1928* (Boston: Beth Israel Hospital in Association with the Francis A. Countway Library of Medicine, 1990).
43. Richard C. Cabot, *Social Service and the Art of Healing,* 2d ed. (New York: Dodd, Mead, 1931), 4–5.
44. Charles Rosenberg, "Social Class and Medical Care in 19th-Century America: The Rise and Fall of the Dispensary," *Journal of the History of Medicine and Allied Sciences* 29 (1974): 32–54.
45. Ibid., 32–34.
46. Michael M. Davis, Jr., and Andrew R. Warner, *Dispensaries, Their Management and Development: A Book for Administrators, Public Health Workers, and All Interested in Better Medical Service for the People* (New York: Macmillan, 1918).
47. Ibid., 347.
48. Ibid., 346.
49. Davis, *Immigrant Health and the Community,* 326–43.
50. Ibid., 328.
51. Ibid., 329.
52. Ibid., 330.
53. Ibid., 342. In 1927, Davis created a handbook for those in the field that included an extensive section on the creation of health centers. Michael M. Davis et al., *Clinics, Hospitals and Health Centers* (New York: Harper & Brothers, 1927).
54. George Rosen, "The First Neighborhood Health Center Movement: Its Rise and Fall," *American Journal of Public Health* 61 (1971): 1620–35. On Wilbur Philips and the National Social Unit Organization, see Patricia Mooney Melvin, *The Organic City: Urban Definition and Community Organization, 1880–1920* (Lexington: University Press of Kentucky, 1987). Also see the review of Melvin's book by Alan M. Kraut in *Bulletin of the History of Medicine* 64 (1990): 123–24.
55. This chart is reprinted in its entirety from Davis, *Immigrant Health and the Community,* 331.

56. Marya Tallarico to Lee Frankel, June 1911, in Health Education files, Metropolitan Life Insurance Archives, New York City.
57. An especially fine discussion of this aspect of home nursing is an article by Karen Buhler-Wilkerson, "False Dawn: The Rise and Decline of Public Health Nursing in America, 1900–1930," in Ellen C. Langemann, ed., *Nursing History: New Perspectives, New Possibilities* (New York: Teachers College Press, 1983), 89–106.
58. Ysabella Waters, *Visiting Nursing in the United States* (New York: Charities Publication Committee, 1909). Waters's data are effectively described and analyzed in Karen Buhler-Wilkerson, "Left Carrying the Bag: Experiments in Visiting Nursing, 1877–1909," *Nursing Research* 36 (Jan./Feb. 1987): 42–46.
59. Buhler-Wilkerson, "Left Carrying the Bag," 45.
60. Davis, *Immigrant Health and the Community*, 285.
61. Ibid., 286.
62. Ibid., 288.
63. Ibid.
64. *Thirteenth Annual Report of the Visiting Nurse Society of Philadelphia, For the Year Ending February 28th, 1899*, 13. By the turn of the century, nativity was regularly included in Philadelphia Visiting Nurse Society reports, with each of four branches reporting its unique ethnic configuration. Next to native-born Americans at 359 and "colored" at 320, the largest number of patients (for whom nativity was known) from the main branch at Lombard Street were Russians (mostly Jews) at 158, 12 percent; Irish, 141, 11 percent; and Italians, 106, 8 percent. The College Settlement branch remained busiest with immigrants. Of the 290 patients for whom nativity data were recorded, 22.7 percent were Italian and 17.9 percent were Russian. *Fifteenth Annual Report of the Visiting Nurse Society of Philadelphia For the Year Ending February 28th, 1901*, 7, 17. The annual reports of the VNS of Philadelphia were located at the Center for the Study of the History of Nursing at the University of Pennsylvania School of Nursing.
65. *Twenty-Fifth Annual Report of the Visiting Nurse Society of Philadelphia For the Year Ending February 28th, 1911*, 13.
66. *Report of the Board of Managers of the Visiting Nurse Society of Philadelphia to the Contributors at their Annual Meeting, held March 6, 1914, Comprising the Report of the Nursing Work Together With the Accounts of the Treasurer, For the Year Ending February 28th, 1914*, 9.
67. *Report of the Board of Managers of the Visiting Nurse Society of Philadelphia For the Year Ending February 29th, 1916*, 8.
68. Ibid.
69. *Report of the Board of Managers of the Visiting Nurse Society of Philadelphia, 1917*, 8.
70. Lillian D. Wald, *The House on Henry Street* (New York: Henry Holt and Company, 1915), 27.
71. There is no definitive biography of Lillian Wald. Her own autobiography,

The House on Henry Street (1915), remains the best published source for insight into her perceptions of her early career. Among recent studies of Wald are Beatrice Siegel, *Lillian Wald of Henry Street* (New York: Macmillan, 1983); Doris Groshen Daniels, *Always a Sister, The Feminism of Lillian D. Wald* (New York: Feminist Press at the City University of New York, 1989). Some useful correspondence is published in Clare Coss, ed., *Lillian D. Wald, Progressive Activist* (New York: Feminist Press of the City University of New York, 1989).

72. Wald, *House on Henry Street,* 28.
73. Ibid.
74. Davis, *Immigrant Health and the Community,* 379.
75. Diane Bronkema Hamilton, "The Metropolitan Life Insurance Company Visiting Nurse Service (1909–1953)," (Ph.D. diss., University of Virginia, 1987), 72.
76. The increasing prestige of physicians in the United States can be demonstrated through the manner in which they were characterized in cultural expressions high and low. See Chauncey D. Leake, "Medical Caricature in the United States," *Bulletin of the Society of Medical History of Chicago* 24 (1928): 1–29.
77. Davis, *Immigrant Health and the Community,* 138.
78. Ibid.
79. Alan M. Kraut, "Healers and Strangers, Immigrant Attitudes Toward the Physician in America—A Relationship in Historical Perspective," *Journal of the American Medical Association* 263 (Apr. 4, 1990): 1807–11.
80. Max Danzis, "The Jew in Medicine," *American Hebrew and Jewish Tribune* (Mar. 23, 1934), photocopy located in Morris Lazaron papers, American Jewish Archives, Cincinnati, Ohio.
81. Morris S. Lazaron, "The Jewish Student in Medicine," draft located in ibid.
82. Burton D. Meyers to Rabbi Morris S. Lazaron, Oct. 30, 1934, in Jews in Medicine Survey of Medical Schools, 1930–1934, collection number 71, file 37/16.
83. C. C. Bass, M.D., to Rabbi Morris S. Lazaron, Mar. 8, 1934, ibid.
84. W. McKim Marriott, D.D., to Everett R. Clinchy, June 10, 1930, ibid.
85. A. C. Bachmeyer to Rabbi Morris S. Lazaron, Feb. 24, 1934, and Arthur C. Curtis, M.D., to Rabbi Morris S. Lazaron, Feb. 20, 1934, ibid.
86. The Association of American Medical Colleges report is quoted by Dan A. Oren, *Joining the Club, A History of Jews and Yale* (New Haven: Yale University Press, 1985), 146.
87. Davis, *Immigrant Health and the Community,* 145.
88. Michael M. Davis, Jr., "Pain Killers and New Americans," *The Survey* 45 (Jan. 29, 1921): 635–36. The Dr. Landis mentioned in the article is very likely Leonard Landes, who was prosecuted in New York State by the Bureau of Industries and Immigration in 1915 and given thirty days and a fine of one hundred dollars. Under the name Leo Landes, he placed advertisements aimed at immigrants of different groups in at least twelve of the ethnic newspapers examined in the Cleveland study.

89. Ibid.
90. Ibid.
91. Davis, *Immigrant Health and the Community,* 148.
92. Ibid., 149.
93. Ibid., 151–52.
94. Ibid., 152.
95. The best volumes on medical quackery and the fight for pure food and drug legislation are by James Harvey Young. See Young, *The Medical Messiahs, A Social History of Health Quackery in Twentieth-Century America* (Princeton, N.J.: Princeton University Press, 1967) and *Pure Food, Securing the Federal Food and Drug Act of 1906* (Princeton, N.J.: Princeton University Press, 1989).
96. File on "Mr. John Smuk" in Report #5,"Fake Doctors," papers of the Immigrants' Protective League, box 4, folder 39, located in the manuscript division of the University of Illinois Library at Chicago Circle.
97. Edward Kremus and George Urdang, *History of Pharmacy,* rev. Glenn Sonnedecker, 4th ed. (Madison, Wis.: American Institute of the History of Pharmacy, 1976; orig. 1940), 290–315.
98. Davis, *Immigrant Health and the Community,* 131–32.
99. Ibid.
100. Ibid., 132.
101. Ibid., 133.
102. "K. S. Ramashauskas," in "Fake Doctors," IPL papers, box 4, file 39, 6.
103. "Drug Stores in Relation to Venereal Disease in Chicago," report of an investigation by the American Social Hygiene (changed to "Health" in 1960) Association made at the request of the Social Hygiene Agencies of Chicago, Mar. 18, 1931, in American Social Health Association papers, box 99 in Social Welfare History Archives, University of Minnesota Library.
104. Ibid., 13–15. See also Allan M. Brandt, *No Magic Bullet, A Social History of Venereal Disease in The United States Since 1880* (New York: Oxford University Press, 1985).

CHAPTER 9: "EAST SIDE PARENTS STORM THE SCHOOLS"

1. United States Immigration Commission, *Abstracts of Reports of the Immigration Commission,* 61st Cong., 3d sess., 1910, S. Doc. 747, vol. 2 (Washington: Government Printing Office, 1911), 17.
2. Ibid.
3. The most blatant expression of the thesis that the schools were little more than a weapon in class warfare is Colin Greer, *The Great School Legend: A Revisionist Interpretation of American Public Education* (New York: Viking, 1972).
4. Irving Fisher, "Report on National Vitality, Its Wastes and Conservation," *Bulletin 30 of the Committee of One Hundred on National Health* (Washington, D.C.: Government Printing Office, 1909), 1.

5. Lillian D. Wald, "Medical Inspection of Public Schools," *Annals of the American Academy of Political and Social Science* 25 (Jan.–June 1905): 290.
6. Diane Ravitch, *The Great School Wars, New York City, 1805–1973: A History of the Public Schools As Battlefields of Social Change* (New York: Basic Books, 1974), 164.
7. *U.S. Immigration Commission Reports*, vol. 2, 56. While New York City had the largest number of foreign-born teachers and teachers with foreign-born fathers of any sample cities surveyed by the U.S. Immigration Commission, other cities had even higher percentages of teachers with foreign-born fathers than New York. For example, Buffalo, 49.4; Chicago, 51.1; Cleveland, 51.3; Meriden, 47.5; Milwaukee, 54.1; San Francisco, 57.0; Scranton, 59.7; Shenandoah, 69.3; and Worcester, 54.6. Cities with higher percentages of foreign-born teachers than New York included Chicago, 8.1; Detroit, 9.5; and Duluth, 10.4.
8. William H. Maxwell as quoted in *Proceedings of the Dedication of the New York State Education Building, October 15, 16, 17, 1912* in New York State Education Department, *Ninth Annual Report* (Albany: New York State Education Department, 1913), 61. Also cited in Stephan F. Brumberg, *Going to America, Going to School: The Jewish Immigrant Public School Encounter in Turn-of-the-Century New York City* (New York: Praeger, 1986), 12–13.
9. Mary Antin, *The Promised Land, The Autobiography of a Russian Immigrant* (Princeton: Princeton University Press, 1969; orig. 1911), 205.
10. Some of the most original insights about culture and predeparture attitudes shaping views on formal education can be found in the published doctoral dissertation of the former principal of Benjamin Franklin High School in New York's East Harlem, Leonard Covello, *The Social Background of the Italo-American School Child, A Study of the Southern Italian Family Mores and Their Effect on the School Situation in Italy and America,* ed. Francesco Cordasco (Towtowa, N.J.: Rowan and Littlefield, 1972; orig. 1967). A brilliant study using Providence, Rhode Island, data is Joel Perlmann, *Ethnic Differences, Schooling and Social Structure Among the Irish, Italians, Jews, and Blacks in an American City, 1880–1935* (New York: Cambridge University Press, 1988). See also, Stephan F. Brumberg, *Going to America, Going to School, The Jewish Immigrant Encounter in Turn-of-the-Century New York City* (New York: Praeger, 1986) and the essays on individual Euroethnic groups in Bernard J. Weiss, ed., *American Education and the European Immigrant: 1840 and 1940* (Urbana: University of Illinois Press, 1982).
11. "East Side Parents Storm the Schools," *New York Times,* June 28, 1906.
12. Ibid.
13. Ibid.
14. Ibid.
15. Ibid. For a discussion of private physicans' complaints about public sector competition, especially public schools and dispensaries, see Paul Starr, *The*

Social Transformation of American Medicine (New York: Basic Books, 1982), 182–83 and 187–89.
16. Ibid.
17. Ibid.
18. *New York Tribune,* June 28, 1906.
19. *New York Times,* June 28, 1906.
20. Ibid.
21. "Throat-Cutting Rumors Revive School Rioting," *New York Times,* June 29, 1906.
22. Ibid.
23. John Duffy, *A History of Public Health in New York City, 1866–1966* (New York: Russell Sage Foundation, 1974), 269–70. Also discussed in Ronald L. Numbers, *Almost Persuaded: American Physicians and Compulsory Health Insurance, 1912–1920* (Baltimore: Johns Hopkins University Press, 1978), 8–9.
24. John Duffy, "School Vaccination: The Precursor to School Medical Inspection," *Journal of the History of Medicine and Allied Sciences* 33 (1978): 344–55.
25. John Duffy, "School Buildings and the Health of American School Children in the Nineteenth Century," in Charles E. Rosenberg, ed., *Healing and History, Essays for George Rosen* (New York: Science History Publications), 161–78.
26. John Duffy, *The Sanitarians, A History of American Public Health* (Urbana: University of Illinois Press, 1990), 182. See also the discussion of health in the schools in Duffy's comprehensive *A History of Public Health in New York City, 1866–1966,* chap. 10.
27. J. A. Larrabee, M.D., "The Schoolroom: A Factor in the Production of Disease," *Journal of the American Medical Association* 11 (Nov. 3, 1888): 613–14.
28. Luther Halsy Gulick and Leonard P. Ayres, *Medical Inspection of Schools* (New York: Charities Publication Committee, 1910), 24.
29. Instructions to Medical Inspector, Department of Health, Borough of Manhattan, reprinted in Gulick and Ayres, *Medical Inspection of Schools,* 36.
30. Ibid., 35–37.
31. S. Josephine Baker, *Fighting for Life* (New York: Macmillan, 1939), 78.
32. Ibid.
33. Ibid., 78–79.
34. Wald, "Medical Inspection of Public Schools," 293.
35. Baker, *Fighting for Life,* 78.
36. Ibid., 80.
37. There is no better contemporary account of school nursing than Lina Rogers Struthers, R.N., *The School Nurse, A Survey of the Duties and Responsibilities of the Nurse in the Maintenance of Health and Physical Perfection and the Prevention of Disease Among School Children* (New York: Putnam's, 1917).

38. Duffy, *History of Public Health in New York City, 1866–1966,* 254.
39. Baker, *Fighting for Life,* 80.
40. Ibid., 81.
41. Gulick and Ayres, *Medical Inspection of Schools,* 66.
42. Ibid., 68–69.
43. Ibid., 80.
44. Jacob Sobel, "Prejudices and Superstitions Met with in Medical Inspection of School Children," *Transactions of the Fourth International Congress on School Hygiene, Buffalo, New York, U.S.A., August 25–30, 1913,* ed. Thomas A. Storey (Buffalo: Courier Company of Buffalo, 1913): 78–88.
45. Ibid., 79.
46. Ibid., 79–80.
47. Ibid., 81.
48. Ibid.
49. Ibid., 87.
50. Ibid., 88.
51. Chart, "Diseases for Which Pupils Are Excluded in Different Cities," in Gulick and Ayres, *Medical Inspection of Schools,* 48. Use of public schools in the war on diphtheria is described in Duffy, *A History of Public Health in New York City, 1866–1966,* 560–61. See also Wade W. Oliver, *The Man Who Lived for Tomorrow, A Biography of William Hallock Park, M.D.* (New York: E. P. Dutton, 1941), 418–38.
52. John Spargo, *The Bitter Cry of the Children* (New York: Johnson Reprint Corporation, 1969; orig. 1906), xiii.
53. Ibid., 67.
54. Ibid.
55. Ibid., 85.
56. Selma Cantor Berrol, *Immigrants at School: New York City, 1898–1914* (New York: Arno Press, 1978), esp. chap. 4 on "The Extension of the School," passim.
57. Ibid., 174. I wish to thank Dr. Selma Berrol for calling to my attention the linkage of pickles and alcoholism that some educators perceived, including Lower East Side Superintendent Julia Richman!
58. Spargo, *Bitter Cry of the Children,* 90
59. Burton Hendrick and Paul Kennedy, "Three Cent Lunches for School Children," *McClure's* 5 (Oct. 1913): 121, 128. New York City Department of Education, "Shall the Schools Serve Lunches?," *Bulletin #10* (1913): 2, 4. Also, New York City Department of Education, "The School Lunch Service in New York City," *Bulletin #3* (1914): 9–10. Kitteridges's efforts are discussed at length in Berrol, *Immigrants at School,* 174–80.
60. Berrol, *Immigrants at School,* 179.
61. Spargo, *Bitter Cry of the Children,* 116.
62. Richard Gambino, *Blood of My Blood, The Dilemma of Italian-Americans* (Garden City, N.Y.: Doubleday, 1974), 21.
63. Baker, *Fighting for Life,* 147.

64. Ibid.
65. Ibid., 149.
66. Duffy, *History of Public Health in New York City, 1866–1966,* 484–85. Baker, *Fighting for Life,* 150–51.
67. Wald, "Medical Inspection of Public Schools," 292–93.
68. Ibid., 296–97.
69. Baker, *Fighting for Life,* 158.
70. Lillian Wald was relieved that in such cases the district superintendent of schools cooperated with the district attorney of Manhattan, who said that he would prosecute parents refusing treatment for a child on grounds that they had violated the compulsory education law. Wald recalled that a test case was brought and a father fined ten dollars. Wald, "Medical Inspection of Schools," 94.
71. Steven L. Schlossman, JoAnne Brown, and Michael Sedlak, *The Public School in American Dentistry* (Santa Monica, Calif.: Rand Corporation, 1986), 6. For this theory of focal infections and their importance, see Frank Billings, "Chronic Focal Infections and their Etiologic Relations to Arthritis and Nephritis," *Archives of Internal Medicine* 9 (1912): 484–98. See also by Frank Billings, *Focal Infection, The Lane Memorial Lectures* (New York: Appleton and Company, 1916); "Focal Infection, Wider Application in the Etiology of General Disease," *Journal of the American Medical Association* 63 (Dec. 5, 1914): 2024–25.
72. Alfred C. Fones, "Report of Five Years of Mouth Hygiene in the Public Schools of Bridgeport, Connecticut," *Dental Cosmos* 61 (July 1919): 608. Also cited by Schlossman et al., *Public School in American Dentistry,* 13.
73. Fones, "Report of Five Years," 615 and Schlossman et al., *Public School in American Dentistry,* 14.
74. "Course of Study in Kindergarten, Music, Physical Training with a Syllabus in Each of These Subjects" (New York: Board of Education, 1903). Also quoted in Brumberg, *Going to America, Going to School,* 78.
75. *Brooklyn Daily Eagle,* Sept. 12, 1897. Also cited in Marilyn Thornton Williams, *Washing "The Great Unwashed," Public Baths in Urban America, 1840–1920* (Columbus: Ohio State University Press, 1991), 1.
76. Jacob Riis, *How the Other Half Lives, Studies Among the Tenements of New York* (New York: Hill and Wang, 1957; orig. 1890), 83–84.
77. Maurice Fishberg, "Health and Sanitation of the Immigrant Jewish Population of New York," *Menorah* 33 (Aug. 1902): 74.
78. Riis, *How the Other Half Lives,* 84.
79. Ibid.
80. John Foster Carr, *Guide for the Immigrant Italian in the United States of America* (New York: Arno Press, 1975; orig. 1911), 47, 49.
81. Antonio Stella, "The Effects of Urban Congestion on Italian Women and Children," *Medical Record* 74 (May 2, 1908): 725.
82. Riis, *How the Other Half Lives,* 84.
83. Ibid.

84. Gulick and Ayres, *Medical Inspection of Schools,* 9.
85. Ibid.
86. Ibid., 10. For a discussion of the larger issue of fitness and morality, see Harvey Green, *Fit for America, Health Fitness, Sport and American Society* (New York: Pantheon, 1986).
87. Cary Goodman, *Choosing Sides, Playground and Street Life on the Lower East Side* (New York: Schocken Books, 1979), 61. See also, Dominick Cavallo, *Muscles and Morals, Organized Playgrounds and Urban Reform, 1880–1920* (Philadelphia: University of Pennsylvania Press, 1981).
88. Berrol, *Immigrants at School,* 239–40.
89. Maurice Fishberg, "Health Problems of the Jewish Poor," a paper read before the Jewish Chautauqua Assembly on Monday, July 27, 1903, at Atlantic City, N.J., reprinted from *The American Hebrew* (New York: Press of Philip Cowen, 1903), 15.
90. The data were supplied to Stella by Dr. William Guilfoy, the New York City registrar of vital statistics, and was reprinted by Stella in "Effects of Urban Congestion on Italian Women and Children," 724.
91. David Nasaw, *Children of the City at Work and at Play* (New York: Oxford University Press, 1985), 37–38.
92. *Jewish Daily Forward,* Aug. 6, 1903. The letter is also cited in Irving Howe and Kenneth Libo, eds., *How We Lived, A Documentary History of Immigrant Jews in America, 1880–1930* (New York: Richard Marek, 1979), 51.
93. Duffy, *History of Public Health in New York City, 1866–1966,* 481–83.
94. John Higham, *Strangers in the Land: Patterns of American Nativism, 1860–1925* (New Brunswick, N.J.: Rutgers University Press, 1955), 131–57. See also, Alan M. Kraut, "Silent Strangers: Germs, Genes and Nativism in John Higham's *Strangers in the Land,*" *American Jewish History* 76 (Dec. 1986): 142–58.
95. Edward Alsworth Ross, *The Old World in the New, The Significance of Past and Present Immigration to the American People* (New York: Century Co., 1914), 286.
96. Daniel J. Kevles, *In the Name of Eugenics, Genetics and the Uses of Human Heredity* (New York: Knopf, 1985). See also, Kenneth Ludmerer, *Genetics and American Society: A Historical Appraisal* (Baltimore: Johns Hopkins University Press, 1972).
97. A collection of publicity and photographs of eugenics fairs is in the Charles B. Davenport Papers at the American Philosophical Society in Philadelphia.

CHAPTER 10: "VIRUSES AND BACTERIA DON'T ASK FOR A GREEN CARD"

1. Jay Pierrepont Moffat Diary, May 25, 1939. Also quoted in Richard Breitman and Alan M. Kraut, *American Refugee Policy and European Jewry,*

1933–1945 (Bloomington: Indiana University Press, 1987), 74.

2. Ibid., 73.
3. Sharon R. Lowenstein, *Token Refuge, The Story of the Jewish Refugee Shelter at Oswego, 1944–46* (Bloomington: Indiana University Press, 1986), 70. Also see, Ruth Gruber, *Haven, The Unknown Story of 1000 World War II Refugees* (New York: Coward-McCann, 1983). The Roosevelt administration sent Gruber to Europe to select the refugees for the Oswego camp. The refugees included 874 Jews, 73 Roman Catholics, 28 Greek and Russian Orthodox, and 7 Protestants.
4. Ibid., 80–86.
5. Some important works on the internment are U.S. Commission on Wartime Relocation and Internment of Civilians, *Personal Justice Denied: Report of the Commission on Wartime Relocation and Internment of Civilians* (Washington, D.C.: Government Printing Office, 1982), John Tateishi, *And Justice for All: An Oral History of the Japanese American Detention Camps* (New York: Random House, 1984). Roger Daniels, *Concentration Camps U.S.A.: Japanese Americans and World War II* (New York: Holt, Rinehart, and Winston, 1972), and Jacobus tenBroek, Edward Barnhart, and Floyd Matson, *Prejudice, War and the Constitution: Causes and Consequences of the Evacuation of the Japanese in World War II* (Berkeley: University of California Press, 1970). Roger Daniels has recently published a fresh overview of the internment. See Daniels, *Prisoners Without Trial, Japanese-Americans in World War II* (New York: Hill and Wang, 1993). Older, but with some useful information on the health and hygienic conditions is Audrie Girdner and Anne Loftis, *The Great Betrayal, The Evacuation of the Japanese-Americans During World War II* (London: Macmillan, 1969).
6. The anecdote is recounted by Girdner and Loftis, *The Great Betrayal,* 148.
7. Ibid., 164.
8. For mention of the gradual airing of the experience by its victims, see Ronald Takaki, *Strangers from a Distant Shore, A History of Asian-Americans* (Boston: Little, Brown, 1989), 484–87.
9. As quoted in Leonard Dinnerstein, *America and the Survivors of the Holocaust* (New York: Columbia University Press, 1982), 63–71.
10. Ibid.
11. The single best book to date on the Holocaust survivors who came to the United States and the matter of adjustment is by sociologist William B. Helmreich, *Against All Odds: Holocaust Survivors and the Successful Lives They Made in America* (New York: Simon and Schuster, 1992). On the succeeding generation, see Helen Epstein, *Children of the Holocaust* (New York: Putnam's, 1979).
12. *New York Times,* Apr. 21, 1990. See also, *New York Post,* Apr. 21, 1990.
13. *New York Times,* Apr. 21, 1990.
14. Lawrence Altman, "Debate Grows on U.S. Listing of Haitians in AIDS Category," *New York Times,* July 31, 1983. Altman's article is also cited in Den-

nis Altman, *AIDS in the Mind of America* (Garden City, N.Y.: Doubleday, 1987), 72.

15. For the CDC's public statement on its classification of the Haitian community, see *New York Times,* Apr. 10, 1985.
16. The fullest discussion of the Haitians and accusations against them as the bearers of AIDS to the United States is Paul Farmer, *AIDS and Accusation, Haiti and the Geography of Blame* (Berkeley: University of California Press, 1992).
17. *Washington Post,* Mar. 9, 1992.
18. Centers for Disease Control, "Tuberculosis Among Foreign-Born Persons Entering the United States, Recommendations of the Advisory Committee for Elimination of Tuberculosis," *Morbidity and Mortality Weekly Report* 39 (Dec. 28, 1990): 1.
19. New York City Department of Health, Bureau of Tuberculosis Control, *Tuberculosis in New York City, 1991 Information Summary,* 3, table 2.
20. *Montgomery Journal,* Oct. 14–15, 1992. I am most grateful to Lynn Frank for the information she shared with me in a telephone conversation about TB data in October 1992.
21. Ibid. I am also indebted to Yvonne Richards for her willingness to share available Montgomery County data with me.
22. United States Department of Health and Human Services, Public Health Service, Centers for Disease Control, Center for Prevention Services, Division of Quarantine, *Technical Instructions for Medical Examination of Aliens,* June 1991, p. I-2. My thanks to Nancy McQueen and others at the CDC for their cooperation and generosity in providing the latest reports and answering my questions.
23. The threat posed by drug-resistant TB strains is explained in Frank D. Ryan, *The Forgotten Plague, How the Battle Against Tuberculosis Was Won and Lost* (Boston: Little, Brown, 1992).
24. Lawrence O. Gostin, "Controlling the Resurgent Tuberculosis Epidemic, A 50-State Survey of TB Statutes and Proposals for Reform," *Journal of the American Medical Association* 269 (Jan. 13, 1992): 255–61. Also, *Washington Post,* Jan. 13, 1993.
25. Department of Health of the City of New York, Board of Health, Notice of Adoption of an Amendment to Section 11:47 of the New York City Health Code, copy obtained from public relations office of Department of Health. Also quoted in *New York Times,* Mar. 10, 1993.
26. Ibid.
27. *New York Times,* Dec. 31, 1992.
28. Ibid.
29. *The New York Task Force on Immigrant Health Bulletin* (Fall 1992).
30. In 1992–93, Suzanne Michael, M.S., chaired the task force's Committee on Family Health and taught the course through the Hunter College Center for the Study of Family Policy that sends student interns into health care facilities as translators.

31. Minutes of New York Task Force on Immigrant Health's Roundtable Forum, "Tuberculosis in the Foreign-Born and in Puerto Ricans, March 24, 1992."
32. Ibid., 3–4.
33. Ibid., 4.
34. Ibid., 5.
35. Ibid., 6.
36. Refugee Act of 1980 (94 Stat. 102).
37. Carolyn L. Williams and Joseph Westermeyer, *Refugee Mental Health in Resettlement Countries* (Washington: Hemisphere Publication Corp., 1986). See also, Wayne H. Holtzman and Thomas H. Bornemann, eds., *Mental Health of Immigrants and Refugees, Proceedings of a Conference Sponsored by the Hogg Foundation for Mental Health and the World Foundation for Mental Health* (Austin, Tex.: Hogg Foundation for Mental Health, 1990). A guide to the rich article literature in this field is Carolyn L. Williams, *An Annotated Bibliography on Refugee Mental Health* (Washington, D.C.: Government Printing Office, 1987).
38. Dr. Arthur Kleinman is a psychiatrist trained in anthropology. His research in Taiwan on cross-cultural medicine offers an illuminating perspective on the clinical challenges posed when cultures of medicine East and West clash at the bedside of a patient. Arthur Kleinman, *Patients and Healers in the Context of Culture: An Exploration of the Borderland between Anthropology, Medicine and Psychiatry* (Berkeley: University of California Press, 1980).
39. Both conditions might have been diagnosed as neurasthenia in an earlier era. As Dr. Arthur Kleinman observes, neurasthenia was understood in the nineteenth and early twentieth centuries as "both weakness of the nerves and nervous exhaustion." According to Kleinman, "A portmanteau term, neurasthenia packed together in the same category a syndrome of chronic fatigue, weakness, and a myriad of associated bodily and emotional complaints with a presumed neurological cause; as was stated then and now, it is a 'real physical disease.'" Called "the American disease" by New York neurologist George Beard, who coined the term *neurasthenia,* the malady is no longer even officially a disease in North America, having been removed from the American Psychiatric Association's *Diagnostic and Statistical Manual,* third edition (*DSM-III*) and has been replaced by depressive disorders, hysteria, and a variety of psychophysiological and psychosomatic labels. See Arthur Kleinman, M.D., *The Illness Narratives, Suffering, Healing, and the Human Condition* (New York: Basic Books, 1988), 101–2. Also, Alejandro Portes and Ruben G. Rumbaut, *Immigrant America, A Portrait* (Berkeley: University of California Press, 1990), esp. chap. 5, "A Foreign World, Immigration, Mental Health and Acculturation."
40. Richard F. Mollica et al., *Repatriation and Disability: A Community Study of Health, Mental Health and Social Functioning of the Khmer Residents of Site Two,* Working Document, Harvard Program in Refugee Trauma, vol. 1,

Khmer Adults (Cambridge, Mass.: Harvard School of Public Health and the World Federation for Mental Health), 46.

41. For a discussion of how the clinical experiences of Western psychiatrists during World War II and culture-bound psychiatric ideology shaped the nosology in the first edition of the *Diagnostic and Statistical Manual: Mental Diseases (DSM-I)* in 1952, see Gerald N. Grob, "Origins of *DSM-I:* A Study in Appearance and Reality," *American Journal of Psychiatry* 148 (Apr. 1991): 421–31.
42. Mollica et al., *Repatriation and Disability,* 49–50.
43. Ibid., 50.
44. Robert Berkow, ed., *The Merck Manual of Diagnosis and Therapy* (Rahway, N.J.: Merck, Sharp & Dohme Research Laboratories, 1987), 1508.
45. Patrick Cooke, "They Cried Until They Could Not See," *New York Times Magazine,* June 23, 1991, 24–25, 45–48.
46. Ibid.
47. Ibid. Other discussions of how social experience and culture affect expressions of pain and coping mechanisms are Kleinman, *Illness Narratives,* passim; Mark Zborowski, *People in Pain* (San Francisco: Jossey-Bass, 1969) and David Mechanic, "Role of Social Factors in Health and Well Being," *Integrative Psychiatry* 4 (1986): 2–11.
48. The example and suggested response is offered by medical anthropologist Geri-Ann Galanti, *Caring for Patients from Different Cultures, Case Studies from American Hospitals* (Philadelphia: University of Pennsylvania Press, 1991), 95–96. The importance of understanding patients' diverse cultural backgrounds to the delivery of nursing care is expertly treated in Rachel E. Spector, *Cultural Diversity in Health and Illness,* 3d ed. (Norwalk, Conn.: Appleton and Lang, 1991).
49. Erica Goode, "The Cultures of Illness," *U.S. News and World Report,* Feb. 15, 1993, 74–76.
50. Deinard as quoted in Goode, "The Cultures of Illness," 75.
51. Amos S. Deinard and Timothy Dunigan, "Hmong Health Care—Reflections on a Six-Year Experience," *International Migration Review* 21 (Fall 1987): 862.
52. *New York Times,* Feb. 26, 1989.
53. *Washington Post,* Feb. 7, 1993.
54. Ibid.
55. Ibid.
56. *Washington Post,* Apr. 12, 1990.
57. *New York Times,* Feb. 18, 1992.
58. Ibid.
59. Ibid.
60. United States Department of Agriculture, *Agricultural Statistics, 1986* (Washington, D.C.: Government Printing Office, 1986). See also, Marion Moses, M.D., "Pesticide-Related Health Problems and Farmworkers," *AAOHN Journal* 37 (Mar. 1989): 115–30.

61. Valerie A. Wilk, "Farmworkers and the Health Risks of Pesticides," *Farmworker Justice News* 4 (Summer 1990): 3–4.
62. Joel S. Meister, "The Health of Migrant Farm Workers," *Occupational Medicine* 6 (July–Sept. 1991): 503–18.
63. *Washington Post,* Apr. 24, 1993.

Credits

During the years when I was researching and writing this volume, I published a number of essays that have served as "working papers," texts that have helped me to sharpen my thoughts or to explore particular topics in some depth. These include: "Silent Strangers: Germs, Genes and Nativism in John Higham's *Strangers in the Land,*" *American Jewish History* 76 (December 1986): 142–58; "Silent Travelers: Germs, Genes, and American Efficiency, 1890–1924," *Social Science History* 12 (Winter 1988): 377–94; "Healers and Strangers: Immigrant Attitudes Toward the Physician in America—A Relationship in Historical Perspective," *Journal of the American Medical Association* 263 (April 4, 1990): 1807–11; "Historical Perspectives on Refugee Movements to North America," in Wayne H. Holtzman and Thomas H. Bornemann, eds., *Mental Health of Immigrants and Refugees* (Austin: Hogg Foundation for Mental Health, University of Texas, 1990): 16–37. In this book, I have used ideas, data, and occasionally, passages from these earlier works. Where they are included, it is with the kind permission of the original copyright holders.

Index